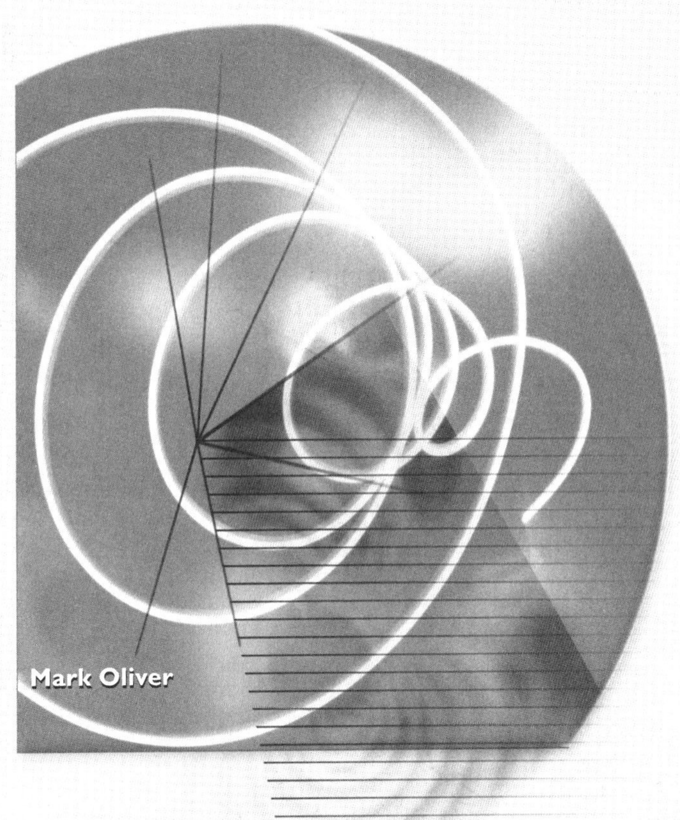

Using AutoCAD Map® 2000

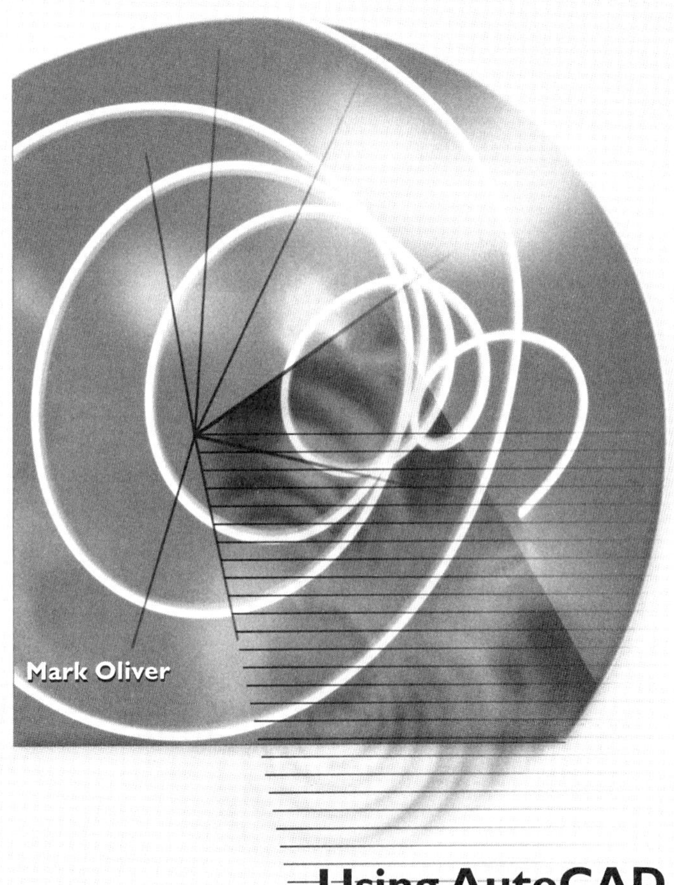

Using AutoCAD Map® 2000

Autodesk Press

Thomson Learning™

Africa • Australia • Canada • Denmark • Japan • Mexico • New Zealand
Philippines • Puerto Rico • Singapore • Spain • United Kingdom • United States

NOTICE TO THE READER

Publisher does not warrant or guarantee any of the products described herein or perform any independent analysis in connection with any of the product information contained herein. Publisher does not assume, and expressly disclaims, any obligation to obtain and include information other than that provided to it by the manufacturer.

The reader is expressly warned to consider and adopt all safety precautions that might be indicated by the activities herein and to avoid all potential hazards. By following the instructions contained herein, the reader willingly assumes all risks in connection with such instructions.

The publisher makes no representation or warranties of any kind, including but not limited to, the warranties of fitness for particular purpose or merchantability, nor are any such representations implied with respect to the material set forth herein, and the publisher takes no responsibility with respect to such material. The publisher shall not be liable for any special, consequential, or exemplary damages resulting, in whole or part, from the readers' use of, or reliance upon, this material. Autodesk does not guarantee the performance of the software and Autodesk assumes no responsibility or liability for the performance of the software or for errors in this manual.

Trademarks

Autodesk, AutoCAD and the AutoCAD logo are registered trademarks of Autodesk, Inc. Thomson Learning uses "Autodesk Press" with permission from Autodesk, Inc. for certain purposes. Windows is a trademark of the Microsoft Corporation. All other product names are acknowledged as trademarks of their respective owners.

Autodesk Press Staff

Executive Director: Alar Elken
Executive Editor: Sandy Clark
Development: Allyson Powell
Executive Marketing Manager: Maura Theriault
Executive Production Manager: Mary Ellen Black
Production Coordinator: Jennifer Gaines
Art and Design Coordinator: Mary Beth Vought
Marketing Coordinator: Paula Collins
Technology Project Manager: Tom Smith

Cover design by Scott Keidong's Image Enterprises

Printed in Canada
1 2 3 4 5 6 7 8 9 10 XXX 04 03 02 01 00 99

For more information contact
Autodesk Press
3 Columbia Circle, Box 15-015
Albany, New York USA 12212-15015;
or find us on the World Wide Web at http://www.autodeskpress.com

Library of Congress Cataloging-in-Publication Data

Oliver, Mark
Using AutoCAD Map 2000 / Mark Oliver.
p. cm.
ISBN 0-7668-0536-0
1. Cartography—Data processing. 2. AutoCAD map. I. Title.
GA102.4.E4044 1999
526'.0285'5369—dc21 99-16965

CONTENTS

Introduction .... **ix**

FEATURES OF THIS EDITION ix
STYLE CONVENTIONS x
WHAT IS GIS? x
HOW TO USE THIS BOOK WITH AUTOCAD 2000 xi
WE WANT TO HEAR FROM YOU xii
ABOUT THE AUTHOR xii
ACKNOWLEDGMENTS xii

SECTION 1 GIS WITH AUTOCAD MAP 2000 .... **1**

CHAPTER 1 Thematic Mapping **3**

WHEN TO USE A THEMATIC DATA QUERY 4
HOW TO SET UP DRAWINGS FOR QUERIES 5
COORDINATE SYSTEMS AND MAP PROJECTIONS 6
ASSIGNING COORDINATE SYSTEMS 7
Tutorial: Assigning Coordinate Systems **8**
LINKING DRAWINGS AND BUILDING A QUERY 12
Tutorial: A Simple Thematic Mapping Exercise **13**
PREPARING DRAWINGS FOR USE IN QUERIES 30
Tutorial: Preparing Drawing for Use in Queries **31**
MULTIPLE DRAWING QUERIES 40
Tutorial: Multiple Drawing Queries **41**
Continuous vs. Discrete Query Functions 64
SUMMARY **65**
REVIEW QUESTIONS **66**

CHAPTER 2 Topology **67**

INTRODUCTION TO TOPOLOGY 67
Tutorial: Drawing Cleanup **70**

Tutorial: Creating Polygon Topology **78**
Tutorial: Creating Network Topology **91**
Tutorial: Creating Node Topology **98**
Tutorial: Network Traces Using Topology **100**
NETWORK FLOOD TRACES 114
Tutorial: Buffer Zone Analysis **115**
Tutorial: Overlay Analysis **121**
SUMMARY **127**
REVIEW QUESTIONS **128**

CHAPTER 3 Raster Images **129**

Tutorial: Viewing Raster Images **130**
SUMMARY **139**
REVIEW QUESTIONS **140**

CHAPTER 4 GIS Techniques **141**

Tutorial: Edge Matching **142**
Tutorial: Rubber Sheeting **152**
Tutorial: Mapbook Printing **157**
Tutorial: Importing Non-DWG Files **174**
SUMMARY **177**
REVIEW QUESTIONS **178**

CHAPTER 5 Generating a Digital Map **179**

Tutorial: Creating Digital Maps Using a Graphics Tablet Using UTM Coordinates **183**
Add a Layer for "Roads" 193
Continuing the Drawing in a Future Work Session 198
Adding Object Data While Digitizing 199
SUMMARY **199**
REVIEW QUESTIONS **200**

CHAPTER 6 External Databases **201**

BACKGROUND INFORMATION AND NOMENCLATURE 201
Tutorial: External Database Connections to dBase3 .... **203**
PART 1 ACTIVATING ODBC 203
PART 2 ACCESSING THE DATA FROM AUTOCAD MAP 2000 205

Tutorial: External Database Connections to Access ... 221
PART 1 ACTIVATING ODBC ... 221
PART 2 ACCESS DATA AND AUTOCAD MAP 2000 ... 224
SUMMARY ... 235
REVIEW QUESTIONS ... 236

SECTION 2 APPLICATIONS IN GIS ... 237

Chapter 7 Civil Engineering Application ... 239
Tutorial: Flood Zone Properties ... 240
Flood Zone Property Exercise ... 244
SUMMARY ... 244

CHAPTER 8 Environmental Application ... 245
ACID RAIN ... 245
Tutorial: Acid Rain ... 246
SUMMARY ... 255

CHAPTER 9 A Tourism Application ... 257
Tutorial: Querying Multiple Drawings ... 258
Tutorial: Establishing Drawings as Data Sets ... 268
Tourism Exercise ... 277
SUMMARY ... 278

CHAPTER 10 Site Selection ... 279
Tutorial: A Simple Site Selection ... 280
SUMMARY ... 285

CHAPTER 11 Facilities Management ... 287
Tutorial: Room Utilization ... 288
Facilities Management Exercise ... 299
Summary ... 299

CHAPTER 12 Optimal Business Delivery Routing ... 301
Tutorial: Route Determination ... 302
Summary ... 327

CHAPTER 13 Architectual Restoration 329

Tutorial: Using Raster Images for Cost Estimates 330
SUMMARY 335

CHAPTER 14 Suggested Projects 337

SUMMARY 338

SECTION 3 RELATED TECHNOLOGIES 339

CHAPTER 15 Autodesk Mapguide Viewer and the Internet 341

Tutorial: Installing/Using Autodesk MapGuide Viewer 342
SUMMARY 347

CHAPTER 16 GPS and MAPS 349

Tutorial: GPS Files to Maps 350
SUMMARY 359

CHAPTER 17 Autodesk World 361

Tutorial: A Brief Introduction to Autodesk World 362
SUMMARY 377

Glossary 379

Index 383

This text is designed to introduce GIS through the use of AutoCAD Map 2000. Novices and experienced users alike will learn how to complete sophisticated applications and techniques including Overlay Analysis, Topological Queries, Edge Matching, attaching External Databases like Access, Excel, dBase, and others. Only minimal AutoCAD experience is required. If you can create layers, draw polylines, and hatch areas, you can master the GIS approaches by following the step by step procedures.

FEATURES OF THIS EDITION

Using AutoCAD Map 2000 is presented in three sections. The first, "GIS with AutoCAD Map 2000," introduces GIS concepts and techniques. This is the only section containing end of chapter review questions. The second component builds on the first through a series of discipline oriented "Applications in GIS." These tutorials give the reader exposure in using AutoCAD Map 2000 in areas as diverse as civil engineering, environmental technology, tourism, industrial site selection, facilities management, delivery routing, and architectural restoration. The final feature focuses on "GIS Related Technologies." These technologies are comprehensive GIS tools that are briefly portrayed in the third section. The purpose is to create a general awareness of these powerful tools.

Field-tested, step-by-step procedures with guiding graphic images, lead you through 30 tutorials, each complete with all the necessary AutoCAD Map 2000 maps and related files. The major GIS topics like Topology, Thematic Mapping, Coordinate Systems, Map Projections, Rubber Sheeting, Digitizing, Raster Images, Buffer Zones, and Queries are explained in clear, concise terminology and activities are provided to reinforce their understanding.

Incorporated into the tutorials are applications related to engineering, tourism, industrial location, facilities management, delivery routing, architecture, and the environment.

Supplementing these hands-on, tutorial-based chapters are related exercises that are designed to further reinforce the user's acquisition of GIS skills.

Also included is a section showcasing related technologies, like mapping on the Internet with Autodesk MapGuide, using GPS units with Autodesk Map 2000, and a introduction to Autodesk World.

With more than 40 sample drawing files on the included CD-ROM and over 450 guiding graphics along with clear, step-by-step directions, *Using AutoCAD Map 2000* will assist you in your application of GIS technology.

"Look into your future and you will find GIS."

STYLE CONVENTIONS

This text uses **Bold** type to indicate a command or menu item selection. Generally, pull-down menu items are utilized and the screen captures reflect this approach to completing a task. Many AutoCAD users are familiar with keyboard commands and prefer that tactic to the mouse-activated menu bar along the top of the screen. Because this text was designated as an introduction to GIS, the menu approach was utilized.

The main objectives are identified at the beginning of the chapter, as are any key terms that might be encountered during the course of the tutorial.

Autodesk MapGuide, Autodesk World, AutoCAD 2000, and AutoCAD Map 2000 are registered trademarks of Autodesk Inc.

Windows and Microsoft Office are registered trademarks of Microsft Corporation.

WHAT IS GIS?

GIS or Geographic Information Systems utilize computer technology to assist with the collection, storage, analysis, and presentation of data that is geographically referenced. There are generally two main components to GIS. The first is the graphical representation of geographical, human-made, or political features on the earth. The second is the data associated with the attributes or features of the map.

The technology has its roots in the early 1960s when a computer-based land information system was developed in Canada. GIS technology became mainstream however, only when the cost of computing power dropped throughout the late 1980s. As the computing power per dollar ratio continues to improve, more and more computer users are finding benefit in using GIS technology to aid them in the decision-making processes associated with their particular applications.

The main benefit derived by the use of GIS is the capability to identify answers to questions like "what is where ?" based on data provided from a wide variety of sources. The result of querying data can be displayed in a manner that invariably improves the level of understanding for the analyst and therefore leads to better decision making. Whether one is trying to determine effective marketing strategies, discover new resource locations, plan new transportation routes, or evaluate the environmental impact of a natural disaster, GIS is a means to an end!

HOW TO USE THIS BOOK WITH AUTOCAD 2000

AutoCAD Map 2000 is an enhanced release of AutoCAD 2000 providing GIS capabilities to AutoCAD 2000. AutoCAD Map 2000 facilitates the utilization of existing drawing files or the creation of new ones for the graphical representation of the earth's features, as well as providing the capability of analyzing block attribute, object, or SQL database-linked data.

This text focuses on the utilization of the added capabilities of AutoCAD Map 2000 and does not attempt to duplicate the contents of the numerous AutoCAD user guides that are available. Rather, this manual has you become familiar with the GIS features and capabilities of Map 2000 by walking you through exercises that show you the power of this technology and the ease with which you can make use of this capability. Additionally, GIS techniques and terminology are presented and explained.

The fundamental purpose of AutoCAD Map 2000 is to enable the user to access multiple drawings and/or data sets simultaneously and then analyze them and display the results. Supplementary to these functions are built-in capabilities that enhance the process of working with drawings and data from a variety of sources and that have been prepared in different software programs. These tasks are beyond the capabilities of AutoCAD 2000.

Before beginning work with this text the user should copy all the tutorial files from the accompanying disk to the users computer. This text suggests that all these files should be placed in a separate directory identified as **Lessons**. These will include several DWG files as well other file formats such as TIF and DBF files. *The exceptions, located in the Lessons 6 directory, are outlined as follows*:

The PROPERTI.DBF and DBASEMAP.DWG files for chapter 6 are to be copied into a subdirectory created inside AutoCAD Map 2000. The subdirectory to be used is **Sample**, and it should be inside **AcadMap4** , which itself is located as a subdirectory of **Autodesk**, which in turn is located inside the **Program Files** directory. The path to the **Sample** directory and properti.dbf file would typically look like:

C:\Program Files\Autodesk\AcadMap4\sample\dbf\properti.dbf

Similarly, the ACCESS1 file and ACESSMAP.DWG files for chapter 6 should be placed in a subdirectory of **Office** called **Samples**. The two files can be copied into that location from the CD. The path to the **Samples** subdirectory and the ACCESS1 file would typically look like:

C:\Program Files\Microsoft Office\Office\Samples\access1.mdb

The CUSTOMER1.mdb used in Chapter 12 should also be placed in the **Office\Samples** subdirectory.

The placement of the files in these specific locations will ensure that the step-by-step instructions and images of the text will match precisely with what the user is experiencing as they go through the procedures. If any of the files are installed into different directories, then accomodation must be made for these changes when referring to the tutorials in the text.

WE WANT TO HEAR FROM YOU

If you have any questions or comments, you can contact us at:

The CADD Team
c/o Autodesk Press
3 Columbia Circle
P.O. Box 15015
Albany, NY 12212-5015

Please visit our Web sites:

www.autodeskpress.com
www.cadd-drafting.com

ABOUT THE AUTHOR

Mark Oliver is the manager of GIS Solutions and is the GIS Coordinator/Head of Geography for Napanee District Secondary School in Ontario, Canada. A teacher for more than 20 years, he has been teaching GIS since 1992. In 1998 he was awarded the Prime Minister of Canada's "Award for Teaching Excellence" in Technology as a result of his work with Geographic Information Systems.

ACKNOWLEDGMENTS

We would like to thank and acknowledge the many professionals who reviewed the manuscript to help us publish this AutoCAD Map 2000 text. A special acknowledgment is due to the following instructors and professionals who reviewed the chapters in detail:

Ray Eisenberg, Autodesk, Inc.
Shelly Fry, Central Piedmont Community College
Herman Gruenwald, University of Oklahoma
Andrew Laudick, ITT
Steven Means, New Mexico Junior College
David Steinhauer, Tidewater Community College

A special thanks goes to Bill Wittreich of Wittreich Associates, who performed a technical edit on the manuscript.

The author would like to acknowledge and thank the following staff members of Delmar Publishers:

Publisher: Alar Elken
Executive Editor: Sandy Clark
Developmental Editor: John Fisher
Production Coordinator: Jennifer Gaines
Art & Design Coordinator: Mary Beth Vought
Marketing Coordinator: Paula Collins
Editorial Assistant: Allyson Powell

The author also would like to acknowledge and thank the following people:

Copyediting: Laura Poole, Archer Editorial Services
Composition: John Shanley, Phoenix Creative Graphics

The author wishes to extend a special thank-you to Dr. Roland Tinline, G.I.S. Lab, Queen's University; Dickson Mansfield, Faculty of Education, Queen's University; Frank Nanfara, Torcomp Group, Terry Moloney, PCI Geomatics, and Gordon McElravy, Autodesk Canada, for their guidance, support, and enthusiasm for G.I.S. education. Most important, the author wishes to thank his wife, Marjorie, and his sons, Colin and Brian, for their patience and encouragement.

GIS with AutoCAD Map 2000

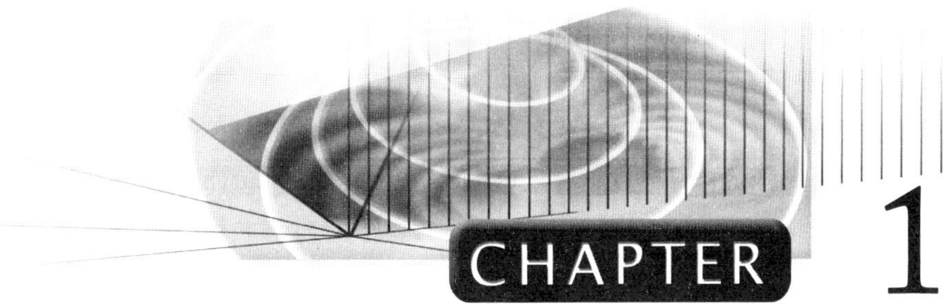

Thematic Mapping

OBJECTIVES

After completing this chapter you will know

- when to utilize a Thematic Data Query
- what Object Data is
- how to set up your drawings for the queries
- what a coordinate system is used for
- how to establish coordinate systems for a work session
- how to conduct a data query involving multiple drawings
- the difference between a Discrete and Continuous query

KEY TERMS

Source drawing
Work session
Base drawing
Project drawing
Thematic map
Equivalent projection
Conformal projection
Global Coordinate System
Object Thematic Query
Object data
Closed polygon
Continous query function
Discrete query function

Thematic maps are generally used to represent spatial data according to numerical values of the data attributes (Figure 1–1). A Thematic map could be utilized, for example, to reveal patterns in average income level by state, unemployment level by county, or daily caloric intake by nation. The patterns revealed by the map illustrate the distribution of the data. It is important to note that non-numeric information can also be categorized or classified and dislayed through Thematic Maps.

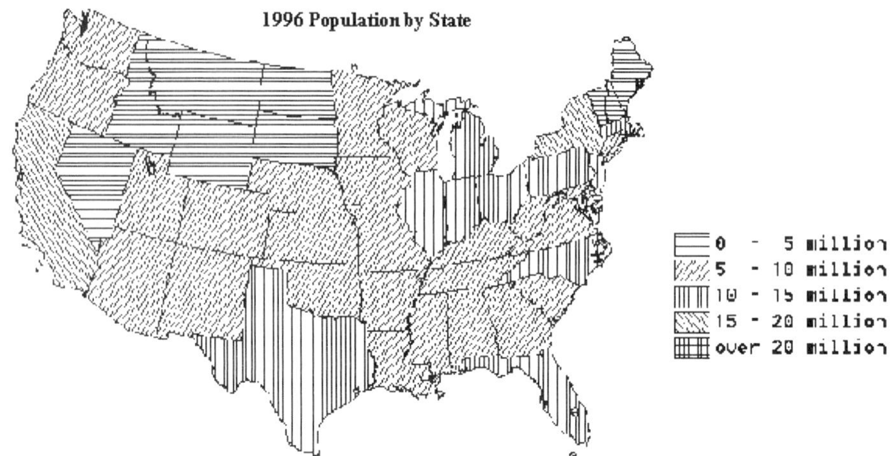

Figure 1–1 *Sample Thematic Map –1996 Population*

WHEN TO USE A THEMATIC DATA QUERY

This approach to visualizing data through a graphic display is likely to be the most common use of AutoCAD Map 2000. You would use thematic mapping of data whenever you wanted to view a pattern associated with the distribution of that data or study the relationships that existed between two or more drawings and the data associated with each. The examples of when this technique might be employed are virtually limitless.

For example, if you were studying a plan for a subdivision, and drawings had been prepared by two different agencies to show high water levels/flood plain zones in one case and storm sewer details in another, these drawings could be viewed simultaneously and queried specifically to determine the location of issues of concern. This example could be extended to include a query that would reveal concerns regarding the location of water and sewage mains, gas pipelines, underground cables, and so on.

Another example of an appropriate use of thematic mapping would be to evaluate the impact of municipal sewage treatment plant discharges on colioform bacteria counts in discharge waterways. Drawings revealing discharge plumes could be displayed concurrently with bacteria count data shown in concentration levels represented by contours. A time-based series of such displays would facilitate the temporal and spatial analysis of the relationship between the two variables. Other examples would include things like the relationship between areas of acid rain deposition and the frequency of sterile lakes or the impact of increasing population density on the extirpation of wildlife.

An example of a simple thematic map would be the illustration of average housing costs by state.

HOW TO SET UP DRAWINGS FOR QUERIES

A thematic mapping query can be utilized to generate a thematic map when you have data on one or more maps and you wish to investigate this data or the relationships that exist within the data and have the results illustrated on yet another map.

If you have a data set that you want to display as a thematic map, you could manually hatch in the regions on your map according to a classification scheme you devise. The thematic mapping query function, however, establishes links with data-enriched drawings, which can be used as part of the mapping process and reused in other mapping queries.

To initiate object thematic mapping, you must first identify which drawings you want to query. These are generally referred to as *source drawings*. Source drawings can be thought of as drawing files that include attribute data that you want to access. This data may be in the form of *block attribute data*, *object data*, or *database-linked data*. Block attribute data is text data that is associated with a block. An example might be the name of the particular block. Object data can be numerical or textual data attached to an object in the drawing. This object could be a node, a line, text, or a polygon. Object data is stored in an internal database and hence can be used only in Autodesk products like AutoCAD Map 2000. It also increases the size of the drawing file, which may be a consideration. Database data is stored in conventional database software packages and is not actually contained in the drawing file itself.

The base drawing on which the results of the query will be displayed must also be identified. The entire collection of drawings you are working with is referred to as a *work session* and are listed in the drive tree section along the left of the main AutoCAD Map 2000 screen while you work. The base drawing is identified as the project file and any source drawings are listed as drawings (see Figure 1–2).

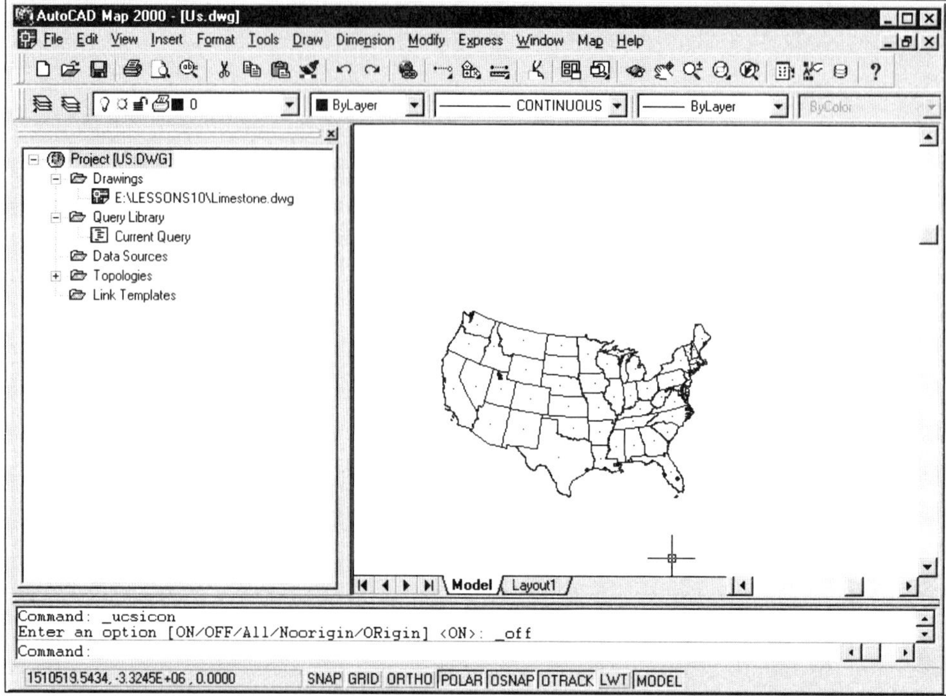

Figure 1–2 *Base/project drawing and source drawing identification*

Once you have identified the base drawing file, the next step in the process requires you to attach the source drawings that contain the data to be queried to your base drawing. You then design and implement a *query*.

A query is a request for specific information from the drawings. To run a query, you determine what you want to find, build an equation or query statement to search for the information, execute the query, and analyze the returned data when it is displayed on the base drawing.

To generate effective query statements, the user needs to know what data is available that relates to the drawings and how that data has been stored. The query statement must be compiled accordingly. If, for example, the data is included as part of an external database, the query process must adopt that strategy and not try to query object data or block attribute data.

COORDINATE SYSTEMS AND MAP PROJECTIONS

If you want to access data from more than one drawing, you must ensure that the drawings all were generated using the same geographical framework or projection.

For example, if the drawings are of a subdivision, then all drawings being accessed in the work session must have the same coordinate system. For simplicity, it is preferable that the outline drawing or base map be the same for all the drawings to be queried and displayed. If the drawings are from different projections, they should be converted to the same projection and coordinate system.

A map projection is a formula designed to minimize distortion and error associated with representing the spherical surface of the earth in a two dimensional model. Depending on the size of the surface of the earth to be represented or the particular purpose of the project, different formulas or projections can be applied.

Equivalent projections maintain area accuracy and *conformal projections* preserve angular values. Mercator projections secure linear directionality.

In AutoCAD Map 2000 you can select from a variety of projections or coordinate systems to work in. If you have drawings that have originated from different systems, you can assign a coordinate system to the entire group of drawings to be used in the work session.

The choice of projections should be based on some fundamentals, however, as it is prudent to use the same coordinate system that is the standard for your organization, local agencies, or others who may want to access your data.

Numerous texts contain highly detailed information on map projections, and the online help section can provide added insight into the wide variety of choices available to the user.

ASSIGNING COORDINATE SYSTEMS

A coordinate system can be assigned to a work session by setting a global coordinate system as the default for the drawings in the work session. AutoCAD Map 2000 provides an extensive list of choices and also offers the flexibility of adding extra projections if required.

The following tutorial illustrates the process of assigning a coordinate system to a work session. In this case, the NAD83 or North American Datum from 1983 is used.

TUTORIAL: ASSIGNING COORDINATE SYSTEMS

Step 1 Start AutoCAD Map 2000 and from the **File** menu select **Open**. Find and open the **USA.DWG** file. Use the **Zoom** and **Extents** commands to adjust the field of view to present the entire USA map, if necessary.

Step 2 From the **Map** menu select **Tools** and then **Assign Global Coordinate System**.

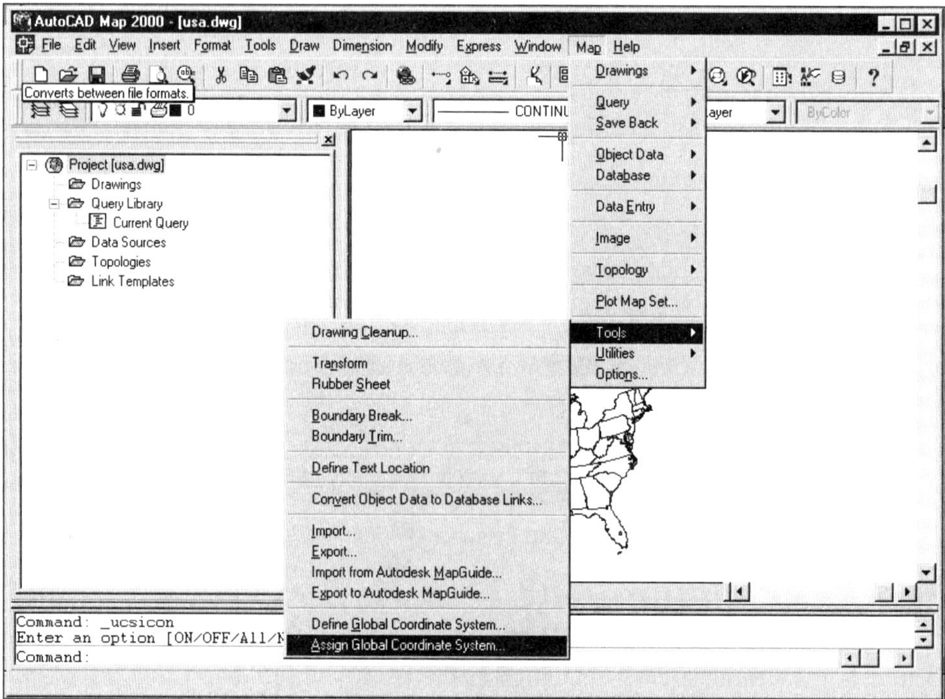

Figure 1–3

The Assign Global Coordinate System dialogue box will appear.

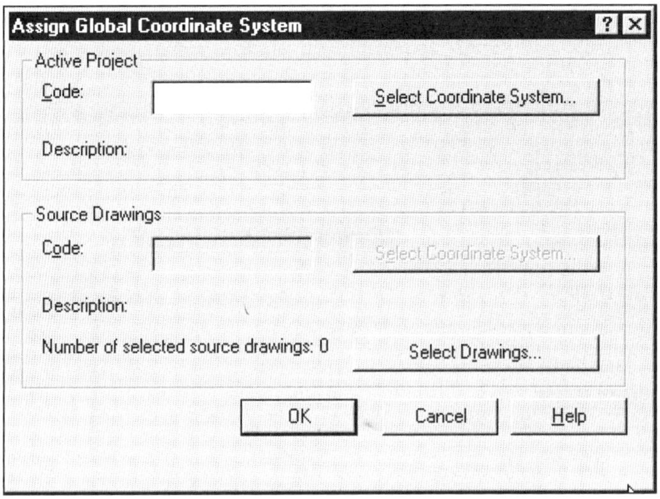

Assign Global Coordinate System
Active Project
Code:
Select Coordinate System...
Description:
Source Drawings
Code:
Select Coordinate System...
Description:
Number of selected source drawings: 0
Select Drawings...
OK
Cancel
Help

Figure 1–4

Step 3 Click on the **Select Coordinate System** button to view the options available. Several available global coordinate systems will be shown as available for the given category, which in the case shown is Lat Longs.

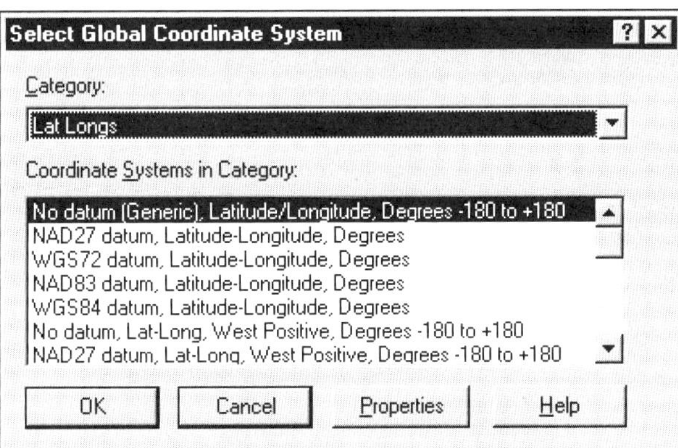

Figure 1–5

Step 4 Click on the downward pointing arrow located to the right of Lat Longs and a selection of categories will be presented. Scroll down the list until you get to the **USA-USGS-Albers 1:2,000,000** item and select it.

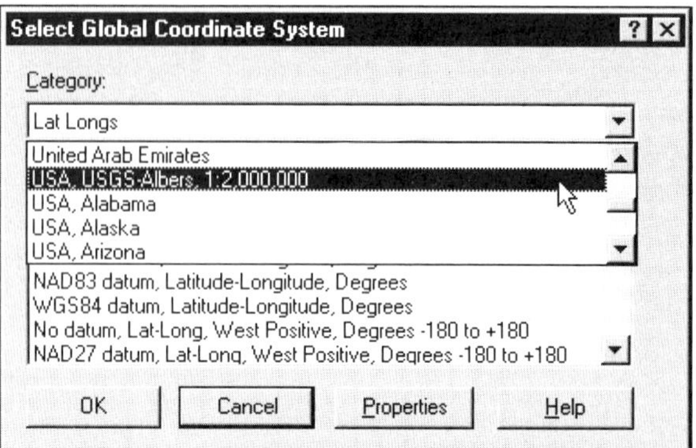

Figure 1–6

AutoCAD Map 2000 will provide a list of several Albers Equal Area projection options. Scroll down the list to the **NAD83 Albers Equal Area** item. The NAD83 or North America Datum 1983 is an ellipsoid that approximates the surface of the earth, which serves as a reference plane for mapping. You will notice that many datums exist covering different parts of the world, and some duplicates exist that cover the same part of the world but were created at different times, such as NAD027 and NAD83. The online help can provide more information about the projection and datum choices.

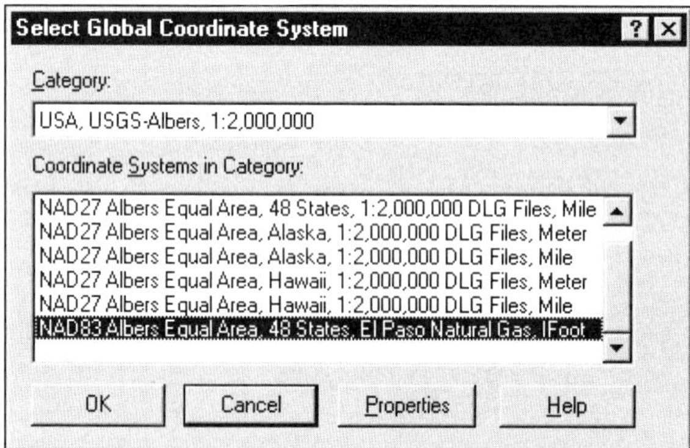

Figure 1–7

Click on **OK**. The Assign Global System Coordinates dialogue box will be presented again.

Step 5 To assign the same coordinate system to drawings that you will be querying, you first identify the drawings you will be working with, then the coordinate system is assigned. Click on the **Select Drawings** button. In the Select Drawings to Assign Coordinate System dialog box highlight the drawings you wish to use. Select the **USATUTOR1** file and then click on **Add** then again on **OK**.

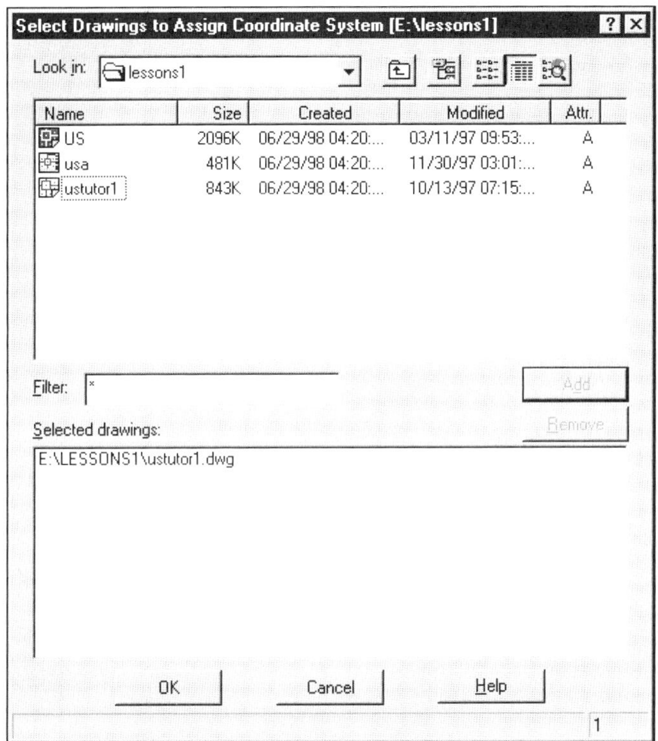

Figure 1–8

Click on the **Select Coordinate System** button for the source drawings. Select the same projections for the source drawings as was selected for the base drawings, **USGS-Albers 1:2,000,000** for the category and **NAD 83 Albers Equal Area** as the coordinate system.

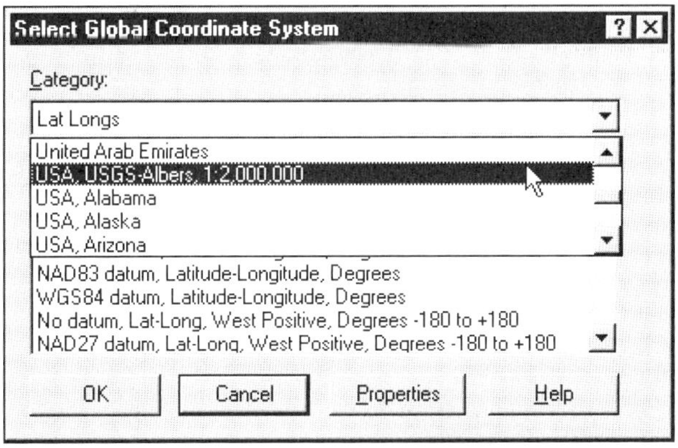

Figure 1–9

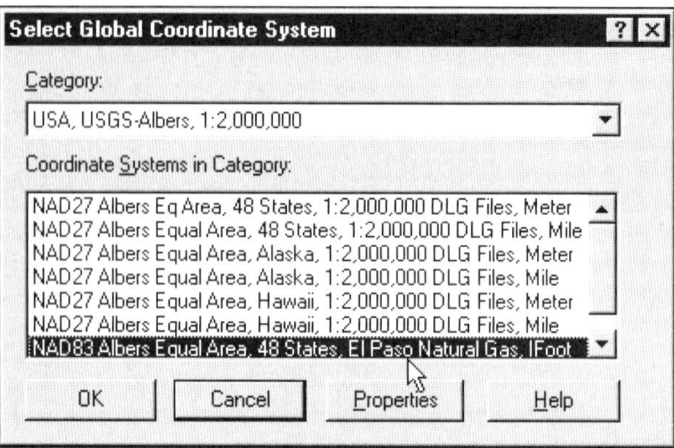

Figure 1–10

Click on **OK** to return to the Assign Global Coordinate System dialog box. Click on **OK** again, and AutoCAD Map 2000 will save a backup copy of the drawings without the coordinate system registered.

The drawings now have matching coordinate systems. The base drawing can be saved with the changes.

LINKING DRAWINGS AND BUILDING A QUERY

Once a set of drawings has been identified as containing data to be used for thematic mapping and the coordinate systems have been set equal, AutoCAD Map 2000 can be used to produce a map that is derived from the data in one or more drawing files.

A thematic map is as the name implies: a graphical representation of data displayed in a spatial context.

The following tutorial will guide you through a simple thematic mapping application.

TUTORIAL: A SIMPLE THEMATIC MAPPING EXERCISE

This tutorial uses an outline of the United States as a base drawing to which a second drawing is attached and queried. For the purposes of this exercise, the second drawing has had numerical data attached to four states in an arbitrary manner; Florida has been assigned a value of 5, New York a value of 10, Texas a value of 15, and California a value of 20. These numbers are examples of object data, data attached to features or objects in the drawing. In this case, the data has been attached to the polygons representing the various states. Object data can be textual or numerical, and in this case it is numerical. You will use the thematic mapping function to identify these four states according to the numerical data associated with each.

Step 1 Use **File** and **Open** to select the drawing entitled USA.DWG from the appropriate directory. In this text the directory is Lessons. (The directory Lessons was created on the hard drive and the tutorial files from the companion CD were copied to this directory)The USA map will serve as the base map. After opening the base map, the next step is to add the data-rich map, which is to be queried, to the work session.

Step 2 Go to the **Map** menu. Select **Drawings** and then click on **Define/Modify Drawing Set**.

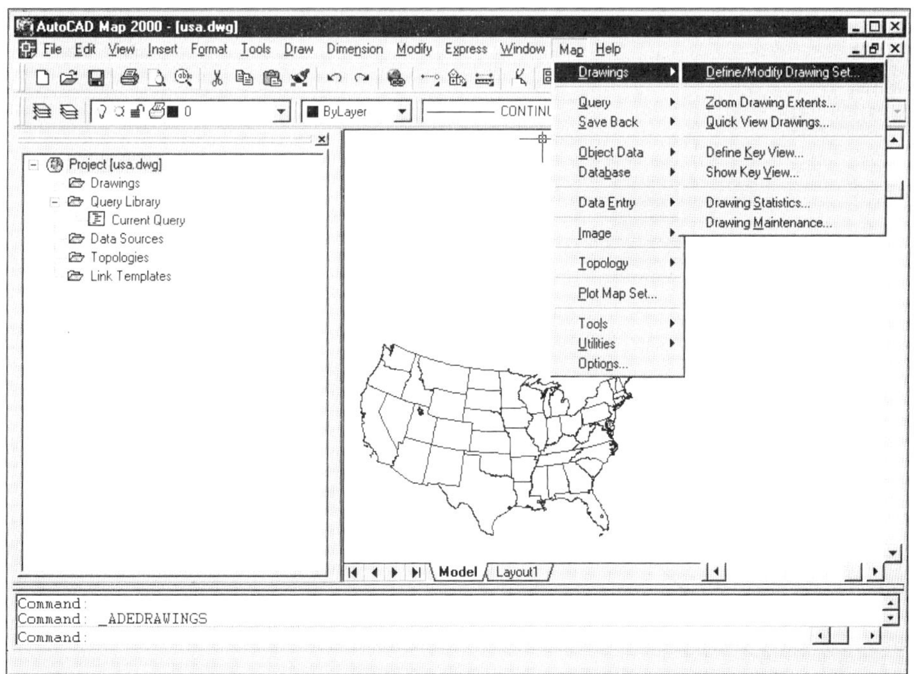

Figure 1–11

A **Define/Modify Drawing Set** dialogue box will appear.

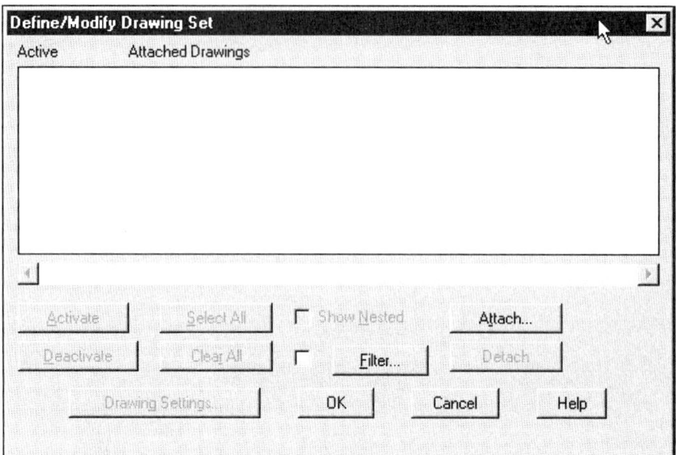

Figure 1–12

Step 3 Click on the **Attach** button and you will be prompted to select a directory and file. Find the Lessons directory under the directories list located in the upper half of the dialog box and double-click on it to see the files located in that directory.

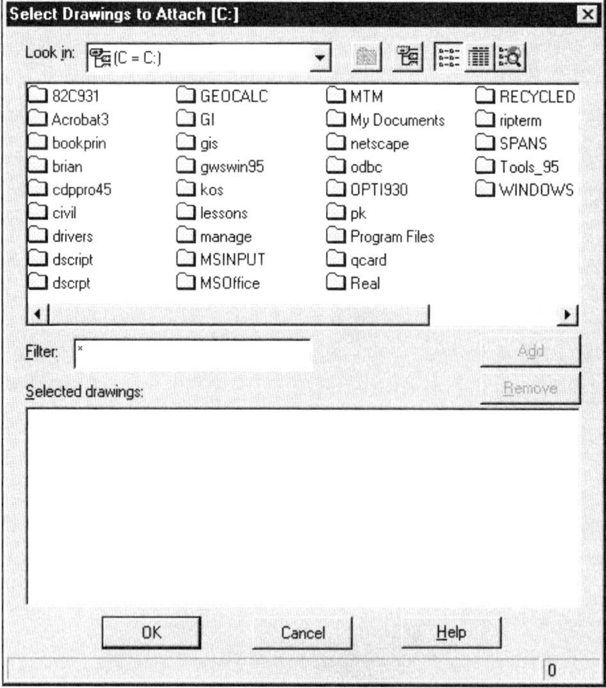

Figure 1–13

As soon as you complete this action, the contents of the directory will be revealed. Locate the file called **USTUTOR1** and click on it.

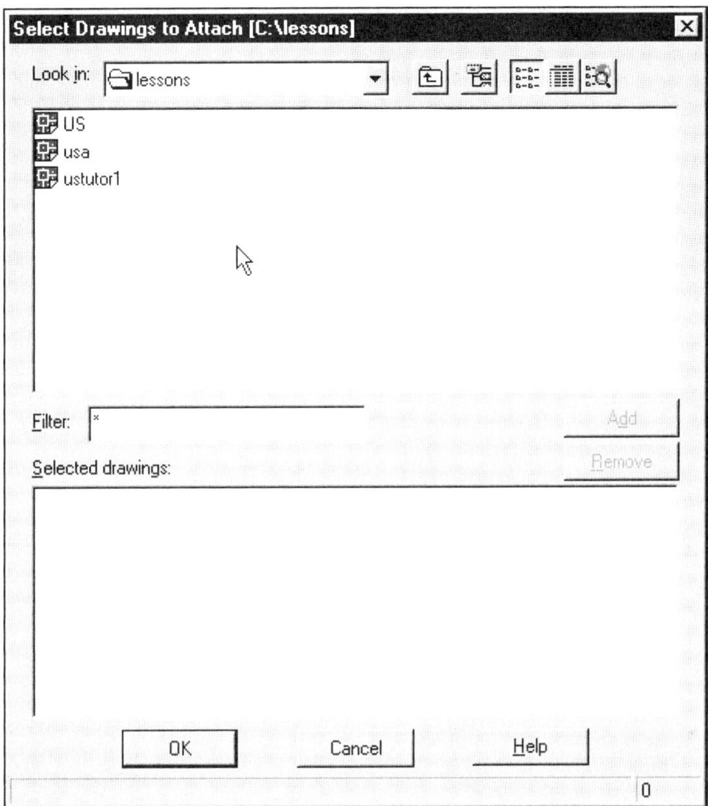

Figure 1–14

When it is highlighted, click on **Add**, and this file will be listed in the bottom half of the dialog box as a file to be attached to the original.

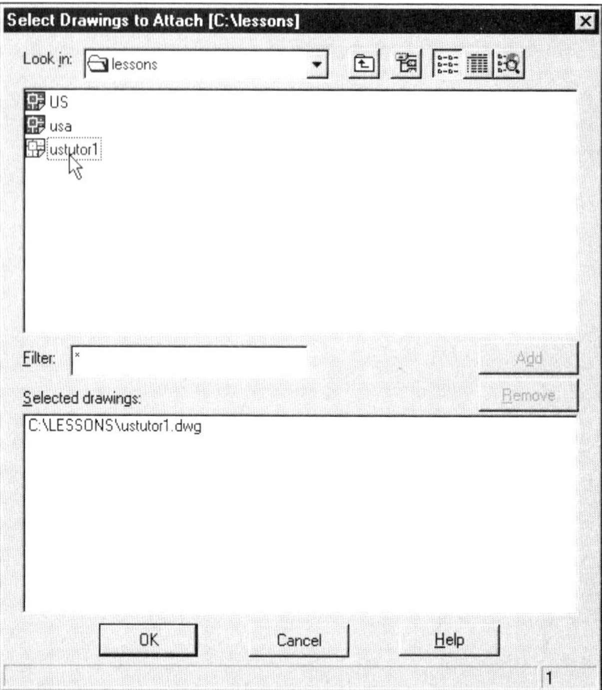

Figure 1–15

Step 4 Click on **USATUTOR1** to highlight it, then click on **OK** again and a dialog box will appear indicating that this file was attached to the original USA drawing. This means that AutoCAD Map 2000 can now make a link to the data contained in the USATUTOR1 drawing. Click on **OK**.

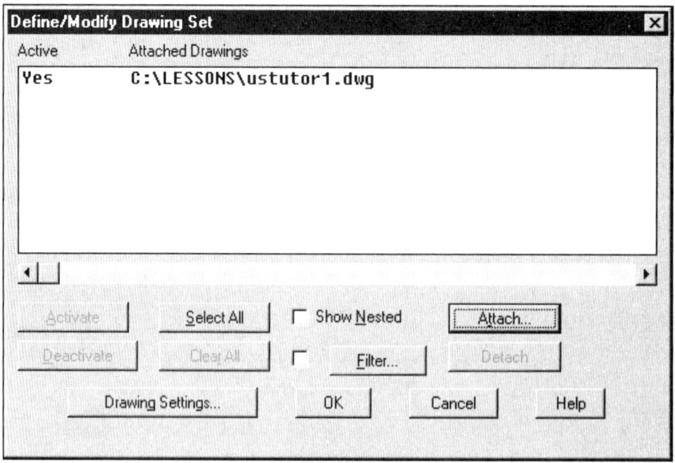

Figure 1–16

AutoCAD Map 2000 offers the capability of conducting different types of queries depending on the data and the nature of the source drawings. In this instance, the data embedded in the USATUTOR1 drawing is integer-based Object Data and the query established in the next steps has AutoCAD Map 2000 search for the numerical information.

Step 5 Go to the **Map** menu, select **Query**, then **Object Thematic Query**.

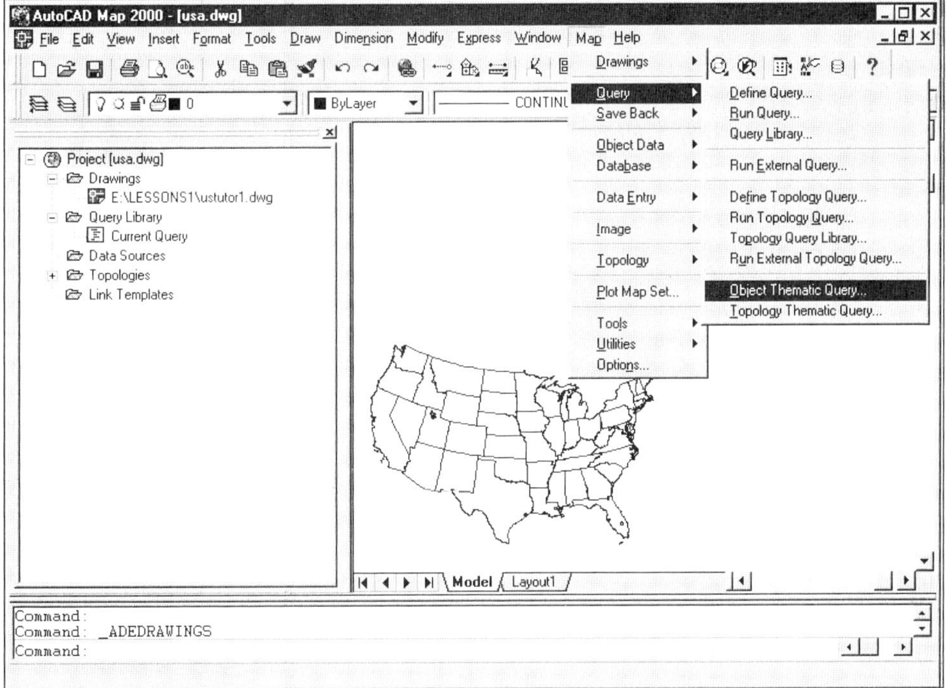

Figure 1–17

The Object Thematic Mapping dialog box will appear. Here you design the query. There are generally three parts to the process of defining a query. The first step is to select a location to be queried (the Objects of Interest section), the second is to determine what objects you want to have queried (the Thematic Expression section), and the third is to select a display style for the results of the query (the Display Parameters section).

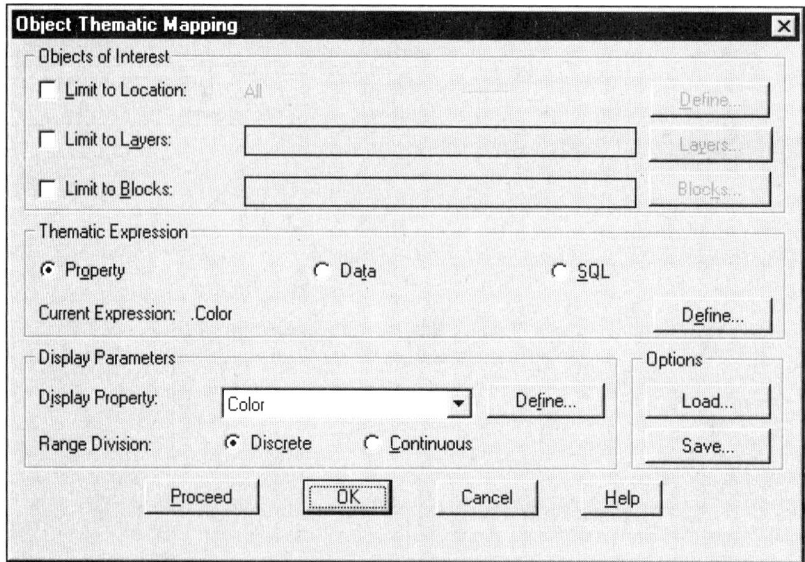

Figure 1–18

Step 6 To complete the selection of the area to be queried select the **Limit to Location checkbox** in the upper portion of the screen. A checkmark will appear in the square when you click on it.

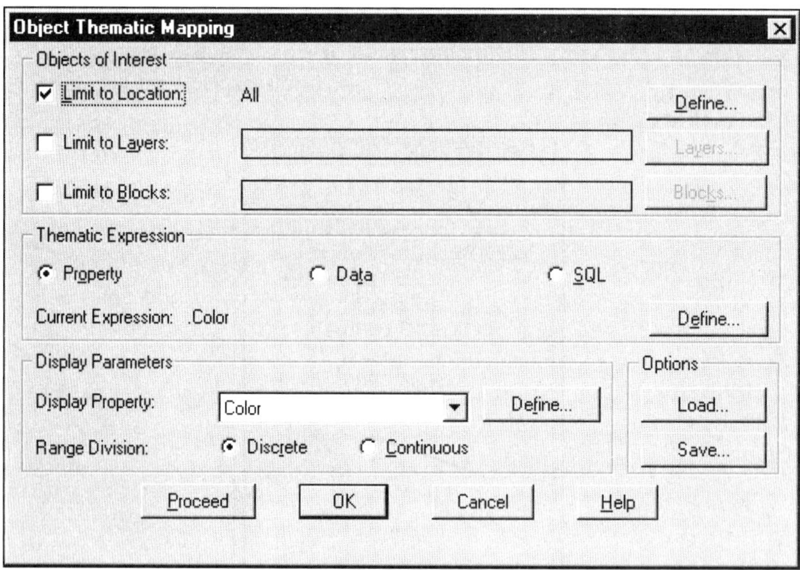

Figure 1–19

Click on **Define**.

Several different approaches can be utilized for determining the area to be queried. Depending on the nature of the drawing files, some boundary definition devices might be easier to use or more appropriate than others. In this exercise, you will be searching for data embedded in some of the states and so it is appropriate to search the entire continental United States; searching inside a circle that encompasses the entire region is one approach. The Inside option selects objects inside the perimeter of the search location, in this case a circle. If Crossing was selected, only objects that intersected the circle would be queried.

For the Boundary Type, click on the **Circle** radio button; for the Selection Type, select **Inside**. Then click on the **Define** button located near the bottom of the screen.

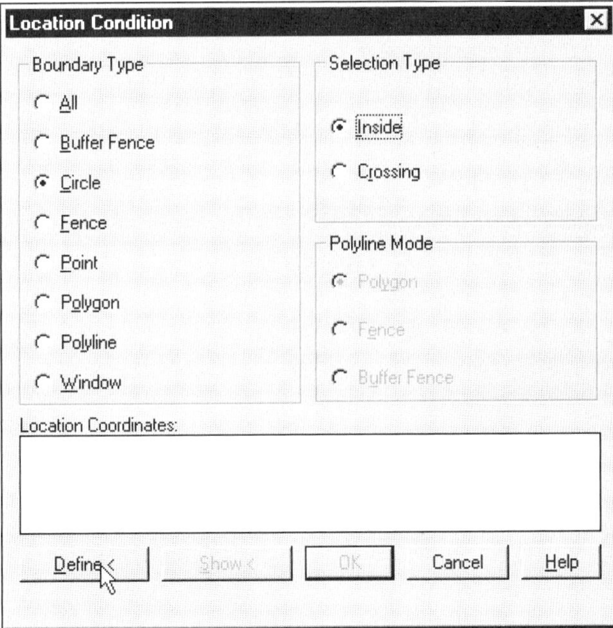

Figure 1–20

The mouse pointer should be placed near the center of the USA, and then click the left mouse button. As the mouse is moved, an outline of a circle will appear. Make sure that the circle encloses the entire USA and then click the left mouse button again.

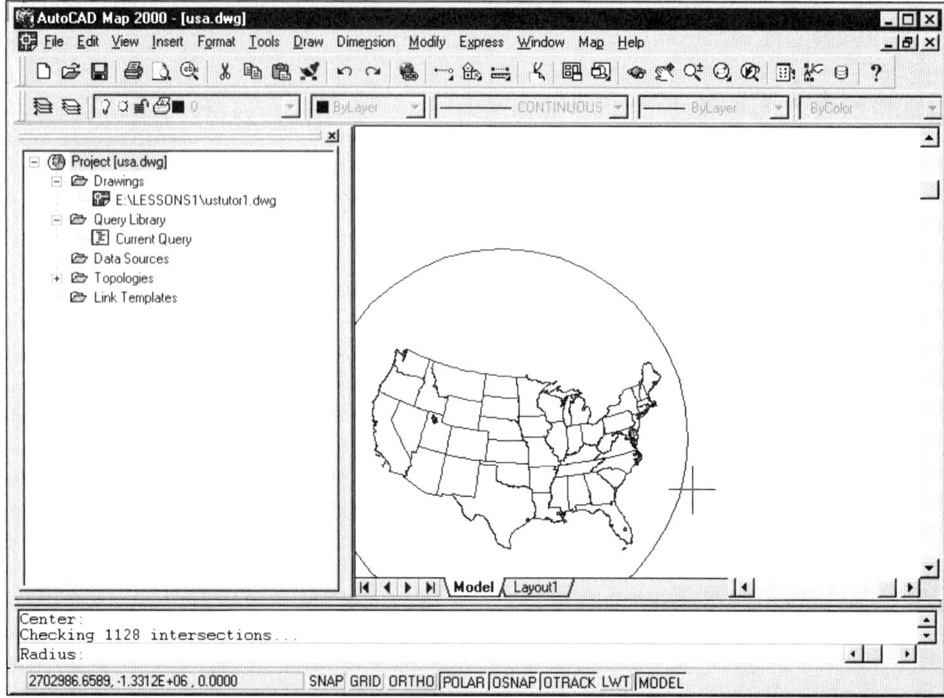

Figure 1–21

Step 7 After the circle creating the boundary of the region to be queried is defined, the Location Condition dialog box will reveal its coordinates. Click on **OK** and you will be returned to the Object Thematic Mapping dialog box.

The central portion of this dialog box identified as the Thematic Expression. The Thematic Expression area allows you to choose the type of data you will query. The options include Property, Data, and SQL. Property covers features like color, layer, text, and topology. Data can be used to select Object Data and SQL can be used to access information in a database. Click on the Data option then on the Define button located to the right. The Data Expression dialog box appears. To the right of the word Tables is a downward-pointing arrow. Use this drop-down menu to locate the table Tutorial 1 from the list available.

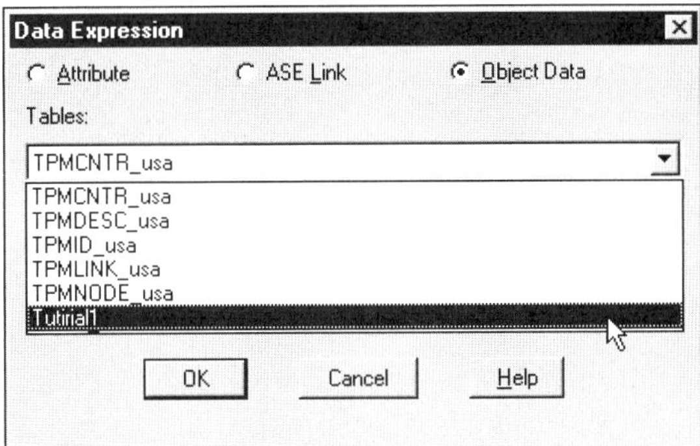

Figure 1–22

Highlight **Tutorial 1** from the list and click on **OK** again. Tutorial 1 will be listed as the Object Data Table and also in the Object Data Fields section of the dialog box.

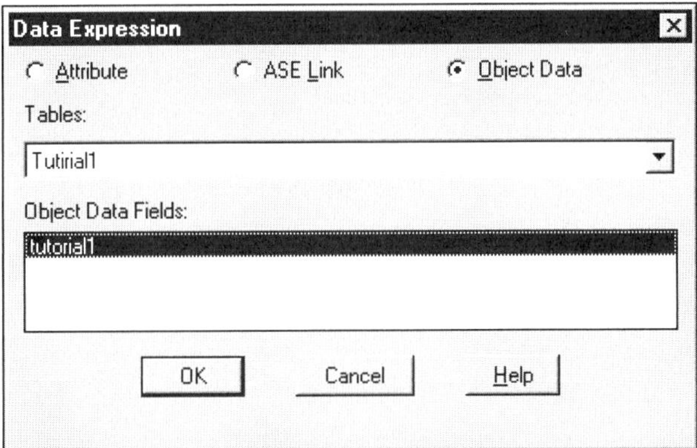

Figure 1–23

Click on **OK.**

Step 8 When you return to the Object Thematic Mapping dialog box, go to the Display Parameters section and activate the drop-down list in the Display Property field.

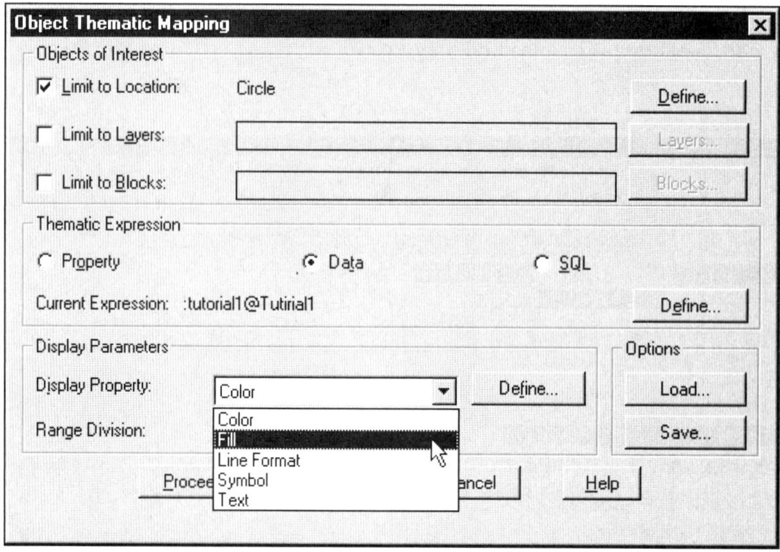

Figure 1–24

Select **Fill** from the list then click on **Define**. You are presented with the Thematic Display Options dialog box. It will be empty because you have yet to define the parameters of the search. Click on **Add** to launch the Add Thematic Range dialog box.

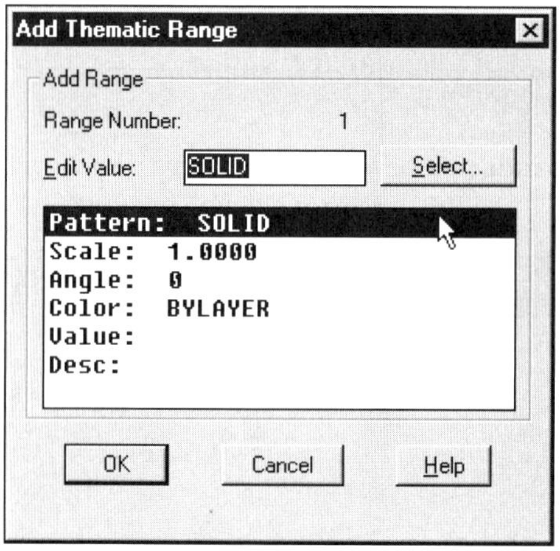

Figure 1–25

In the Add Thematic Range dialog box, set the colors and fill patterns, as well as identify the value to be queried and any description you might want placed in the legend or key of your finished project.

Step 9 For this exercise you will assign different display patterns to each of the four states. Click on the word **Pattern**, then on **Select**, then scroll down the list until you see the **Line** option.

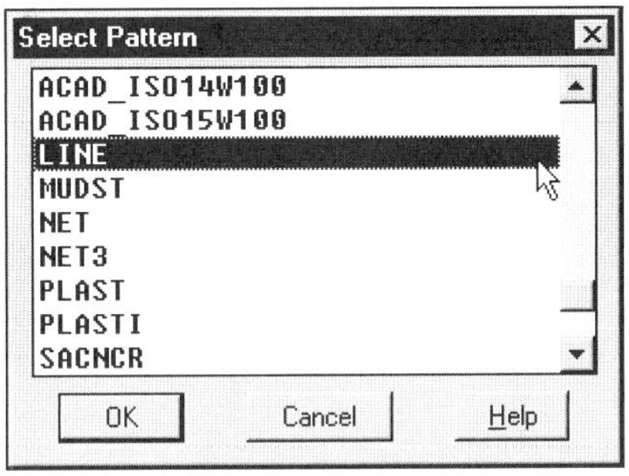

Select Pattern
ACAD_ISO14W100
ACAD_ISO15W100
LINE
MUDST
NET
NET3
PLAST
PLASTI
SACNCR
OK Cancel Help

Figure1–26

Highlight the **Line** option and click on **OK**, then register the change by clicking on the **Pattern** button again. A variety of hatching patterns may be used, but more ornate patterns produce a larger drawing file; in the interest of expediency, the Line pattern is utilized in most of these tutorials. Click on the **Scale** line in the Add Thematic Range dialog box to activate it. In the Edit Value edit box, set the scale of the line spacing at **40000** and register the setting by clicking on the word **Scale** again.

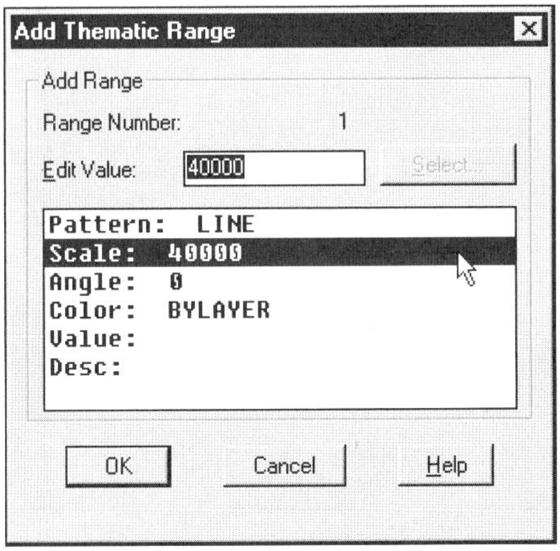

Add Thematic Range
Add Range
Range Number: 1
Edit Value: 40000 Select...
Pattern: LINE
Scale: 40000
Angle: 0
Color: BYLAYER
Value:
Desc:
OK Cancel Help

Figure 1–27

The value of 40,000 establishes a line spacing in meters, making the distance between the lines 40 km, or approximately 25 miles. A smaller value than 40,000 would result in the lines being spaced closer together. The Angle can be left at **0**. Set the color to **white** by clicking on the Color item and then again on the **Select** button.

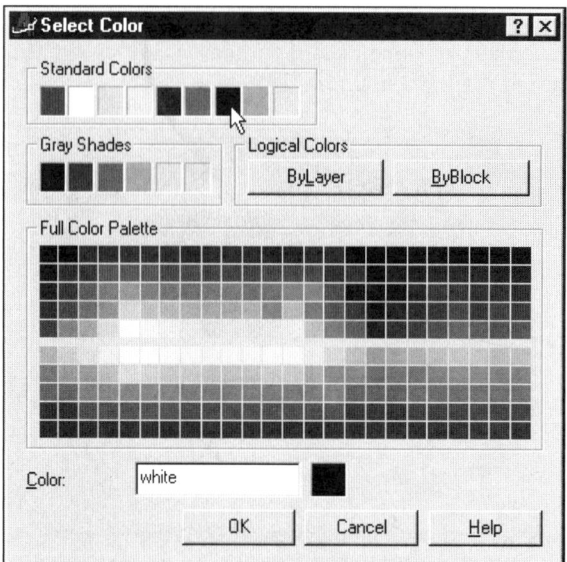

Figure 1–28

Select **white** by clicking on the **black** cell in the upper horizontal grouping of standard colors.(The black is named white because the default color for the AutoCAD Map 2000 screen is black and so white is used to represent black on such a screen. On a white screen background, the color black will show up black, but still be named white. In either case it prints black on color printers. If you are using a customized screen background color then you may wish to utilize your own choice of color). White will be identified with a numerical value of 7. Click on **OK**.

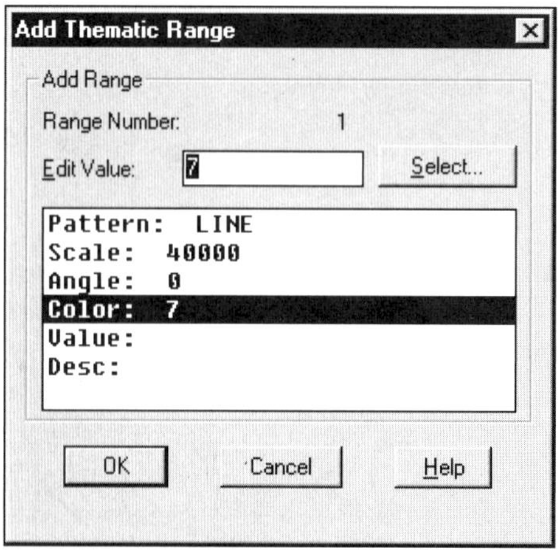

Add Thematic Range

Add Range

Range Number: 1

Edit Value: 7 Select...

Pattern: LINE
Scale: 40000
Angle: 0
Color: 7
Value:
Desc:

OK Cancel Help

Figure 1–29

The entire query process depends on having AutoCAD Map 2000 search for the data embedded in the attached source drawings. This search is based on the Value entered in the next step.

Step 10 In the Add Thematic Range dialog box, click on **Value** and in the Edit Value box type in **5** and then register your entry by clicking on **Value** again. Enter a description name by highlighting **Desc** then typing in **Florida** and clicking on **Desc** again.

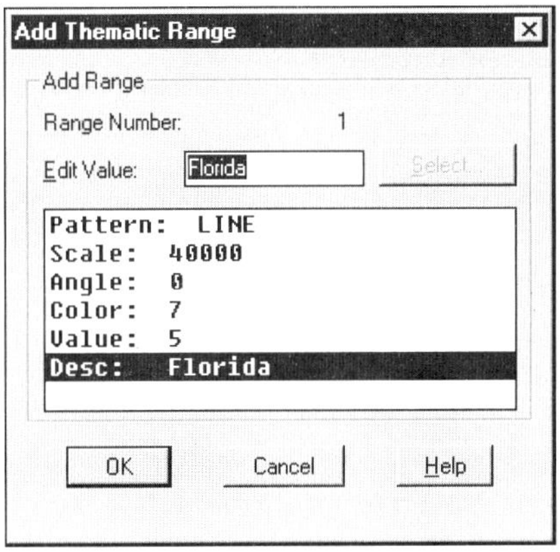

Add Thematic Range
Add Range
Range Number: 1
Edit Value: Florida
Select...
Pattern: LINE
Scale: 40000
Angle: 0
Color: 7
Value: 5
Desc: Florida
OK
Cancel
Help

Figure 1–30

Step 11 Click on **OK** and you will be returned to the Thematic Display Options dialog box, only now you will see that one item has been added.

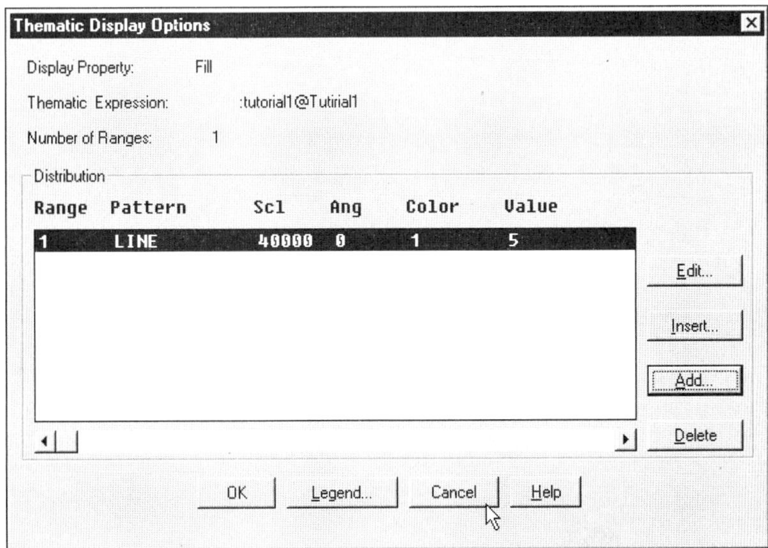

Figure 1–31

Step 12 This process must be repeated for the other three states. Use the following guide to assist you in finishing the establishment of the ranges.

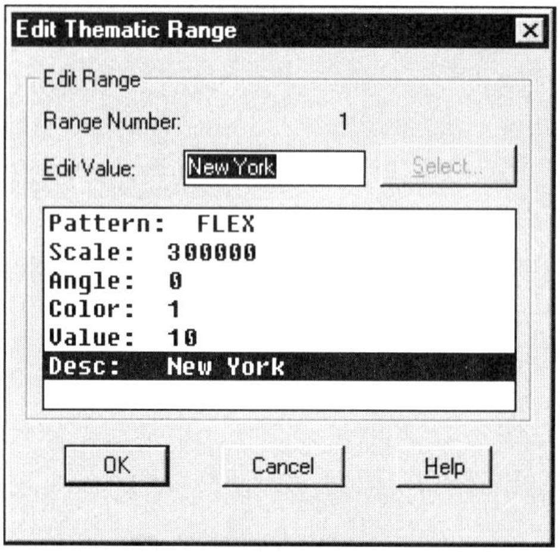

Figure 1–32

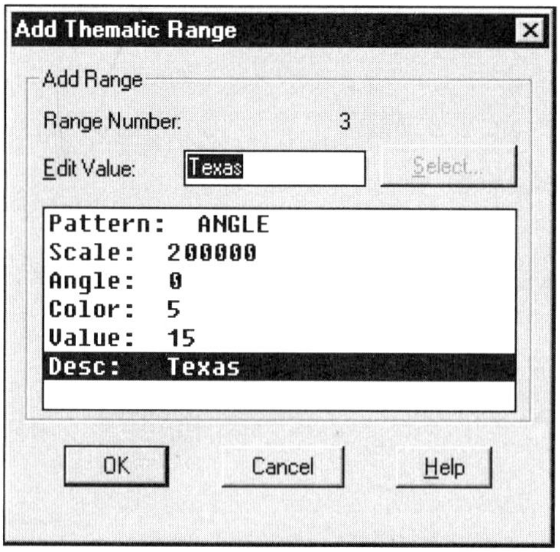

Figure 1–33

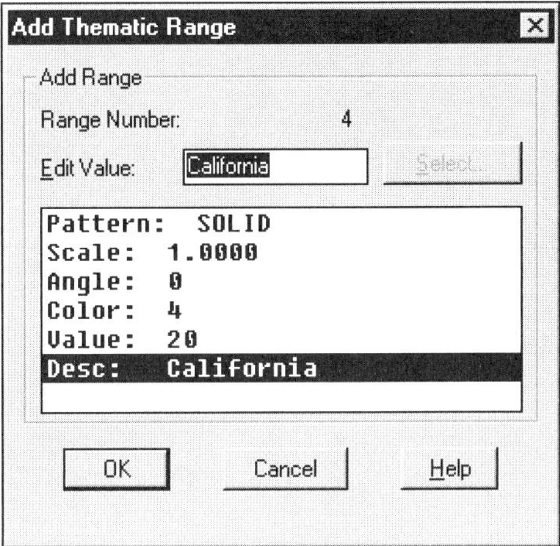

Figure 1–34

Step 13 When you have completed these range classifications, the Thematic Display options screen will appear.

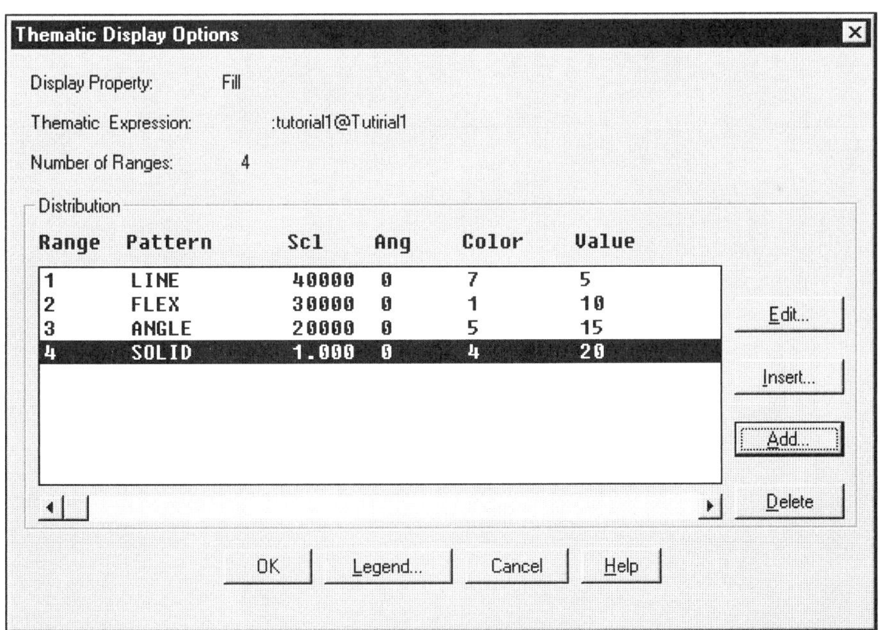

Figure 1–35

Notice that the order of range building is from smallest to largest by value. This is an important consideration if the software is to complete the query properly. Click on **OK.**

Step 14 To have AutoCAD Map 2000 place a legend on a drawing, click on the **Legend** button. The Thematic Legend Design dialog box will appear. You must confirm by clicking on the **Create Legend** box in the upper left corner. A checkmark will appear and the entire dialog box will become active.

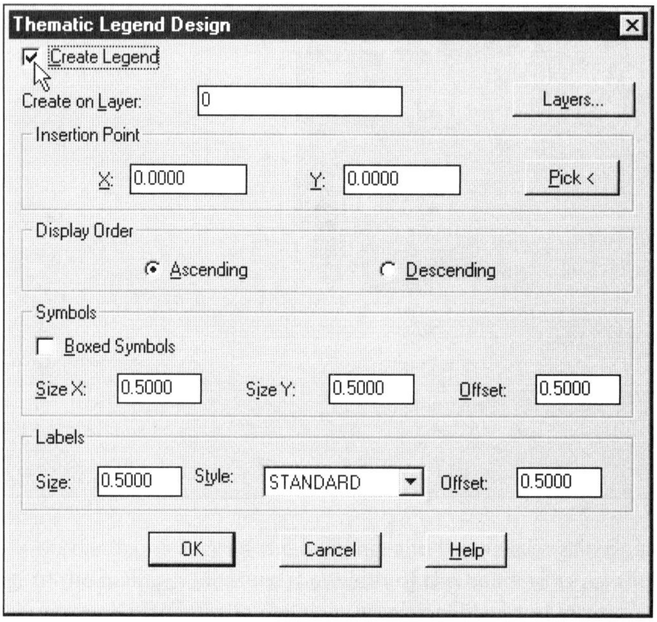

Figure 1–36

Step 15 Select a location on your drawing where you would like to have the legend placed by clicking on the **Pick** button. As you click on this button, the dialog box will disappear. Use the mouse to select a location for the legend. The place where you click will be the top left corner of your legend.

You can decide which order you want to have the legend displayed, but for this exercise leave the setting at ascending.

Step 16 Click on the **Boxed Symbols** checkbox to select it. This option builds a frame around the color patches used to shade in the drawing.

Step 17 In the text field located to the right of Size X, type in a value of **200000**. Enter this same value in the Size Y box by simply clicking the left mouse button in that box. At the same time label sizes will be adjusted automatically.

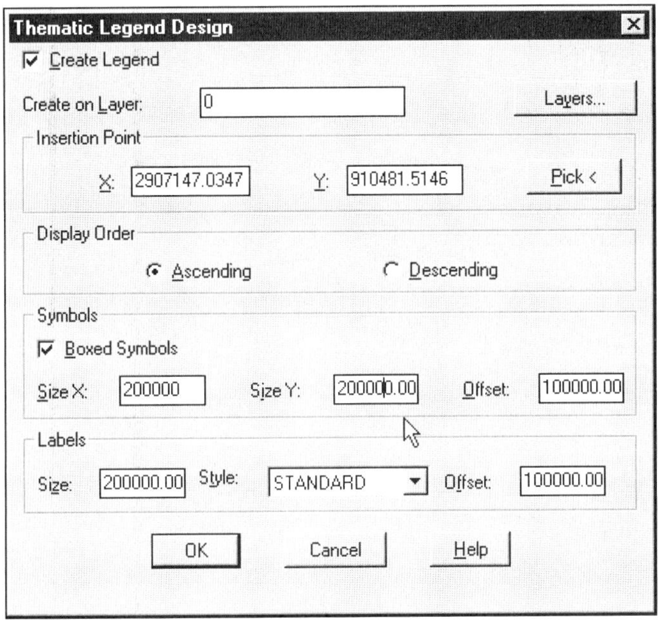

Figure 1–37

Click on **OK**. This takes you back to the Thematic Display Options screen. Click on **OK** again. This will return you to the Object Thematic Mapping screen. Leave the Range Division setting at **Discrete.** (A later section will deal with the difference between continuous and discrete.) Click on **Proceed**. In the command line area, AutoCAD Map 2000 will provide information pertaining to the progress of the query. In a few moments the data will be presented on screen according to the parameters you specified in this exercise. You may want to refresh the screen after the query is completed and the thematic map is generated on the screen.

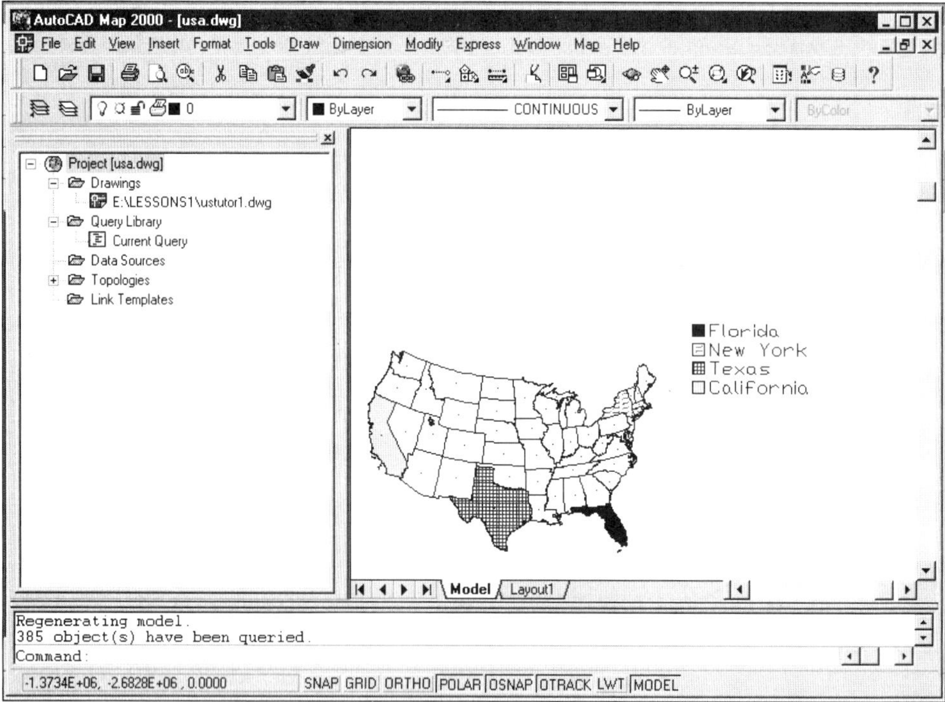

Figure 1–38

The resulting image can be printed or saved as desired.

Either continue working through the tutorials or close AutoCAD Map 2000 by going to File then down to Exit. This file does not need to be saved for any future tutorial.

PREPARING DRAWINGS FOR USE IN QUERIES

The tutorial below will show you how to prepare a drawing to be used in a query. This drawing will then be saved and used in a later exercise in this chapter.

For the purpose of this exercise the base map of the United States called USA.DWG will be used. The premise of this exercise is to identify areas in the United States that may be associated with two different levels of sulfur dioxide emissions representing the potential production of a certain level of air pollution. The base map will be used to build two more drawing files each with attached object data which will be in turn, used in a later tutorial.

TUTORIAL: PREPARING DRAWING FOR USE IN QUERIES

Step 1 Open the drawing file identified as **USA.DWG** included in the lessons directory. You will use this drawing to build a new map. The information used to compose the map is based on general knowledge and information and is for demonstration purposes only.

Step 2 Use the image provided below (Figure 1–39) as a guide to draw a closed polygon on your map that approximates the one shown. Use the typed **pline** command or go to the **Draw** menu and choose the **Polyline** option to activate the polyline drawing function, followed by the **Close** command to draw a closed polygon similar to that shown (Figure 1–39). This polygon will be tagged with object data that will be used to identify this area as a region of sulfur dioxide emissions.

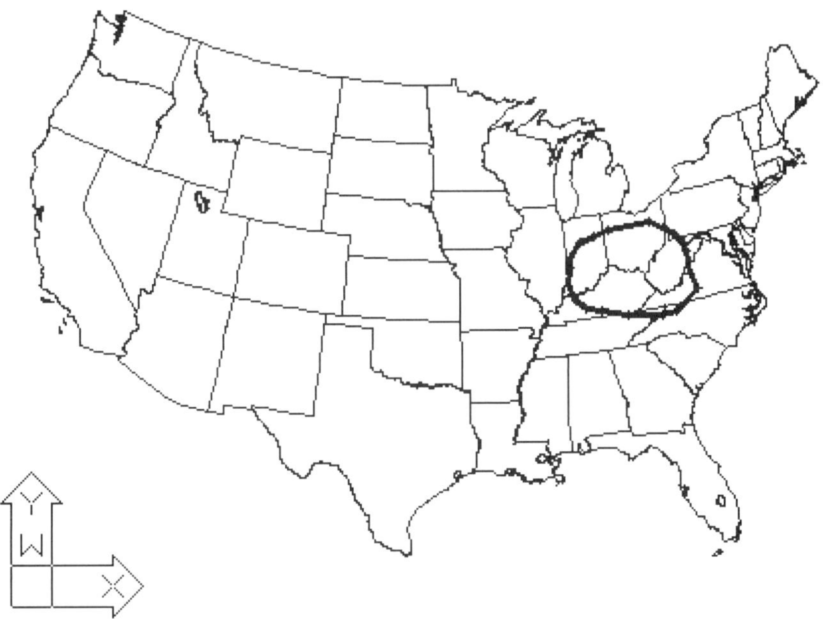

Figure 1–39 *Area of high sulfur dioxide emissions*

Step 3 Use **File** and **Save As** to save this changed version of the map as **USHIGAS.DWG**.

Step 4 Go to the **Map** menu and select **Object Data** then **Define Object Data**.

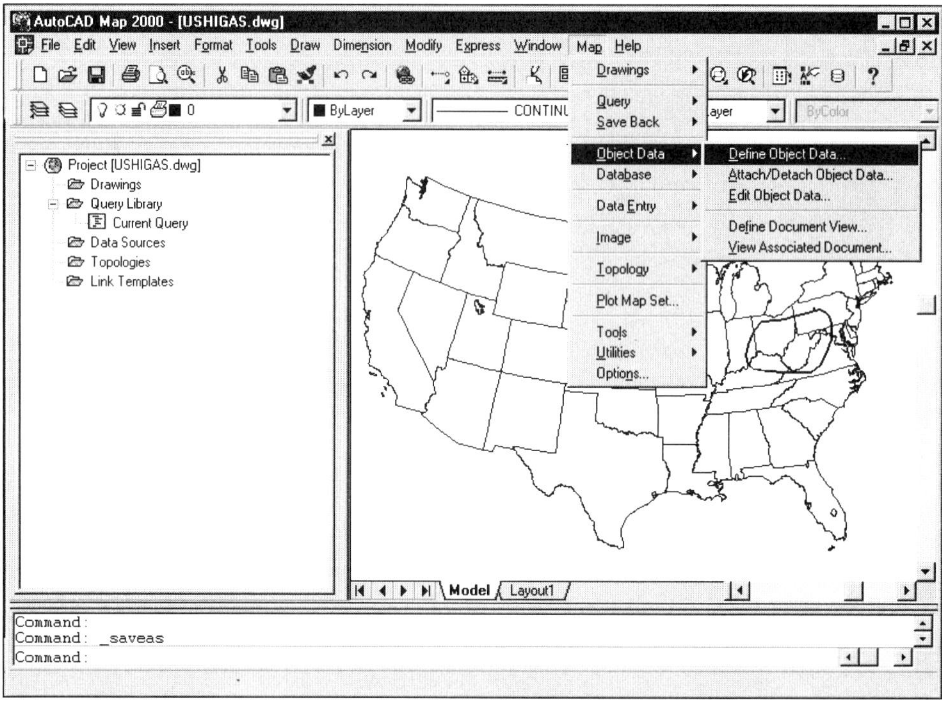

Figure 1–40

The first step in this process is to define a table within which to place the data. This might be likened to an internal spreadsheet for AutoCAD Map 2000. In this process you name the table, specify the type of data you will place in it (character, integer, and so on), and identify the fields or columns of data you want to create.

Step 5 At the Define New Object Data Table dialog box, click in the **New Table** edit box. For the New Table name type in **higas**.

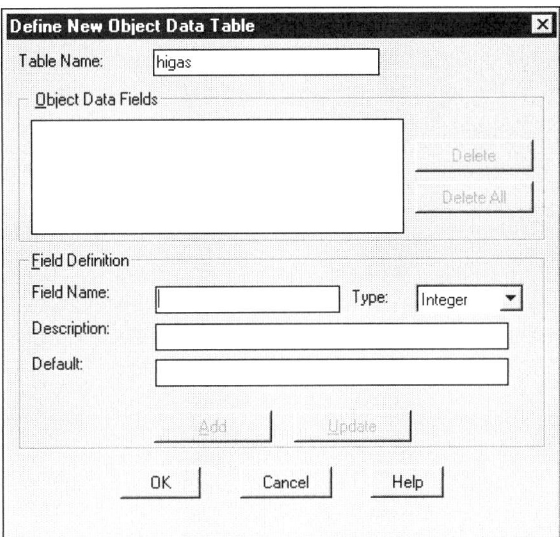

Figure 1–41

Step 6 In the Field Name edit box type in **high** and ensure that the data type is set at **integer**. (The type determines and reflects the nature of the data that you are attaching to the drawing and could be, for example, character- or real number–based.)

In the Description box type in **2 tons per day** then click on **Add.**

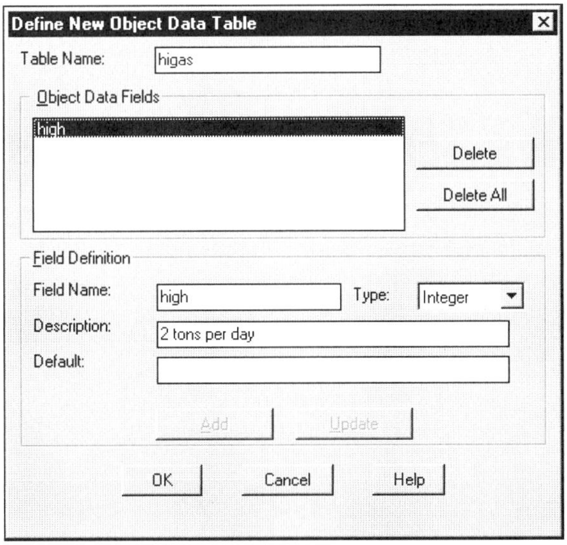

Define New Object Data Table
Table Name: higas
Object Data Fields
high
Delete
Delete All
Field Definition
Field Name: high
Type: Integer
Description: 2 tons per day
Default:
Add
Update
OK
Cancel
Help

Figure 1–42

The **high** data field will be listed under object data fields. The description is used to inform the user about the data table when the data is being accessed in other stages of the query. It is especially useful when one has not worked with a particular data set for a period of time and one can rely on the descriptions for a reminder about the data.

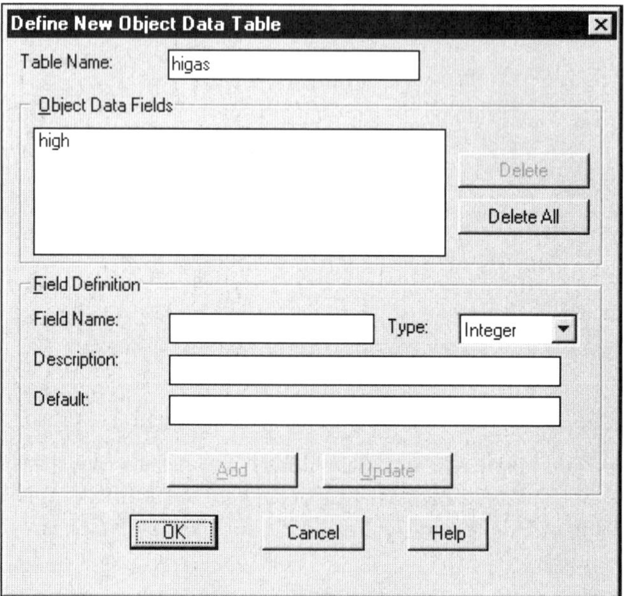

Figure 1–43

Click on **OK.** You will be returned to the Define Object Data screen. Click on **Close.**

Step 7 Go to the **Map** menu and select **Object Data** then **Attach/Detach Object Data**.

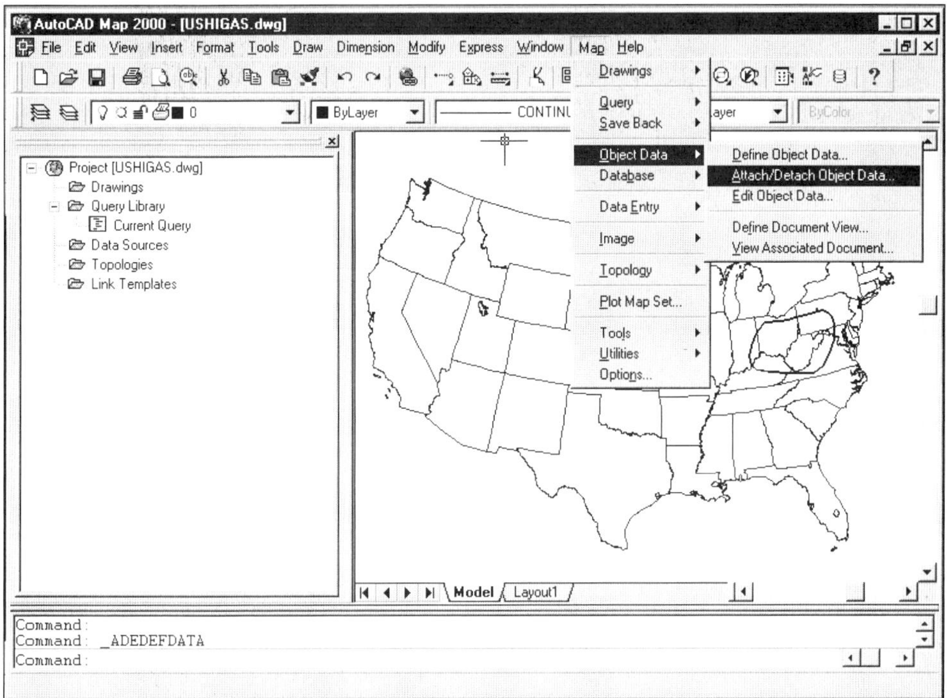

Figure 1–44

In the Attach/Detach Object Data click on **2 tons per day** and the **value** box will become active. Type in a value of **2** and click on **Attach to Objects**. Place the pick or locator box, controlled by the mouse on the line that you drew to indicate the area of high gas levels and click once.

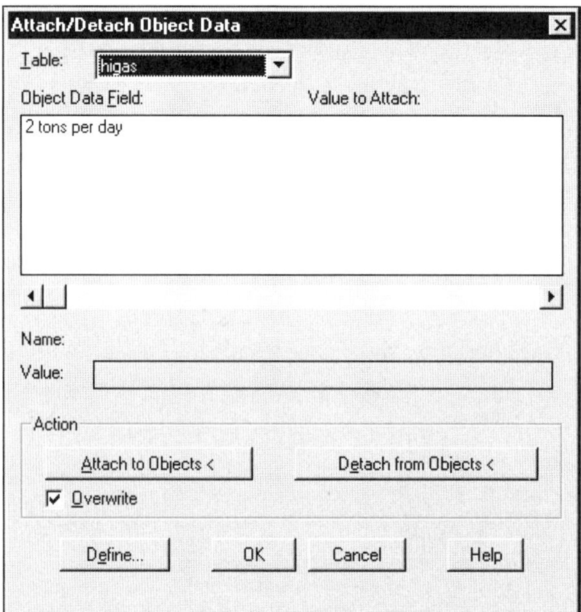

Figure 1–45

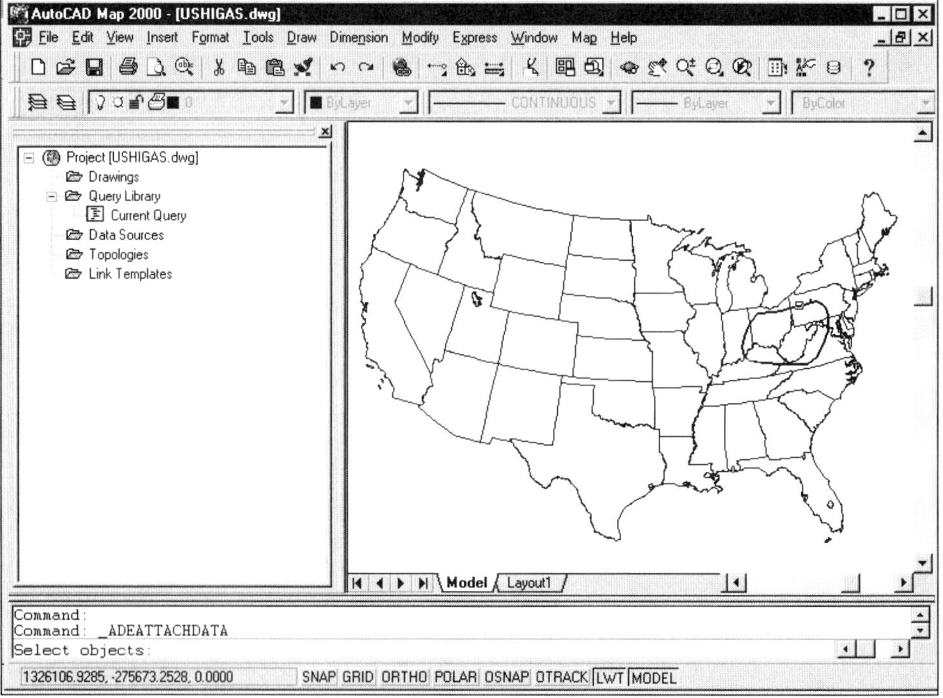

Figure 1–46

A message will be presented in the command line area of the screen informing you that one object was found. Click the right mouse button to end the object selection (or press ENTER) and a message will indicate that data was attached to one object and the selection process is completed.

Go to **File** then down to **Save.** This saves the attached data to the drawing.

Take a moment to observe carefully where you have drawn the region boundary. In the next stages of this exercise you will be asked to draw another region inside this one.

Step 8 Reopen the drawing file identified as **USA.DWG** included in the lessons directory. You will use this drawing to build another new map.

Step 9 Use the **pline** command or **Draw** then **Polyline** again followed by the **Close** command to draw a closed polygon around a region as shown below (Figure 1–47).

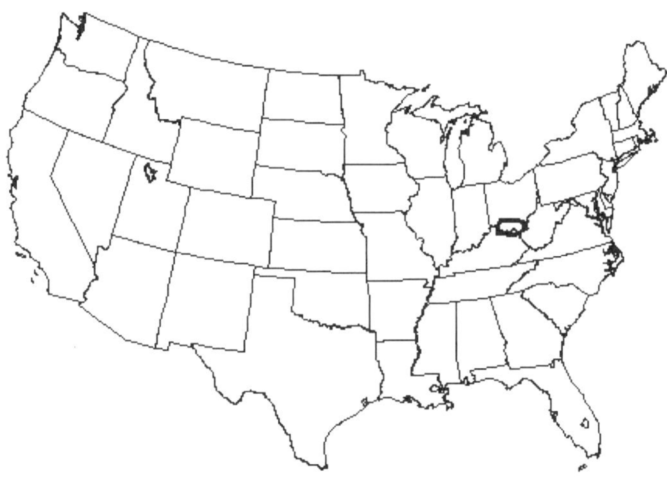

Figure 1–47

This polygon will be used to have additional object data attached to it.

Step 10 Save this map as **HOTSPOT.DWG.**

Step 11 Go to the **Map** menu and choose **Object Data** then select **Define Object Data.** Click on **New Table** and type in the name **HOTSPOT**. For the Field Name, type in **HIGHSO2** and for the **Description** type in **5 tons per day**. Make sure the data type is set at **integer.**

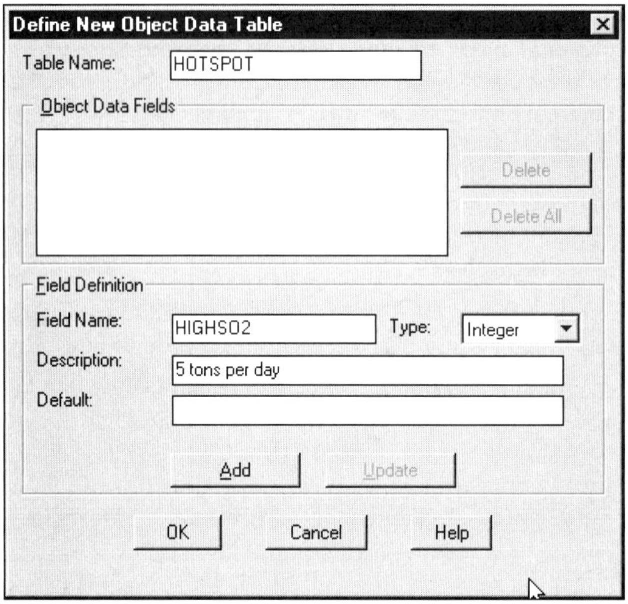

Figure 1–48

Click on **Add** and HIGHSO2 will be added to the box containing the list of object data fields. Click on **OK** and you will be returned to the Define Object Data screen and then click on **Close**. Data can be attached to objects in the drawing by identifying the table and field for the data, then by keying the data into the appropriate text box and picking the object to which the data is to be attached.

Step 12 Go to the **Map** menu and select **Object Data** then **Attach/Detach Object Data** and the Attach/Detach Object Data dialog box will appear. Click on the 5 tons per day entry; it will become highlighted and the Value edit box will become active. In the Value edit box type in **5** and click on the **Attach to Objects** button. Place the mouse-controlled pick box directly over the perimeter of the region you drew and click the left mouse button.

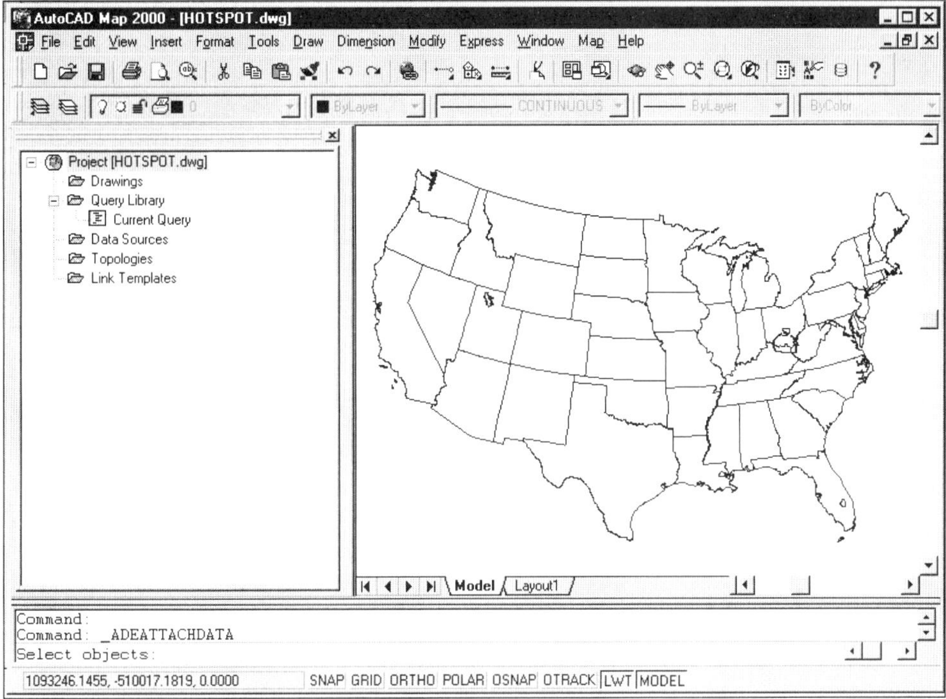

Figure 1–49

The region you clicked on will become highlighted. Click the right mouse button to end the selection process and a message will be presented in the command line area indicating that data was attached to one object.

Step 13 Save the map with the attached data embedded in it.

This tutorial has now led to the creation of two maps with data attached that will be accessed in a query in the next section.

MULTIPLE DRAWING QUERIES

The tutorial below will teach you how to access data that has been embedded in drawings that are not shown on the screen. In this example, a base drawing will be opened and data from two other drawing files will be accessed and presented on the original map.

In an earlier section of this text the assignation of a coordinate system to a work session was discussed. In this case, the two data maps were created using the same map outline and are of the same projection.

The first step, after opening a base drawing, is to attach the source drawings that contain the data to the current work session. In this example two queries will be executed, one for each of the source drawings.

TUTORIAL: MULTIPLE DRAWING QUERIES

Step 1 Open the **USA.DWG** file. With the outline of the United States on screen, go to the **Map** menu and select **Drawings** then **Define/Modify Drawing Set**.

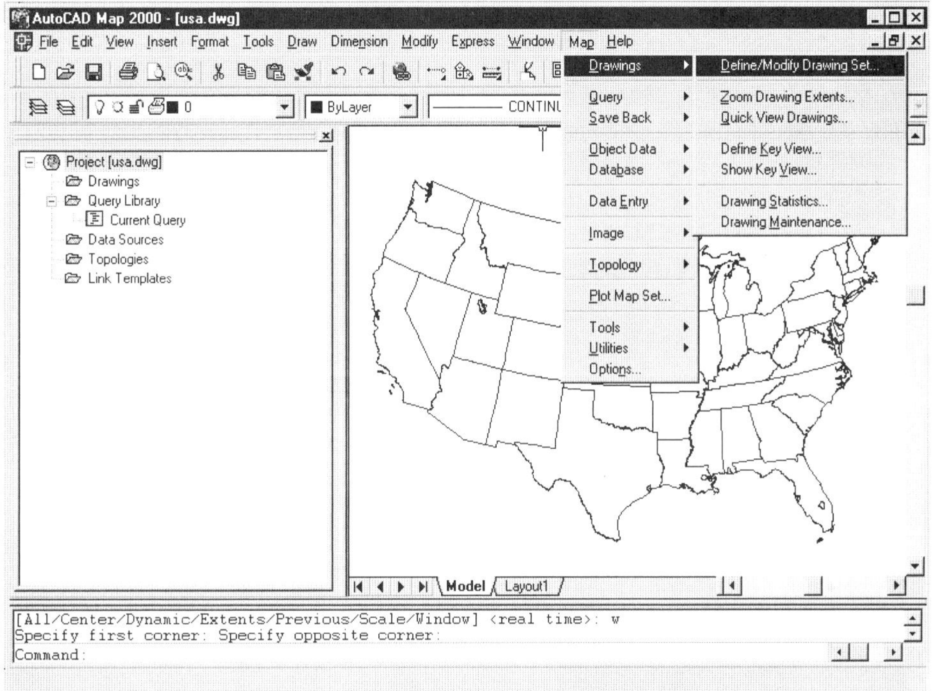

Figure 1–50

In the Define/Modify Drawing Set dialog box click on **Attach**. The Select Drawings to Attach dialog box will appear.

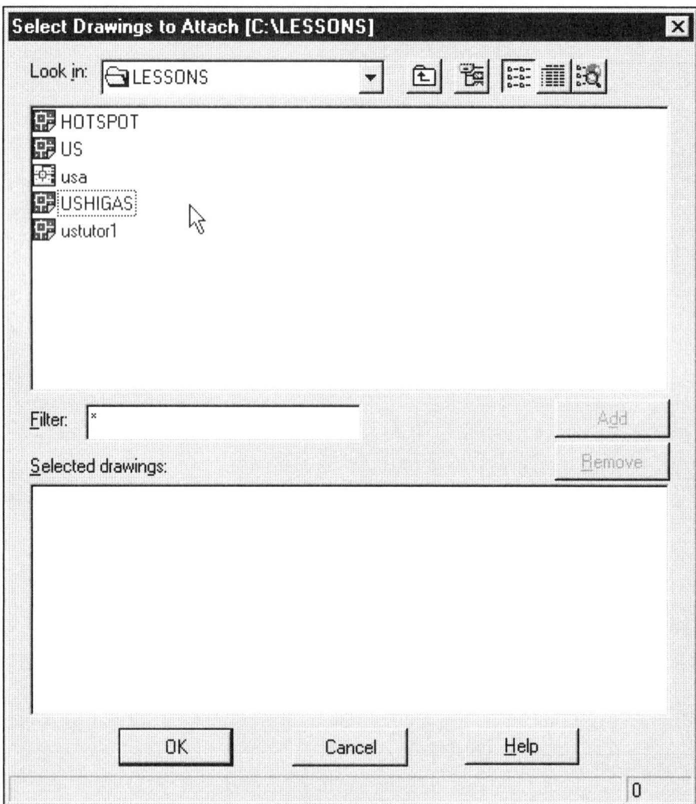

Figure 1–51

Step 2 Scroll through the directory list to the location of the drawing files and select the **USHIGAS** file, then click on **Add.**

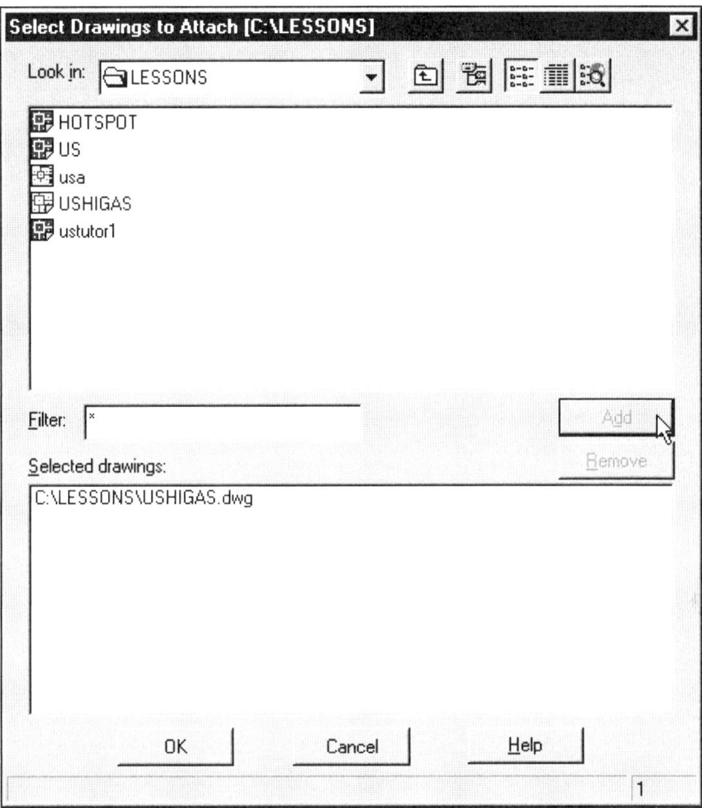

Figure 1–52

The file will be added to the lower portion of the dialog box. Click on **OK**. The Define/Modify Drawing Set dialog box will appear and will now reflect that the USHIGAS drawing was attached and is now active.

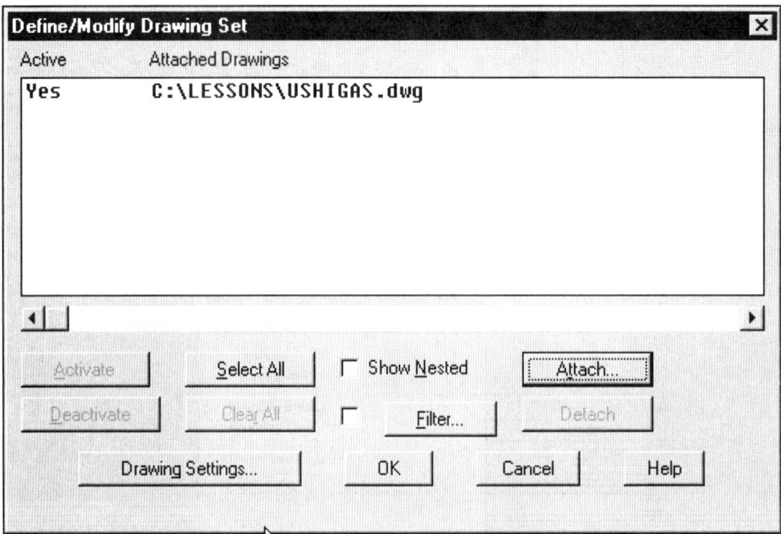

Figure 1–53

Click on **OK.**

The next step is to build a query. In this case you are searching for areas of sulfur dioxide concentrations.

Step 3 Go to the **Map** menu and select **Query** then **Object Thematic Query.**

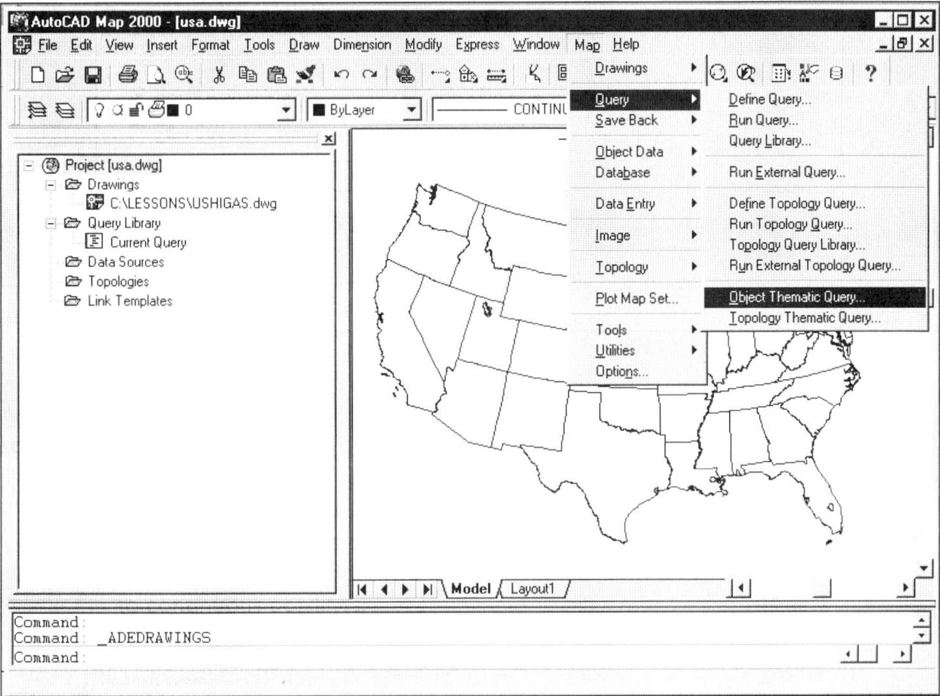

Figure 1–54

The Object Thematic Mapping dialog box will appear.

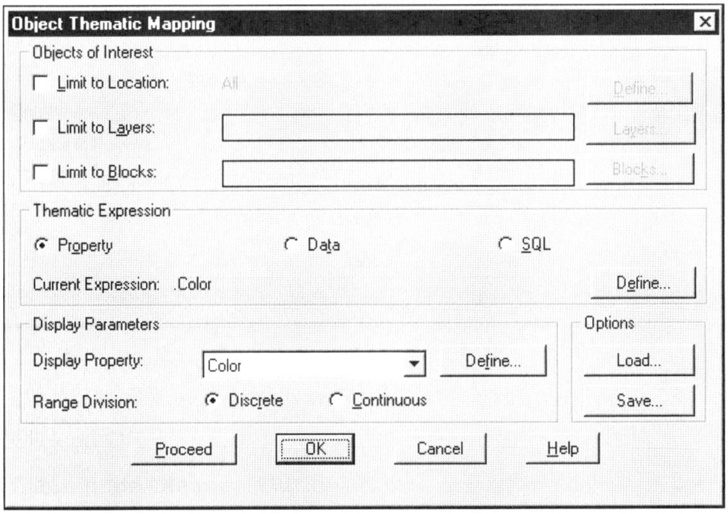

Object Thematic Mapping
Objects of Interest
Limit to Location: All Define...
Limit to Layers: Layers...
Limit to Blocks: Blocks...
Thematic Expression
Property Data SQL
Current Expression: .Color Define...
Display Parameters
Display Property: Color Define...
Range Division: Discrete Continuous
Options
Load...
Save...
Proceed OK Cancel Help

Figure 1–55

Step 4 In the upper section of this dialog box under Objects of Interest, select the **Limit to Location** checkbox and then click the **Define** button. Several options will be presented for determining a region for querying. For the purpose of this tutorial select **Circle** and for the selection type click on **Inside.** This will focus the query on the area and objects inside a circle that you will establish. Click on **Define** again.

Place the mouse pointer near the center of the region you had earlier identified in the USHIGAS as a region with high levels of emissions, and click the left mouse button. Draw a circle that encompasses the region.

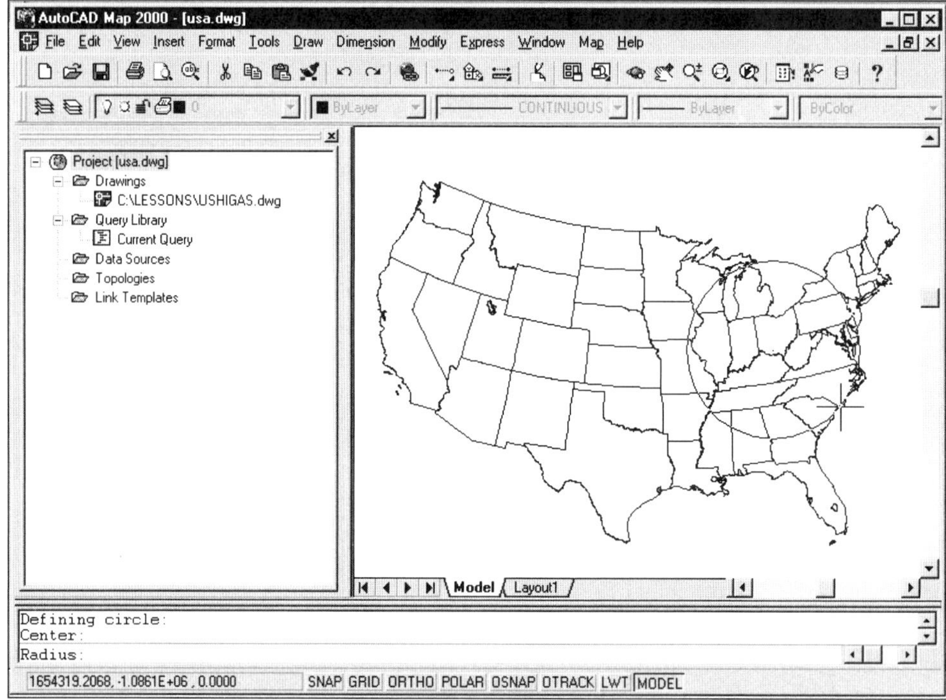

Figure 1–56

After the circle has been defined, click the left mouse button again and you will be returned to the Location Condition dialog box, which provides a numerical description of the circular region being queried. Click on **OK**.

Step 5 You are returned to the Object Thematic Mapping dialog box.

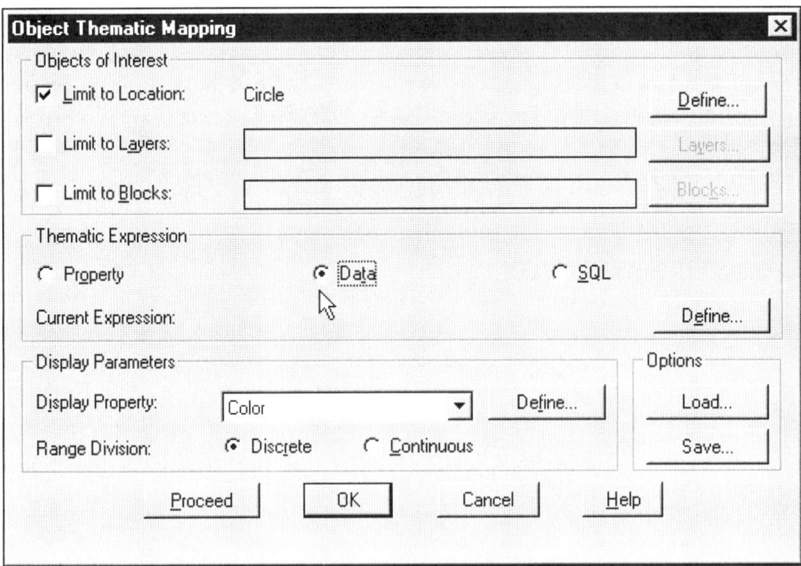

Figure 1–57

In the central portion of the screen identified at the Thematic Expression section, click on the **Data** option, then on **Define**.

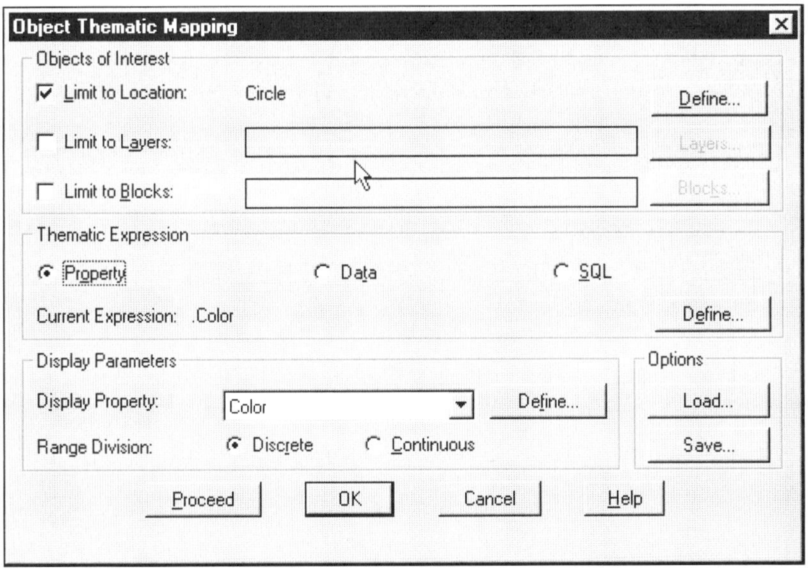

Figure 1–58

Step 6 In the Data Expression dialog box, make sure that the **Object Data** radio button is selected. The search is set to Object Data because that is the nature of the data in the source drawings. In the Tables edit box, select **HIGHGAS** as the table and **HIGH** as the Object Data Field.

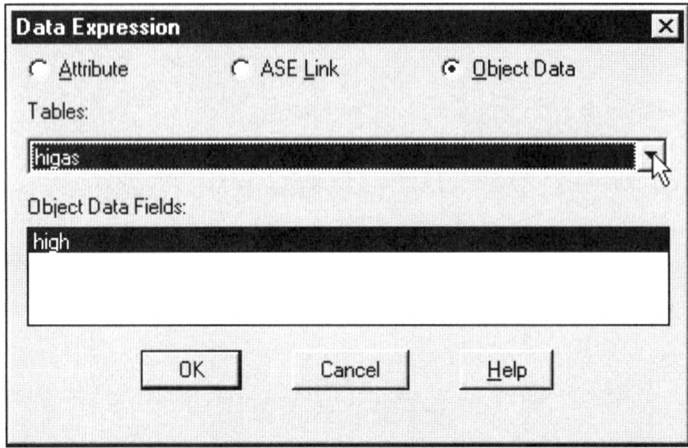

Figure 1–59

Click on **OK**.

Step 7 You will now determine how to have the results of the query displayed. In the Display Parameters section of the screen, click on the **Display Property** arrow and select **Fill** from the list then click on **Define.**

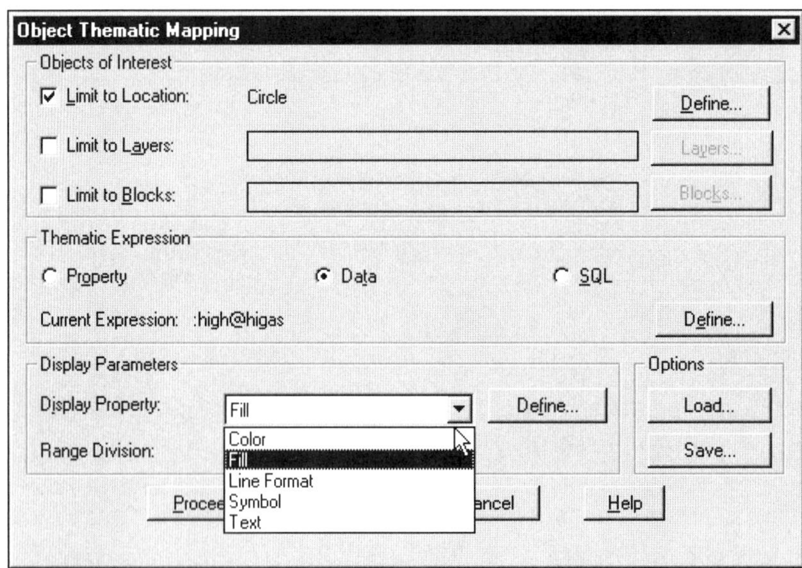

Figure 1–60

Step 8 At the Thematic Display Options dialog box click on **Add**. The Add Thematic Range dialog box will be presented for you to customize the presentation of the query.

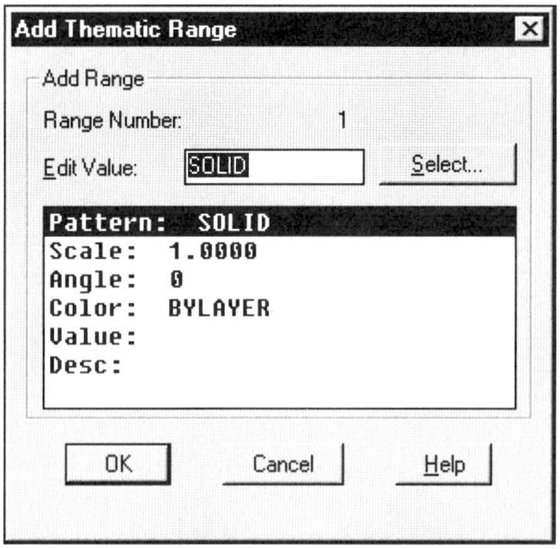

Add Thematic Range
Add Range
Range Number: 1
Edit Value: SOLID Select...
Pattern: SOLID
Scale: 1.0000
Angle: 0
Color: BYLAYER
Value:
Desc:
OK Cancel Help

Figure 1–61

With the **Pattern** item highlighted, click on **Select**. A list of hatching patterns will be presented. Scroll down the list until you see the **Line** option.

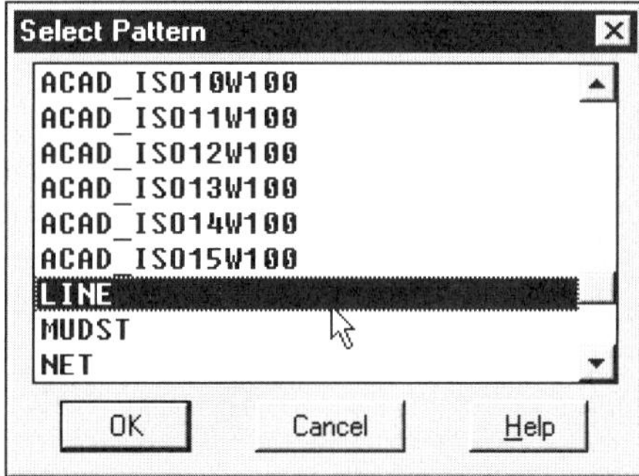

Figure 1–62

Click on **OK** and then click on **Pattern** to register the selection of **Line**.

Step 9 Click on the **Scale** option and change the spacing to **150,000** by highlighting the Edit Value box and replacing 1.000 with 150000. Click on the **Scale** item again to register the change. (As an alternative to selecting Scale a second time to register the selection, you can simply press ENTER.)

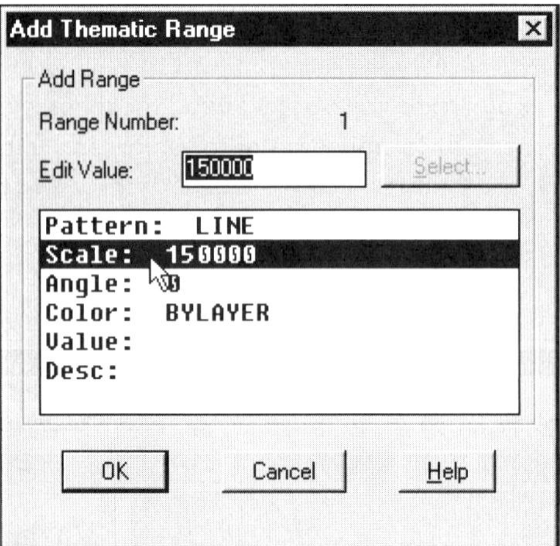

Figure 1–63

The angle for the lines can be left at **0**, which is horizontal.

Step 10 Click on **Color** and then on **Select**. From the standard color strip, select, in this case, **Red** or color **1.**

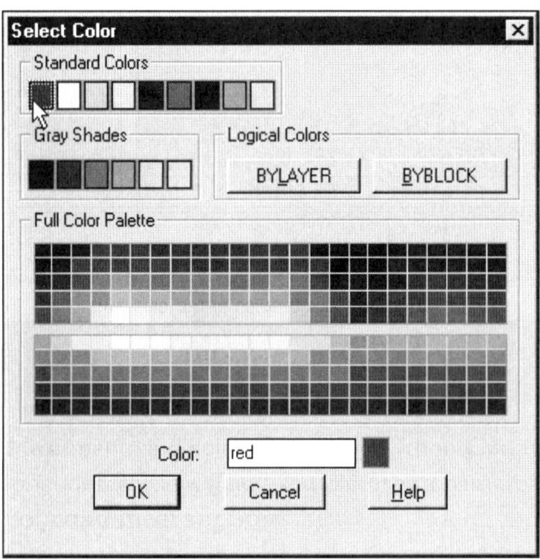

Figure 1–64

This selection will be indicated in the color box at the bottom of the screen. Click on **OK** and then at the Add Thematic Range screen select the **Color** item again to register the color selection.

Edit Thematic Range
Edit Range
Range Number: 1
Edit Value: 1
Select...
Pattern: LINE
Scale: 150000
Angle: 0
Color: 1
Value: 0
Desc: 0
OK
Cancel
Help

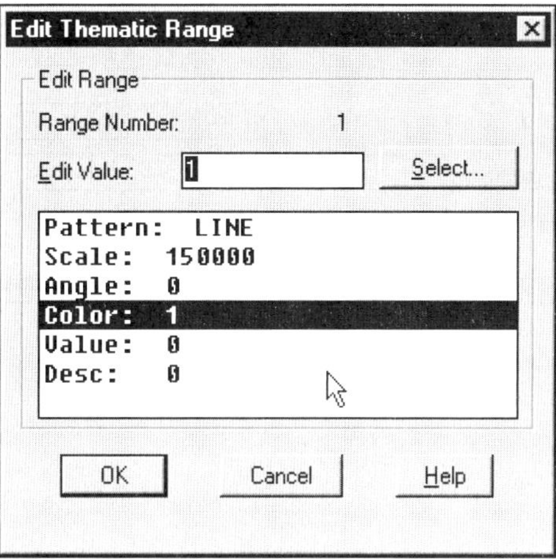

Figure 1–65

Step 11 The Value box is where you specify the data value to be queried and presented. In the **Edit Value** box type in **2**, representing 2 tons per day and register the entry by clicking on **Value** again.

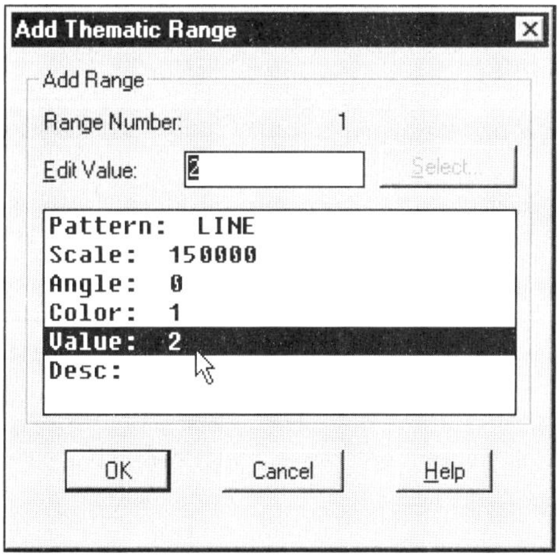

Add Thematic Range
Add Range
Range Number: 1
Edit Value: 2
Select...
Pattern: LINE
Scale: 150000
Angle: 0
Color: 1
Value: 2
Desc:
OK
Cancel
Help

Figure 1–66

Type in a **Description** of **2 tons per day** and register this entry by picking the **Desc** item again. Click on **OK**.

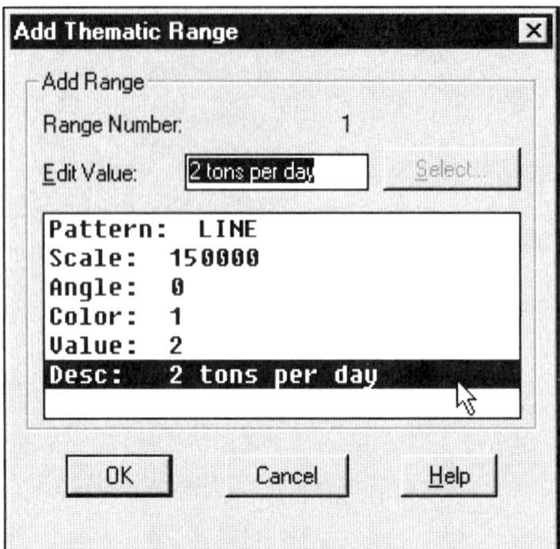

Figure 1–67

Step 12 At the Thematic Display Options dialog box, click on **Legend**. The Thematic Legend Design dialog box will appear. Select the **Create Legend** option by clicking in the small checkbox. The rest of the screen will then become active.

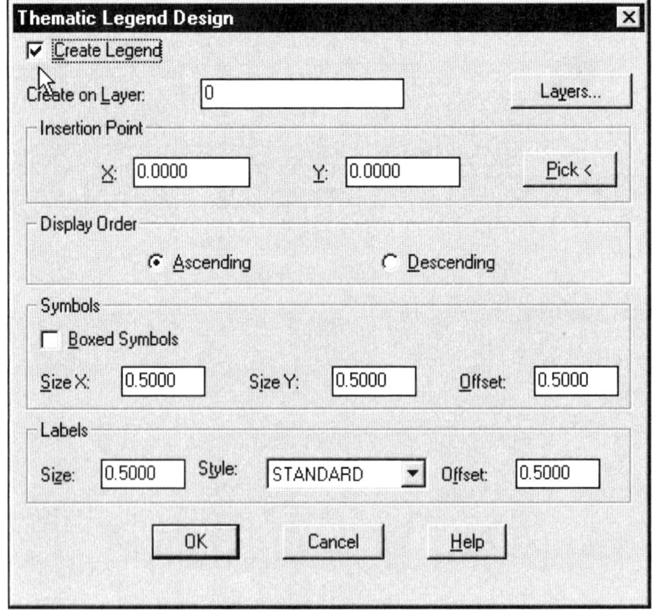

Figure 1–68

Step 13 To place the legend on the drawing at a particular point, you can either specify the *X* and *Y* coordinates or pick an insertion point on the drawing. To use the latter option click on the **Pick** button. Place the pointer just off the East Coast and click the left button. In this case, the order of the legend does not apply because only one item is being displayed.

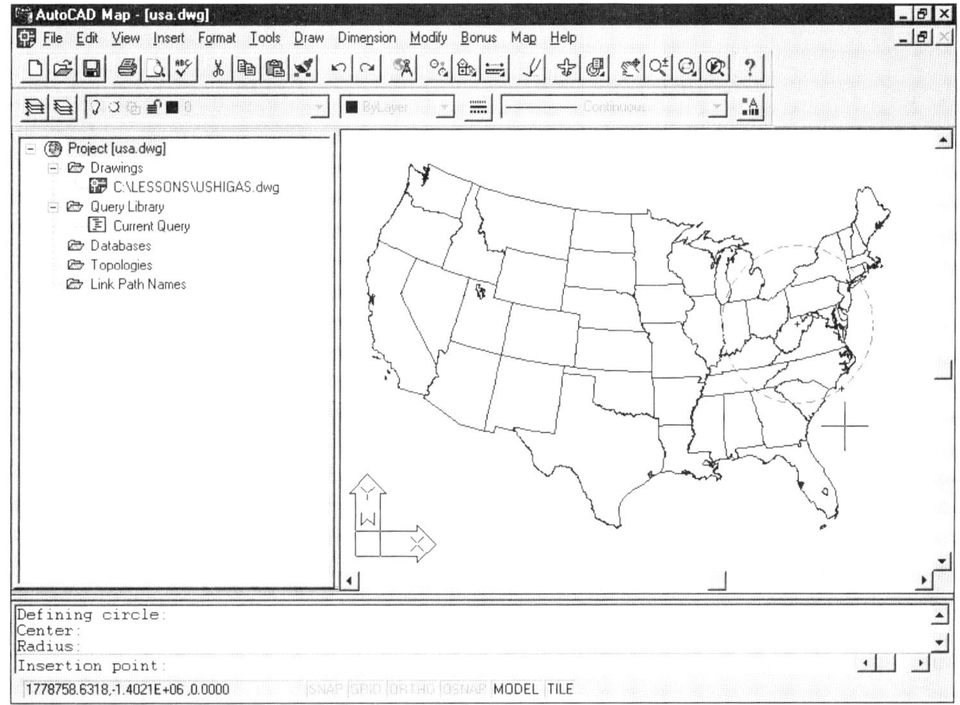

Figure 1–69

Step 14 To have boxed symbols in the legend, select that option by clicking in the checkbox labeled **Boxed Symbols**. The size of the legend items is directly related to the scale of your drawing. Enter a value in the **Size X** box of **200,000.**

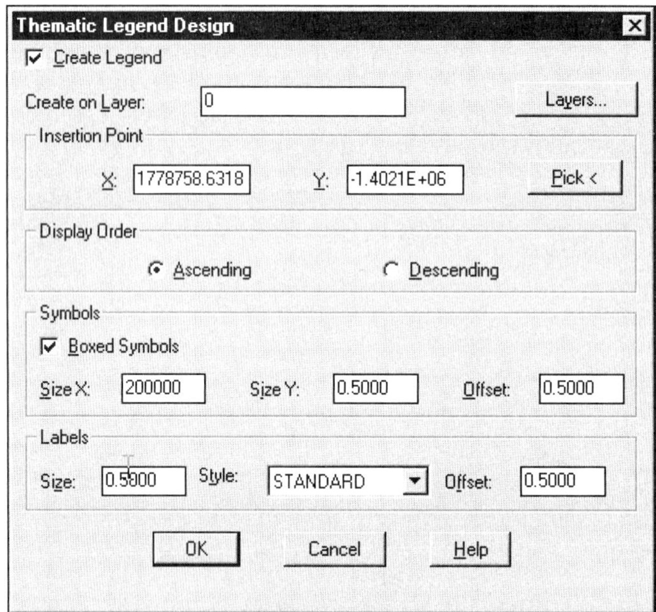

Figure 1–70

Click in the **Size Y** box, and the same value will automatically be placed there and the label height will be set as well. Click on **OK** and then on **OK** again at the Thematic Display Option screen.

Step 15 In the Object Thematic Mapping dialog box click on **Proceed.** The query will be executed and the results drawn on the screen. (If during the process a message appears along the bottom of the screen indicating that no index has been found, ignore it. This implies that AutoCAD Map 2000 is searching for a stored index of a query to run and did not find one.)

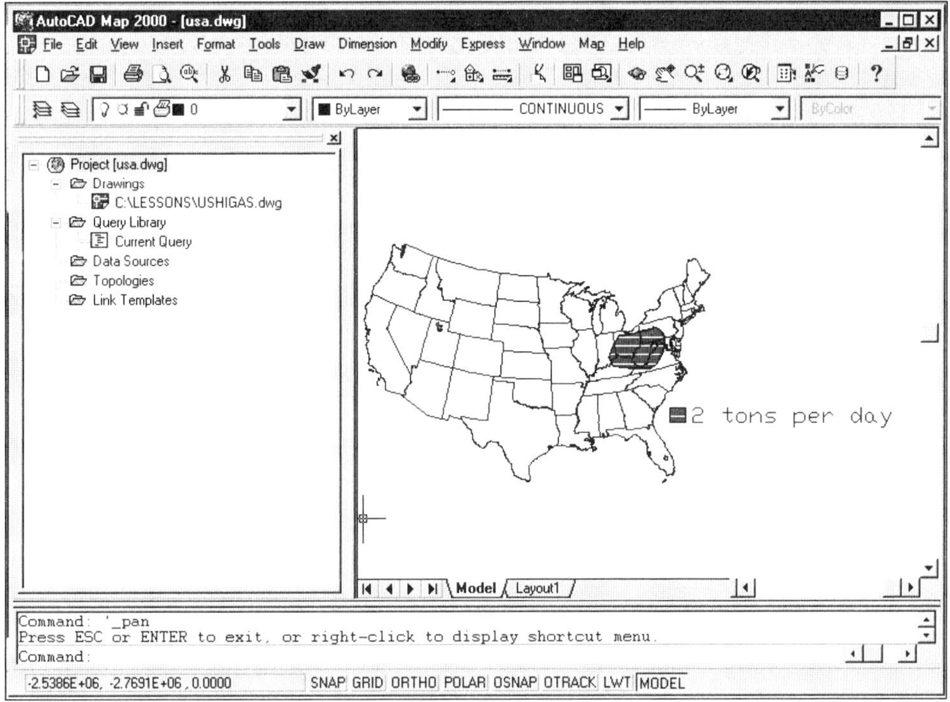

Figure 1–71

Step 16 You will now attach a second source drawing and execute another query to reveal the area on the map with higher emissions.

Go to the **Map** menu and select **Drawings** and **Define Modify Drawing Set**. The dialog box will indicate that the HIGHGAS drawing is still attached.

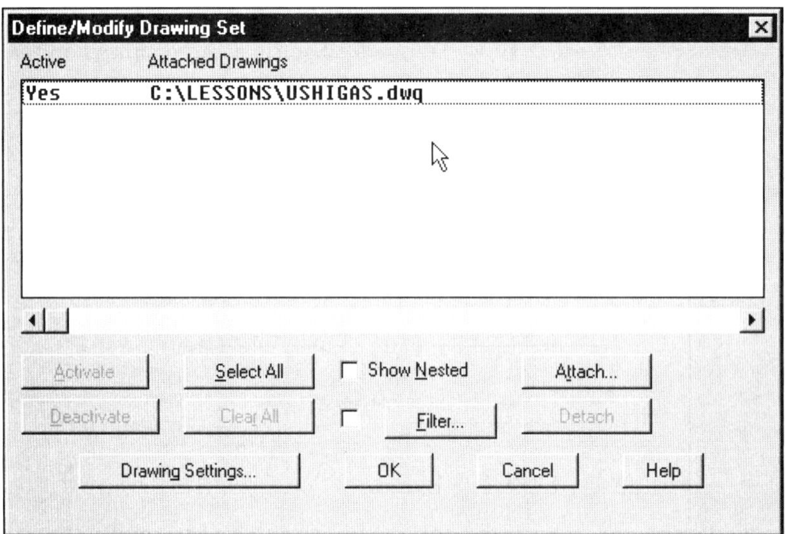

Figure 1–72

Step 17 Click on the **Attach** button. Select the second map you created in the previous tutorial, **HOTSPOT**, as the file then click on **Add**. The file will be listed in the lower portion of the screen. Highlight the HOTSPOT file and click on **OK.** This drawing will be accessed and a dialogue box will appear informing you of the status of your work session files. In this case you will have a base drawing, the USA map open, and have two drawing files attached.

Select Drawings to Attach [C:\LESSONS]
Look in:
LESSONS
HOTSPOT
USHIGAS
Filter:
*
Add
Remove
Selected drawings:
C:\LESSONS\HOTSPOT.dwg
OK
Cancel
Help
1

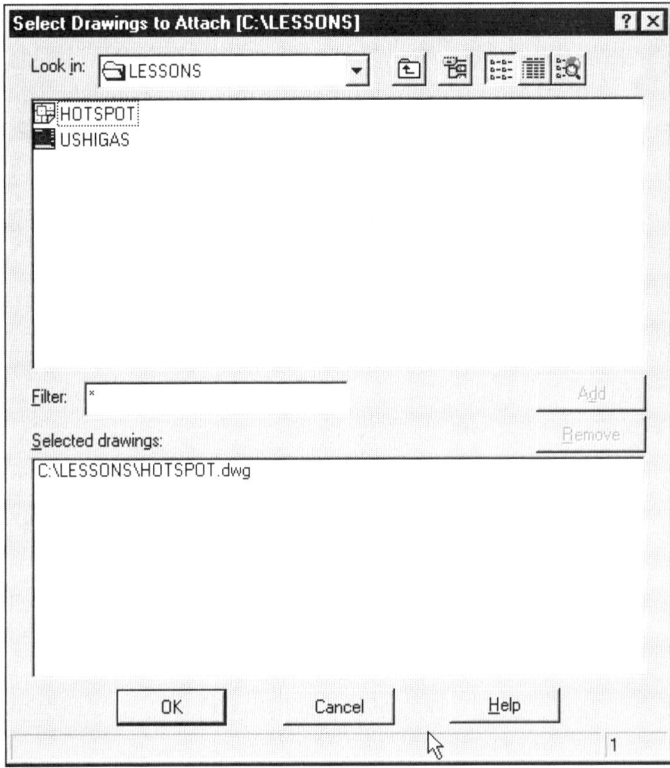

Figure 1–73

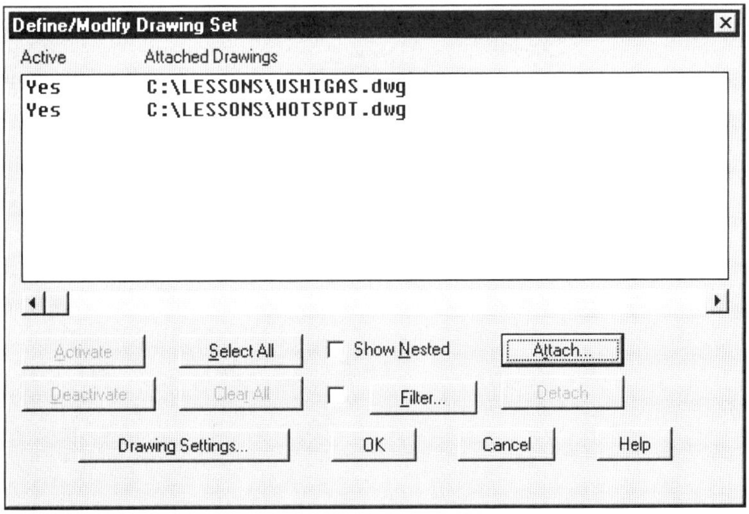

Define/Modify Drawing Set
Active
Attached Drawings
Yes C:\LESSONS\USHIGAS.dwg
Yes C:\LESSONS\HOTSPOT.dwg
Activate
Select All
Show Nested
Attach...
Deactivate
Clear All
Filter...
Detach
Drawing Settings...
OK
Cancel
Help

Figure 1–74

Click on **OK** and the work session drawings will be identified along the left side of the AutoCAD Map 2000 screen. This section of the screen is called the project workspace.

Step 18 Go to **Map** menu and select **Query** and **Object Thematic Query**.

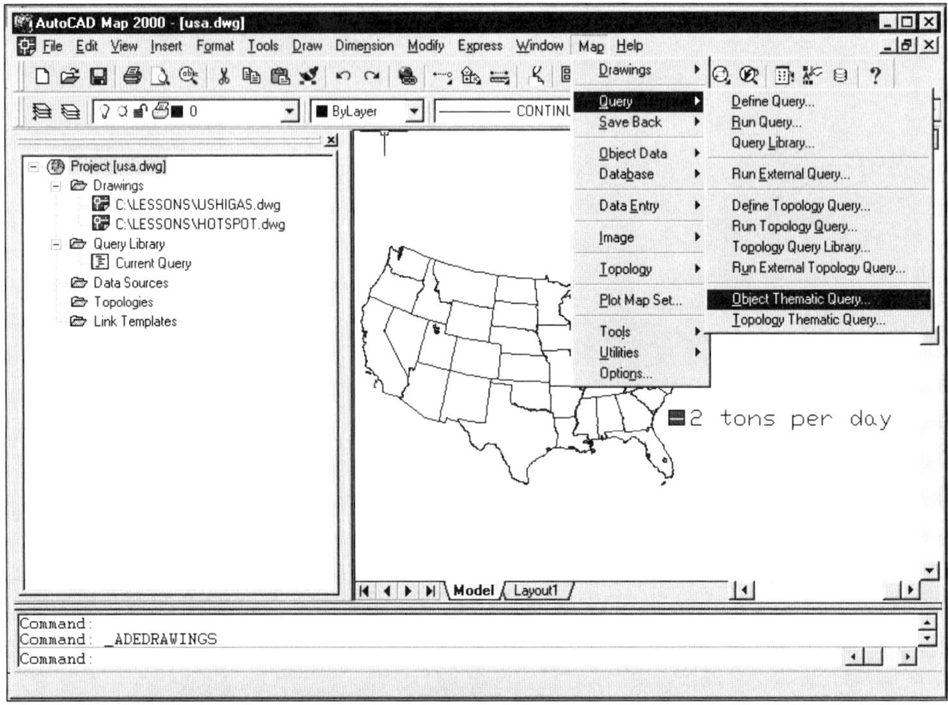

Figure 1–75

Step 19 At the Location Condition dialog box, the previous definition will have been retained, that being a region inside a circle defined by the location drawn previously. If necessary, a new circle could be produced by using Step 4 from above. Click on **OK**.

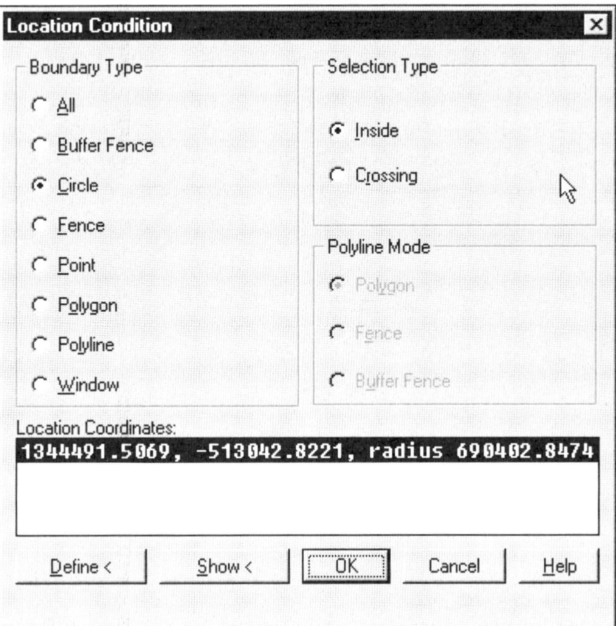

Figure 1–76

Step 20 For a Thematic Expression make sure that the **Data** option is selected and click on **Define.**

Step 21 At the Data Expression dialog box, click on **Object Data**. For the table, select the **HOTSPOT** table, and in the Object Data Field choose **HIGHSO2.**

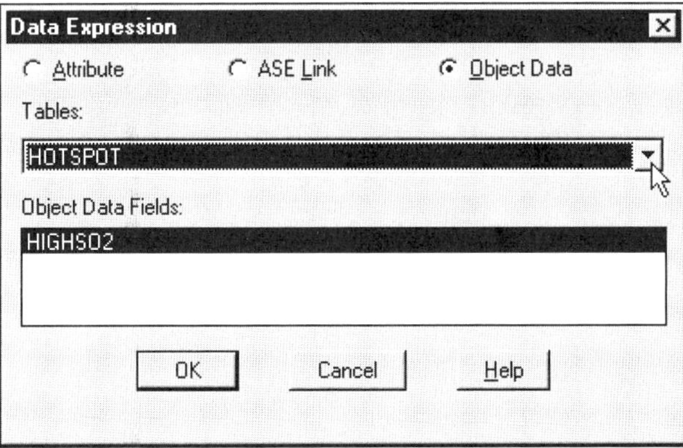

Figure 1–77

Click on **OK**.

Step 22 At the Object Thematic Mapping dialog box, set the Display Property to **Fill** then click on **Define**.

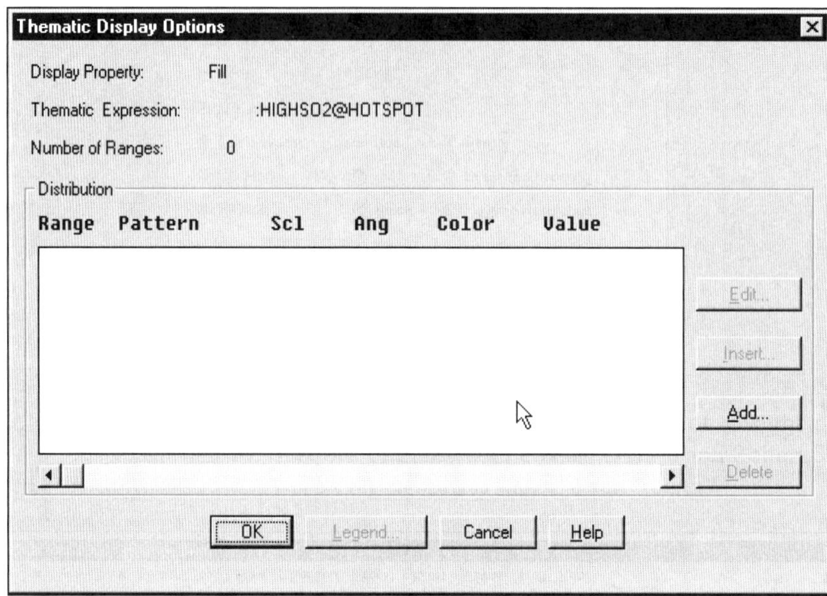

Figure 1–78

At the Thematic Display Options dialog box, click on **Add**. The Add Thematic Range dialog box will pop up. Select a Pattern of **Line**, and set a scale for the line spacing of **125000.**

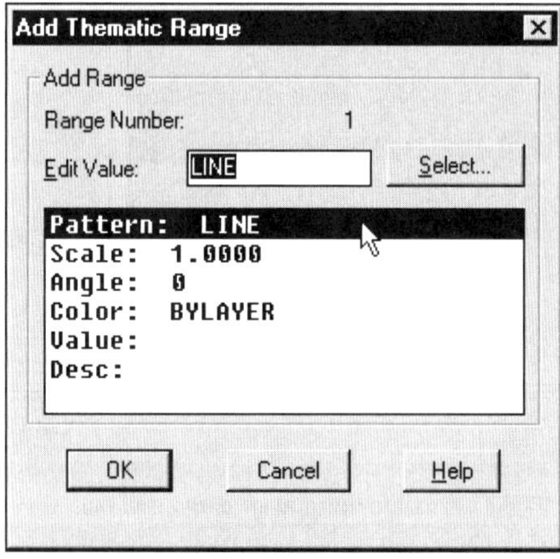

Figure 1–79

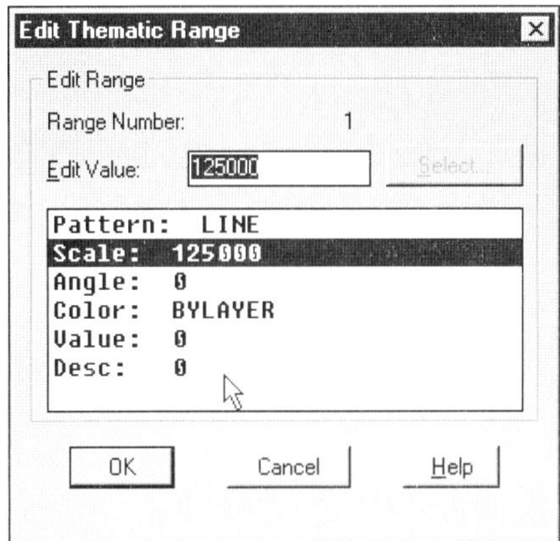

Figure 1–80

Establish an angle for the lines of **90** degrees then pick a color **blue** or **5** .

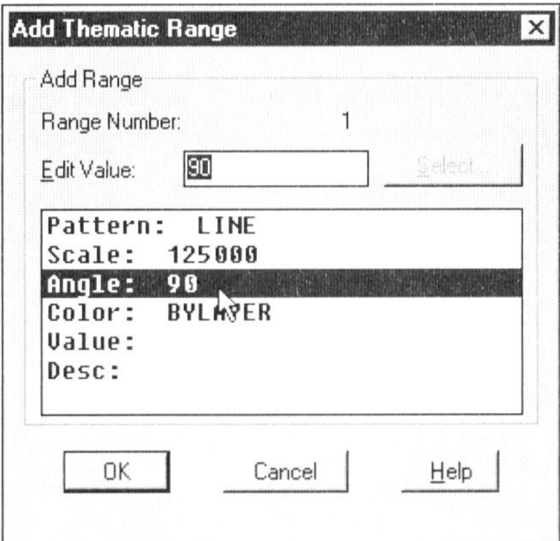

Figure 1–81

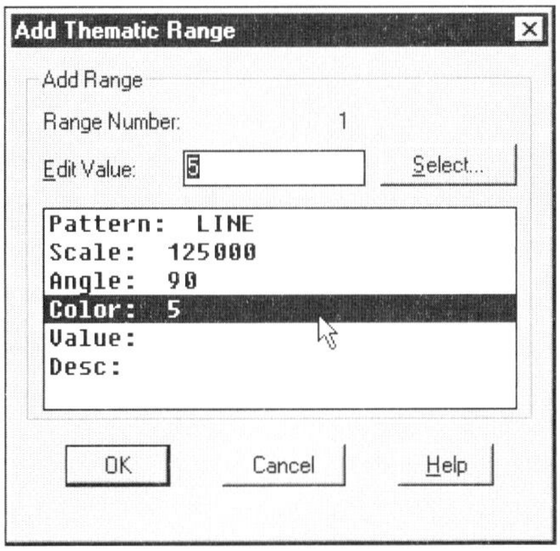

Figure 1–82

Set the value to search for at **5** and enter a description of **5 tons per day.** Click on **OK.**

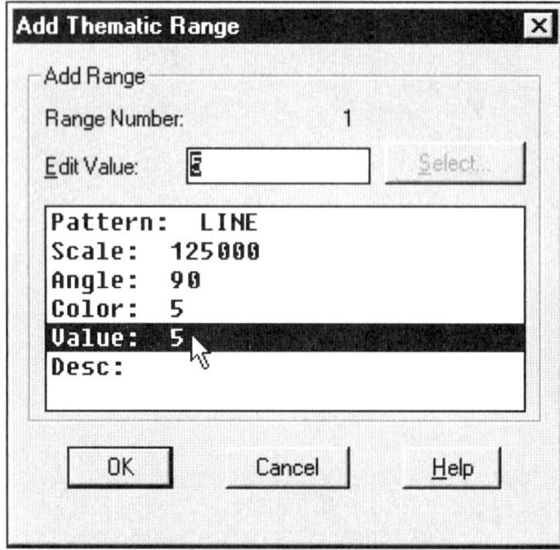

Figure 1–83

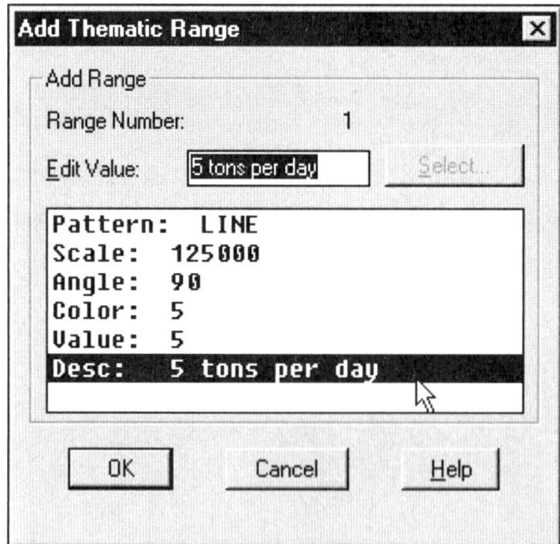

Figure 1–84

Step 23 Select the **Legend** and **Create Legend** option. Use the **Pick** button to select a location for the legend that is, for example, below the legend already existing on the drawing.

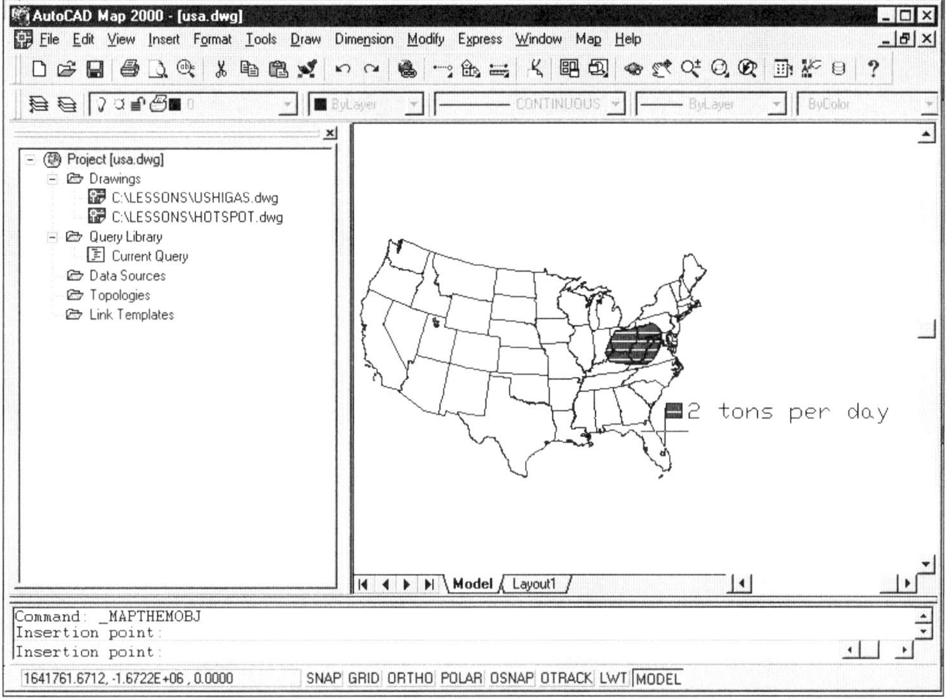

Figure 1–85

The dimensions for the Size X and Size Y should be **200,000** again to match the legend already in existence. Click on **OK**, then **OK** again at the Thematic Display Options screen.

Step 24 At the Object Thematic Mapping dialog box click on **Proceed**. The inner area will be identified on the screen and a legend item will be added.

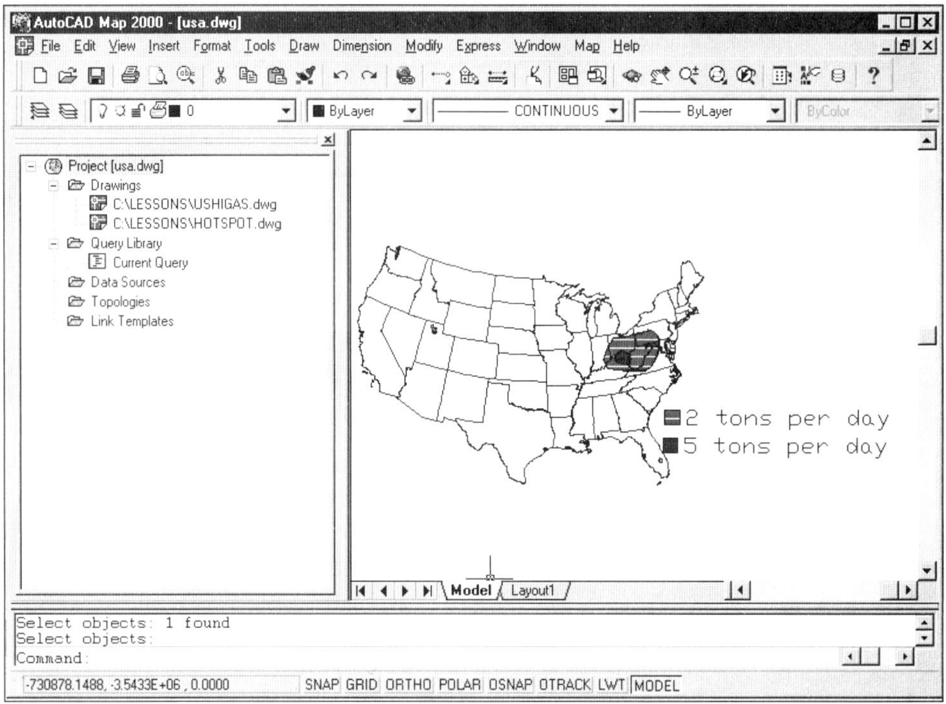

Figure 1–86

The individual legend items can be moved to suit specific needs. Annotations can be added as required.

The drawing can either be saved or discarded as you proceed through the tutorials or exit AutoCAD Map 2000.

CONTINUOUS VS. DISCRETE QUERY FUNCTIONS

The discrete or continuous options (see Figure 1–87) are selected at the Object Thematic Mapping dialog screen based on the nature of the data being queried.

If the data being presented is numeric and is being classified into distinct categories that can be considered to be a continuum of data, then a continuous thematic map setting would be used. If the data is non-numeric, such as land use information, or if it is numeric and not continuous, a discrete setting would be used.

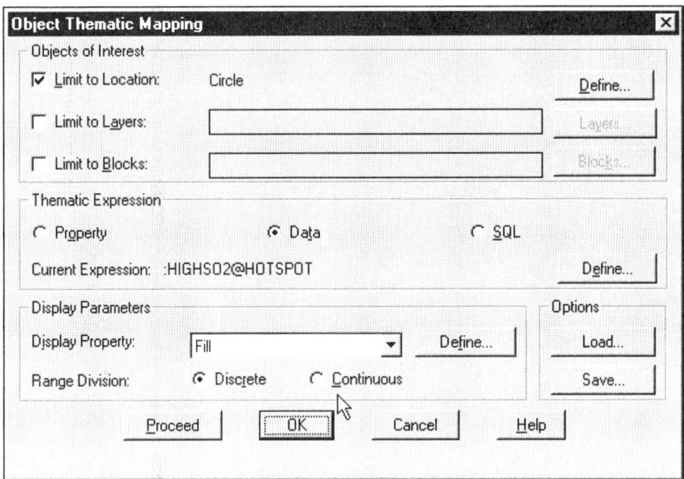

Figure 1–87

Continuous data could be exemplified by population data for each state, which could be classified into categories such as:

0–5 million

5–10 million

10+ million

Discrete data could be exemplified by the prediction of election outcome by classifying each state as either Republican or Democrat based on voter registration.

SUMMARY

Chapter 1 introduced GIS and some of the fundamental concepts associated with mapping such as coordinate systems and projections. If multiple drawings are to be used for any analysis covering a common region, then the map drawings should possess identical coordinate systems and map projections. One strategy for resolving this problem in situations where several people in an organization might be producing maps, is to select a common projection prior to the start of a project. The process of embedding object data in a drawing is one that is quick, requires no other database software, can be easily done in the field, and facilitates the use of queries. While in the tutorial, object data was attached as an integer to a closed polyline, this data can be in the form of text or real numbers and can be attached to features like blocks and text. To successfully generate thematic maps from queries of object data, the attached data rich maps are queried with the results presented on a base drawing which is itself, not queried.

REVIEW QUESTIONS

1. When should thematic mapping be used?

2 What is a critical factor for the base drawing if maps are being prepared by different sources to be used later in thematic mapping?

3. What is the difference between setting the Defining Location for a circular area to Inside compared to Crossing?

4. What is the difference between the Thematic Expression and the Data Expression?

5. What are three variables that are set at the Add Thematic Range dialog box?

6. What are three forms of Object Data that can be used to construct a field within a new table?

7. What is the difference in the screen presentation between a Fill and a Color setting for the Display Property?

8. Why is it important to know the approximate size of the drawing units used to create the base map?

9. After a query has been completed, what function or command can be used to remove the Location Condition from the screen?

10. What is the difference between a Discrete and a Continuous query?

CHAPTER 2

Topology

OBJECTIVES

At the completion of this chapter you will be able to

- create Network, Polygon, and Node-based topology
- complete automated drawing cleanups
- generate Network Traces
- generate Buffer Zones
- conduct Overlay Analysis

KEY TERMS

Topology	**Buffer Zone**
Node	**Overlay**
Network	**Intersection**
Polygon	**Direction**
Centroid	**Resistance**

INTRODUCTION TO TOPOLOGY

Topology is the property that describes how items in a drawing relate to each other spatially. Topology identifies what is adjacent to other features, what is surrounded by other features, and the distance between features. The features of a map are the items that have been drawn as points, lines, or areas and would include all the components of the drawing, such as roads, boundary lines, wells, rivers, lots, or parcels.

AutoCAD Map 2000 uses *nodes*, *networks,* and *polygons* as the topology building blocks.

Nodes can be thought of as points that mark significant locations. Such locations could include intersections, ends of streets, or an isolated point exemplified by a well. A node consists of a single coordinate ordered pair.

Networks are a series of connected lines or polylines. Generally, they are related in purpose, such as a road or drainage system. Because a line or polyline is made up of several ordered pairs of points, a network consists of a multitude of ordered pairs.

Polygons are made from lines or polylines that form boundaries or closed shapes and, hence, regions of interest exemplified by lots or counties and states. Polygons are identified by the coordinates of the vertices of the boundary lines that make up the region.

Figure 2–1 represents a spatial layout for some nodes, networks, and polygons. Nodes are identified by a capital N and a number, links or network components by a capital L and a number, and polygons by a the letter A, B, C, or D.

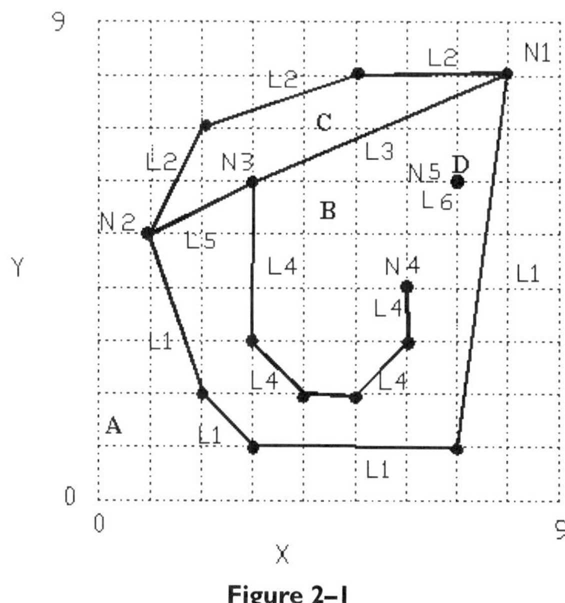

Figure 2–1

All of the node topological information can be stored in a table similar to that shown below.

In this case a sample of the node topology for N1 would be the identification of the lines that link to N1 which are L1, L2 and L3. N2 would be identified as the node to which L2 and L5 linked.

Node Topology

Node	Network Lines
N1	L1, L2, L3
N2	L1, L2, L5
N3	L3, L5
N4	L4
N5	L6

The network topology for L1 would be identified by the start point N2 and the end-point N1 with the polygon to the left and right sides being identified as B and A, respectively. This data, along with the coordinates of any points registered to the line, can be stored in a table exemplified by the one below.

Network (Line) Topology

Line	**Start Node**	**Start (X,Y)**	**Contained Points**	**End Node**	**End (X,Y)**	**Left Polygon**	**Right Polygon**
L1	N2	1,5	2,2 3,1 6,1	N1	7,8	B	A
L2	N1	7,8	5,8 2,7	N2	1,5	C	A
L3	N3	3,6		N1	7,8	C	B
L4	N3	3,6	3,3 4,2 5,2 6,3	N4	6,4	B	B
L5	N2	1,5		N3	3,6	C	B
L6	N5	7,6		N5	7,6	B	B

The polygon topology for polygon B would be identified by the lines that comprise the polygon, L1 and L3. The polygon identified as A is outside the map area, whereas node N5 can be assigned a polygon identifier to give it location with respect to the other polygons as shown in the table below.

Polygon Topology

Polygon	**Network Lines**
A	Outside map area
B	L1, L3, L5
C	L2, L3, L5
D	L6

The presence of topology in a drawing allows the software to perform calculations using the topology data rather than or in conjunction with the coordinate data. This facilitates spatial operations being done quickly.

In the following tutorials, you will be guided through the processes that facilitate the generation of topology and introduced to some of the analysis that can be performed with the spatial drawings.

TUTORIAL: DRAWING CLEANUP

Often, drawing files contain small errors, which require correction before they can be used for useful work. Typically these problems are associated with digitizing errors that occurred during the original creation of the map.

The drawing used in this tutorial has several such errors and the Drawing Cleanup utilities will be used to correct the difficulties. Prior to AutoCAD Map 2000, such errors would have to be corrected manually.

Step 1 Start AutoCAD Map 2000, and **open** the **ROADNET.DWG** drawing. **Zoom** in on the area identified by the circle.

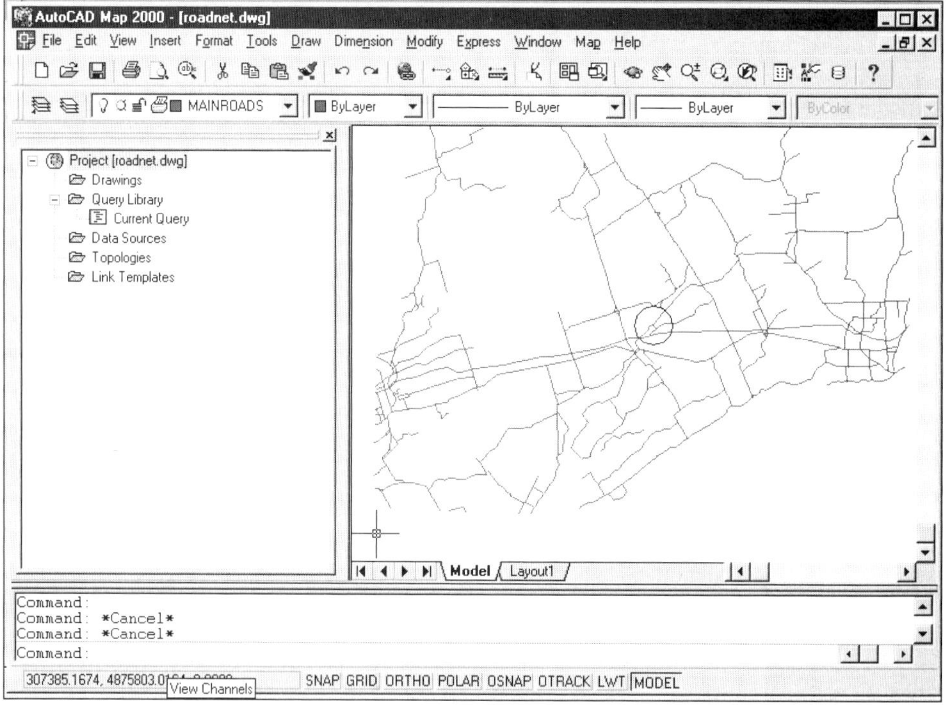

Figure 2–2

You will notice several errors as shown on the next graphic.

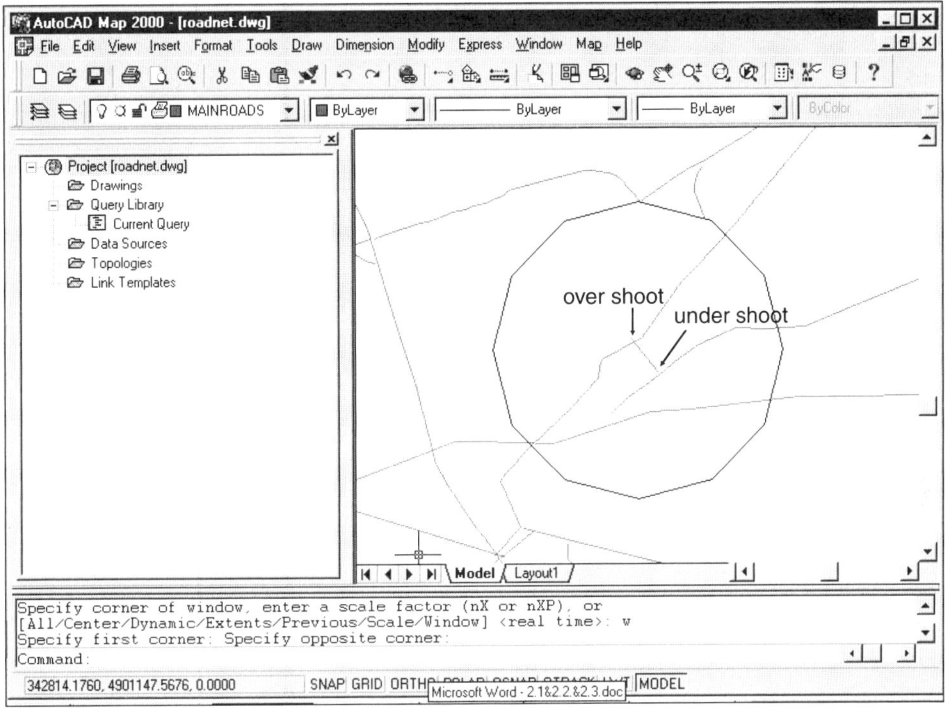

Figure 2–3

Step 2 Go to the **Map** menu select **Tools** then **Drawing Cleanup**.

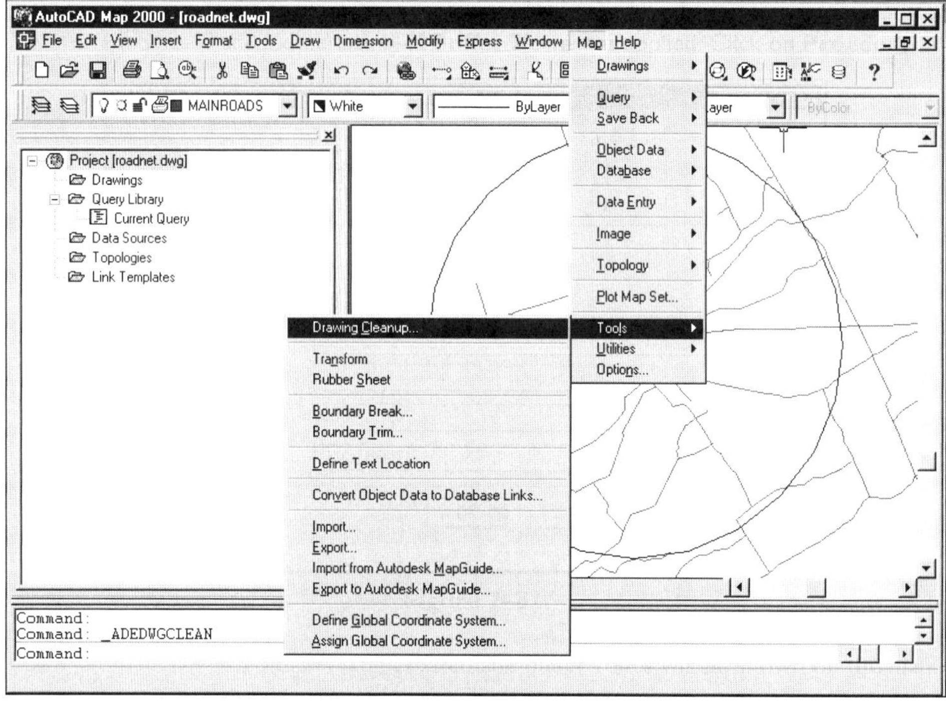

Figure 2–4

Step 3 At the Drawing Cleanup dialog box, click on the **Object Selection** button.

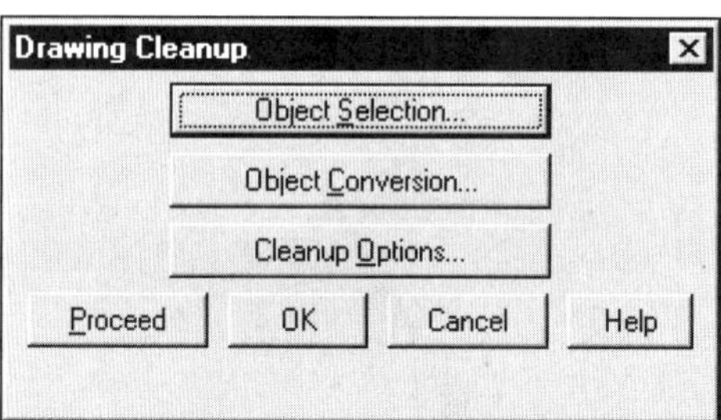

Figure 2–5

Choose the **Select Automatically** radio button from the Object Selection screen and click on **OK**.

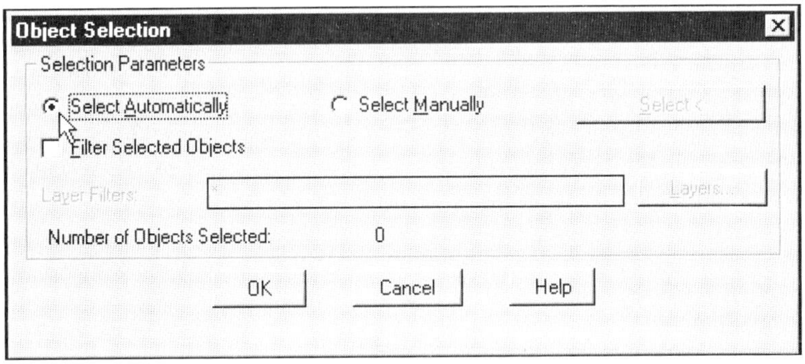

Figure 2–6

This option allows the software to automatically and mathematically select all objects in the drawing that meet the drawing cleanup parameters. Now choose the **Object Conversion** option and select the **Modify Original Object** as the method, the **Line to Polyline** option, then click on **OK**.

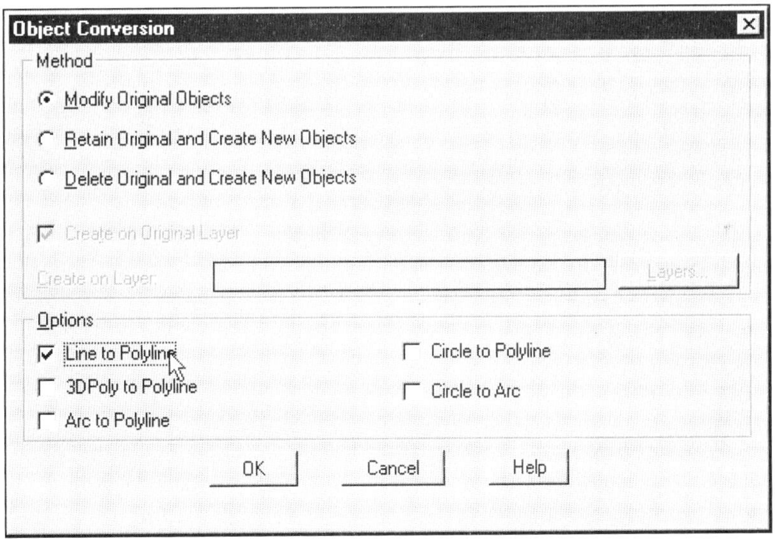

Figure 2–7

This converts existing drawing features rather than creating new features.

Step 4 Select **Cleanup Options** and from the Edit Geometry section highlight the **Delete Duplicate Objects** and **Extend Undershoots** items.

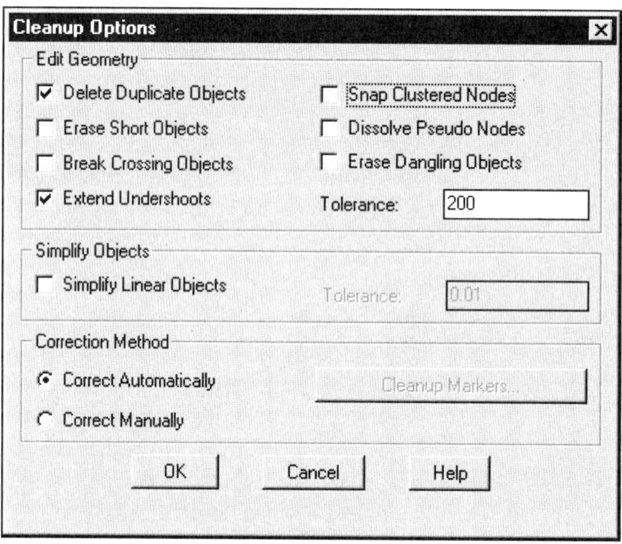

Figure 2–8

Deselect the other default options. The tolerance should be set at **200** to represent 200 meters. This value can be established by using the **Tools** and **Inquiry** function for **Distance** in order to estimate the distance between sections of the drawing that required correction. Set the **Correction Method** to **Correct Automatically.** Click on **OK** then at the **Drawing Cleanup** screen select **Proceed.** In the command prompt area a progress report will be provided and updated indicating objects are being found, modified etc. This process will extend any undershoots that are within 200 meters of another line segment.

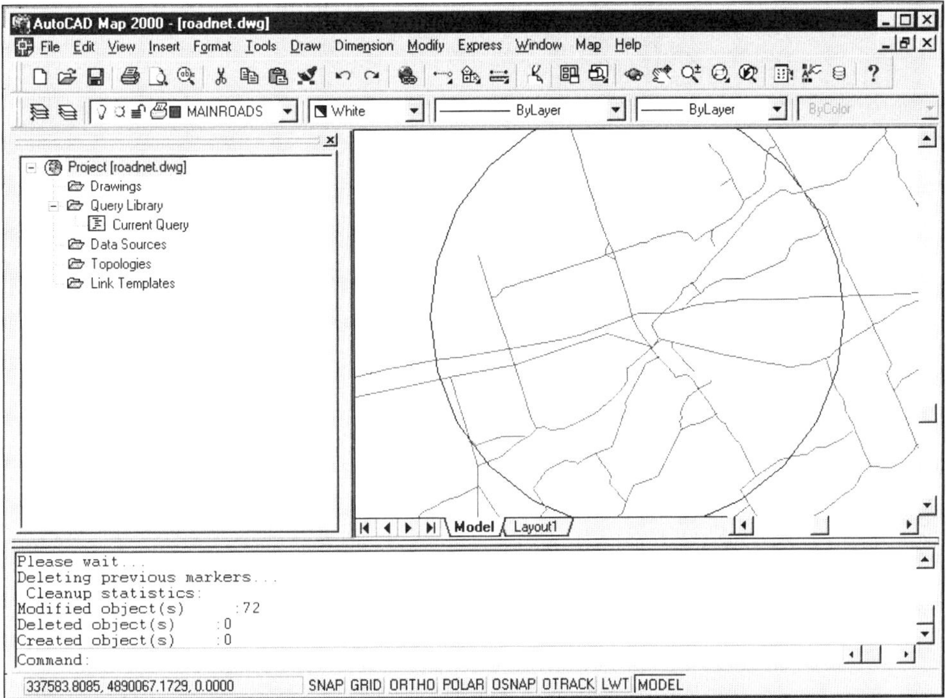

Figure 2–9

Step 5 To correct the overshoots go back to **Map,** select **Tools** and **Drawing Cleanup**. For Object Selection, choose the **Select Automatically** option, then click on **OK**.

Object Selection

Selection Parameters

Select Automatically   Select Manually   Select <

Filter Selected Objects

Layer Filters: Layers...

Number of Objects Selected: 0

OK   Cancel   Help

Figure 2–10

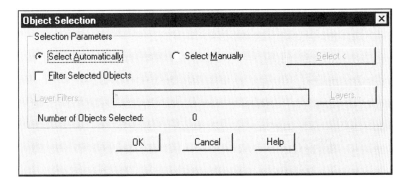

Pick **Object Conversion**, and choose **the Modify Original Objects** and **Line to Polyline** options then click on **OK**.

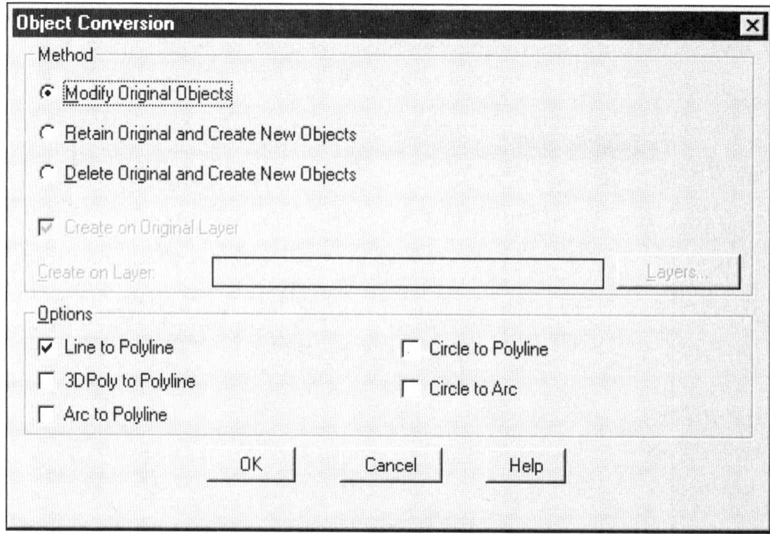

Figure 2–11

For the Cleanup Options highlight **Break Crossing Objects** and **Erase Dangling Objects** and set the tolerance to **150** meters.

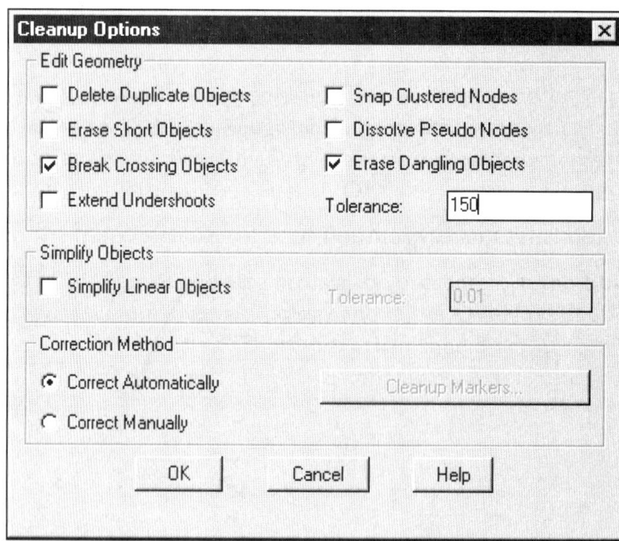

Figure 2–12

These settings will modify any line segment that crosses another line or extends beyond the line being crossed, forming an overshoot. The tolerance distance, in this case 150 meters, can be determined by identifying some of the overshoots and estimating their lengths through the Inquiry and Distance functions. It is important to note that too large a tolerance can result in a loss of real data, such as a short street. Click on **OK** and then again on **Proceed** at the Drawing Cleanup dialog box. The short overshoots will be deleted.

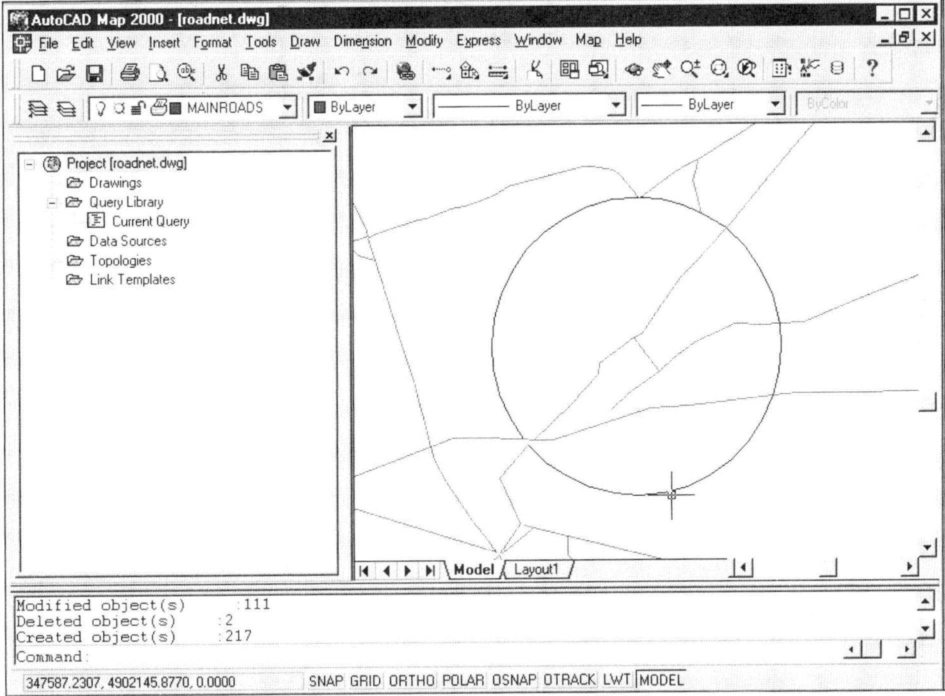

Figure 2–13

This file may be saved with a new name or closed without saving changes.

This process can save countless hours of manual map editing and correction. Some trial and error may have to be used to arrive at the appropriate tolerance levels for a drawing, and sometimes the cleanup procedure should be run more than once on a drawing using increasing tolerance levels.

TUTORIAL: CREATING POLYGON TOPOLOGY

You create polygon topology when you want to attach data to regions of a drawing that were not created as closed polygons. In this example, a drawing of Canada will be used to demonstrate the difference between a drawing with and without polygon topology.

Step 1 Open the **CANADA.DWG** drawing file (illustrated in Figure 2–14) that has been provided.

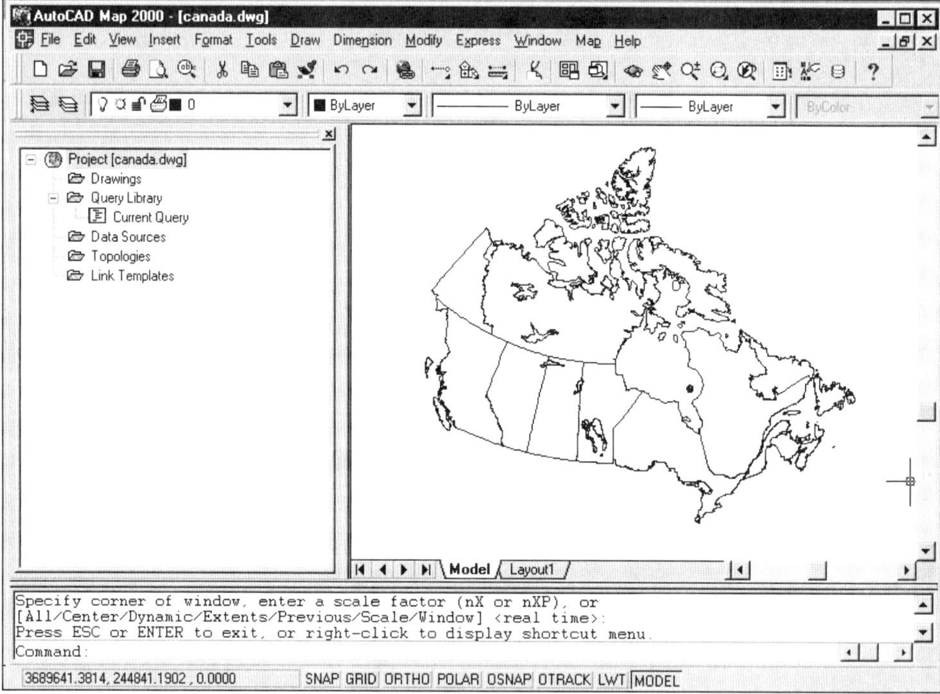

Figure 2–14

In this example, you will design the hatch pattern manually instead of using predefined hatch patterns, utilizing a simple line pattern with a spacing between the lines set in recognition of the scale of the drawing. Type the command **hatch** and press ENTER. Type in **u** for user defined, then press ENTER.

```
AutoCAD - Command Line - canada.dwg
Command: *Cancel*
Command: *Cancel*
Command: *Cancel*
Command: *Cancel*
Command: hatch
Enter a pattern name or [?/Solid/User defined] <ANSI31>: u
```

Figure 2–15

When prompted to determine the **angle** for the hatching lines select the default value of **0** by pressing ENTER or tapping the right mouse button.

```
AutoCAD - Command Line - canada.dwg
Specify angle for crosshatch lines <0>:
Specify spacing between the lines <1.0000>: *Cancel*
Command: *Cancel*
Command: hatch
Enter a pattern name or [?/Solid/User defined] <ANSI31>: u
Specify angle for crosshatch lines <0>: 0
```

Figure 2–16

When prompted for a line spacing, type in **40000**, which represents a distance of 40 km between each line, and press ENTER.

```
AutoCAD - Command Line - canada.dwg
Specify spacing between the lines <1.0000>: *Cancel*
Command: *Cancel*
Command: hatch
Enter a pattern name or [?/Solid/User defined] <ANSI31>: u
Specify angle for crosshatch lines <0>: 0
Specify spacing between the lines <1.0000>: 40000
```

Figure 2–17

At the prompt for double hatching, which allows you to generate sets of intersecting hatch lines, select the default value of **No** by pressing ENTER again.

```
AutoCAD - Command Line - canada.dwg
Command: *Cancel*
Command: hatch
Enter a pattern name or [?/Solid/User defined] <ANSI31>: u
Specify angle for crosshatch lines <0>: 0
Specify spacing between the lines <1.0000>: 40000
Double hatch area? [Yes/No] <N>: N
```

Figure 2–18

Step 2 To select a region for hatching, place the pick box over one of the boundary lines for the province of Ontario. In the example below this box has been placed on the northern shore of Lake Ontario.

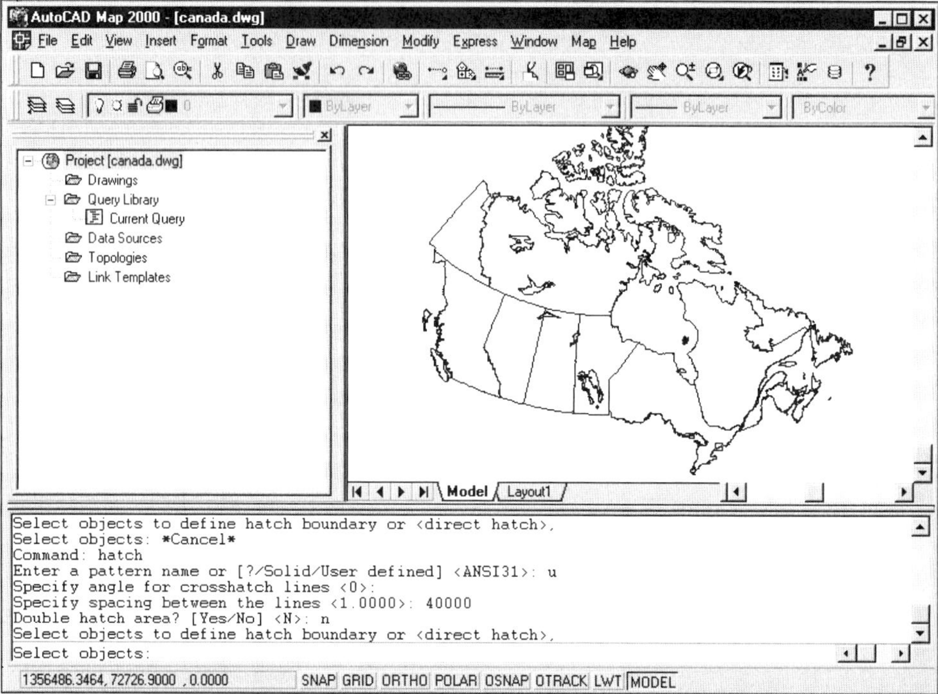

Figure 2–19

When you click on this boundary, part of the border of Ontario will become highlighted.

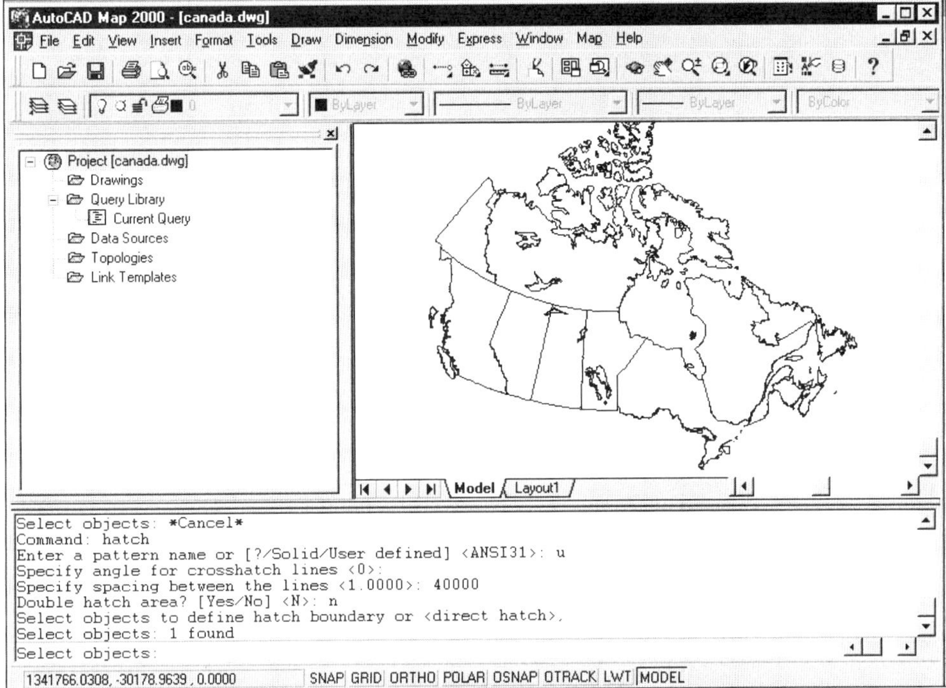

Figure 2–20

Tap **Esc** to cancel the **hatch** command. If the province was a closed polygon, the entire boundary would have been highlighted, but such was not the case as this boundary consists of a collection of individual lines. The creation of polygon topology will permit the creation of closed polygons.

Step 3 Go to the **Map** menu, then select **Topology** and **Create.**

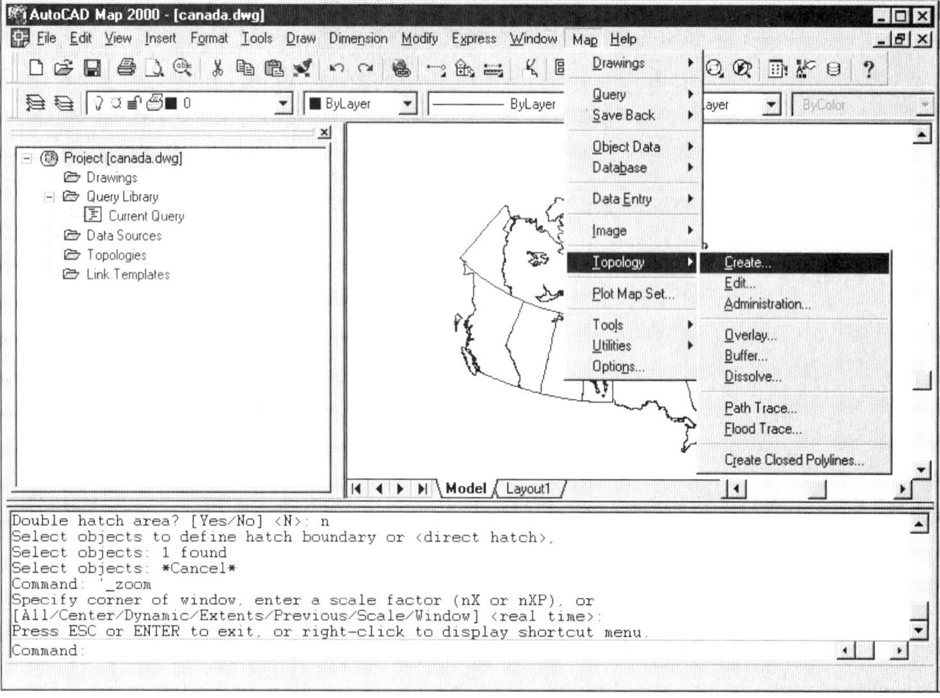

Figure 2–21

Step 4 The Create Topology dialog box will appear.

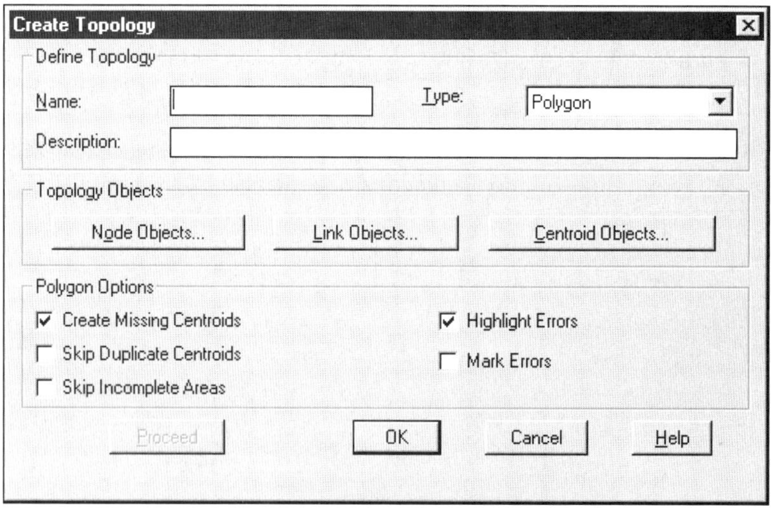

Create Topology
Define Topology
Name:
Type: Polygon
Description:
Topology Objects
Node Objects... Link Objects... Centroid Objects...
Polygon Options
☑ Create Missing Centroids ☑ Highlight Errors
☐ Skip Duplicate Centroids ☐ Mark Errors
☐ Skip Incomplete Areas
Proceed OK Cancel Help

Figure 2–22

In the upper right corner of this dialog box, you select the type of topology to be created. In this exercise choose **Polygon** topology. Create a name for the topology that is to be created. In the example below, **regions** is the name used and an optional description, **Provincial Boundaries** was also added.

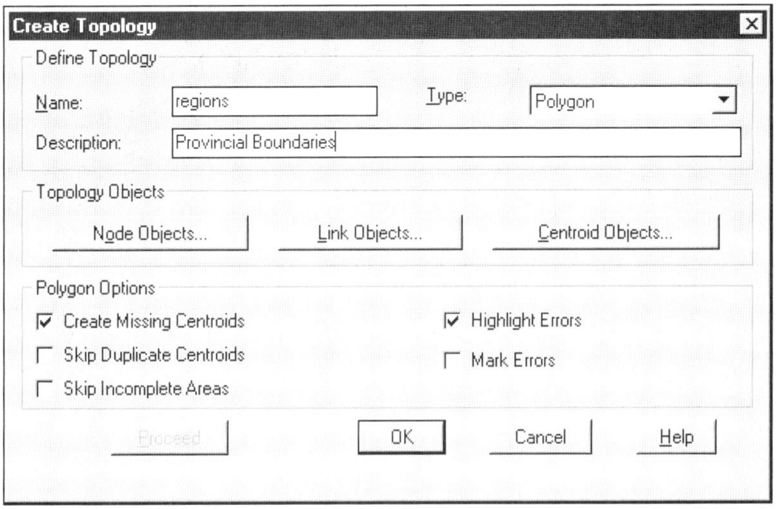

Figure 2–23

As mentioned previously, topology can be created through the selection of nodes, links, and centroids. Nodes are identified as intersections of lines, whereas links or link objects are lines that are shared between two or more polygons. A centroid is an independent point located inside a polygon. It might not be located at the exact center of the polygon or even on the same layer as the polygon. It is located away from any separate features or islands inside the polygon. It serves as an identification point for the polygon it represents and is generally presented as a point.

Step 5 Click on the **Node Objects** button and the Node Objects dialog box will present itself. In the upper portion, under Object Selection, click on the **Select Automatically** radio button; in the lower portion of the screen select **the Create Node Objects** checkbox.

This generates an automatic search for nodes and will result in the creation of nodes (registered ordered pairs of coordinates) where lines intersect or end at places where nodes were not created during the drawing of the map.

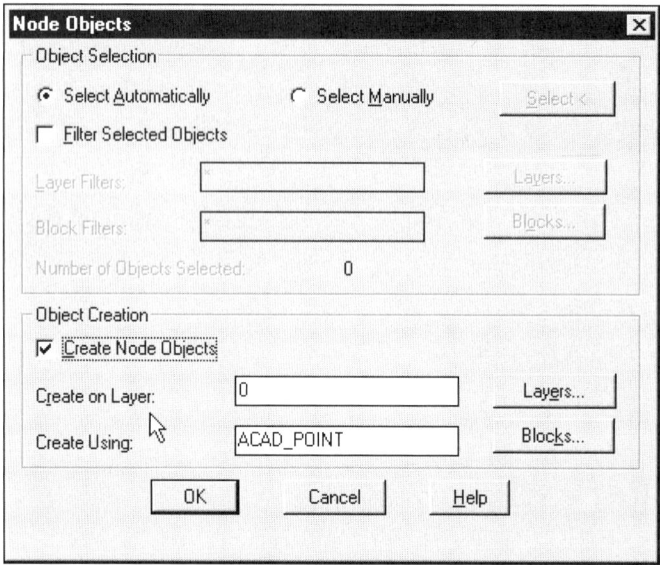

Figure 2–24

Click on **OK**. You will be returned to the Create Topology dialog box.

Step 6 Click on the **Link Objects** button. In the Link Objects dialog box highlight the **Select Automatically** option and then click on **OK**.

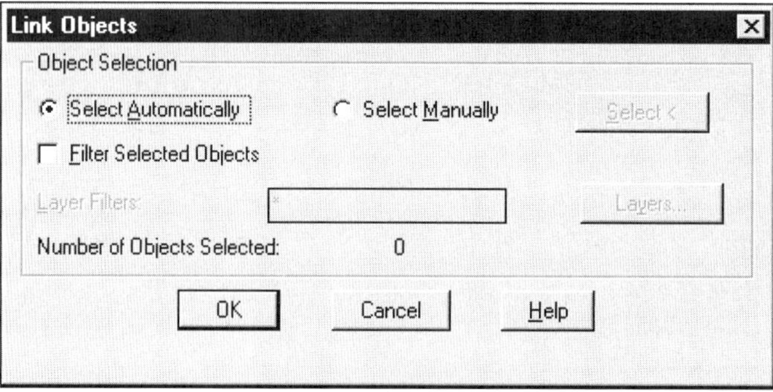

Figure 2–25

This cues AutoCAD Map 2000 to find the linking lines for the Nodes and Polygons.

You will be returned to the **Create Topology** screen.

Step 7 Click on the **Centroid Objects** button and in the Centroid Objects dialog box click on the **Select Automatically** radio button then choose **OK**. This setting directs the software to generate centroids for each polygon detected.

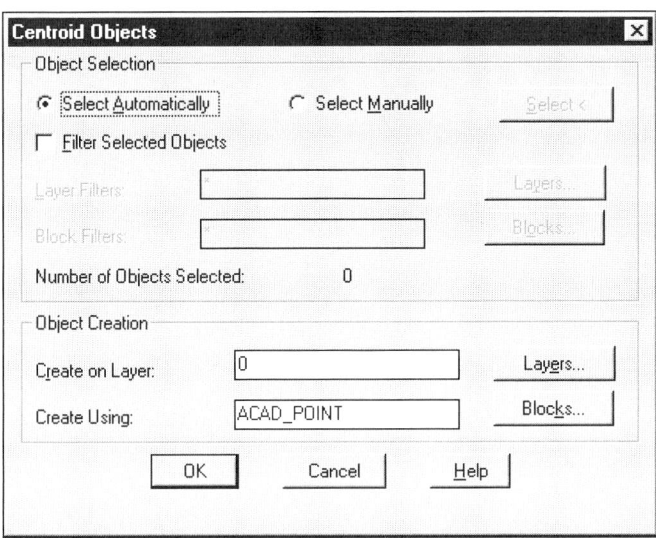

Figure 2–26

Step 8 After you have set the three components required for topology creation, you will again be returned to the Create Topology dialog box. Make sure that the **Create Missing Centroids** option is selected then click on **Proceed.**

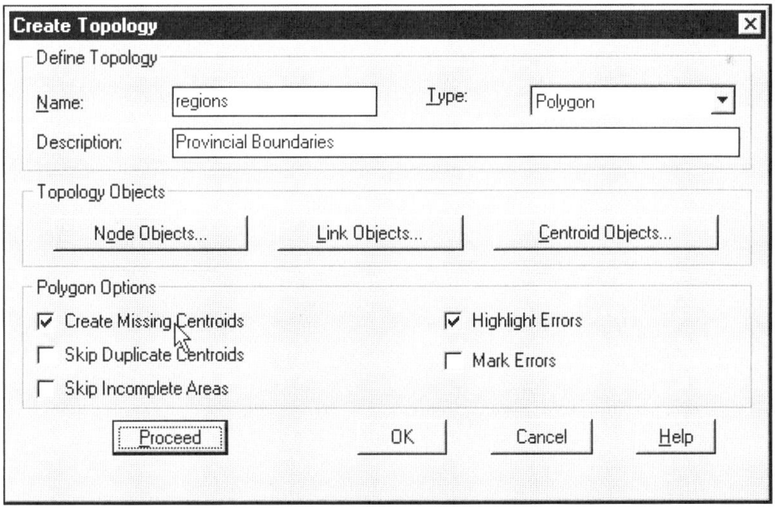

Figure 2–27

The topology will be created in a few seconds. You will see small dots appear inside the polygons. These are the centroid points.

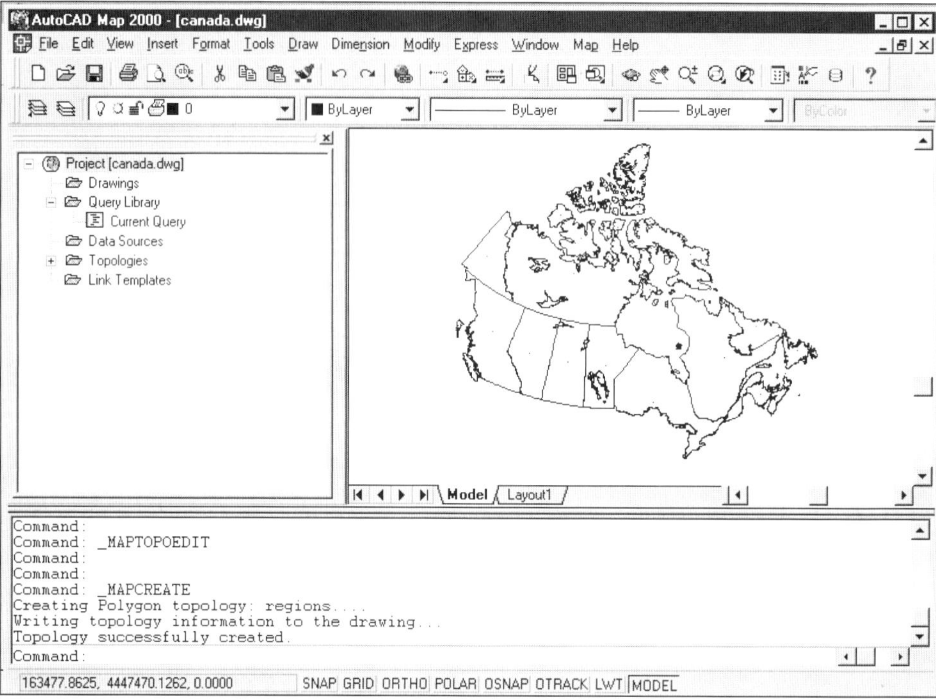

Figure 2–28

Step 9 The polygon-based topology will be used to create closed polygons in the drawing. Go to the **Map** menu and select **Topology** then **Create Closed Polylines**.

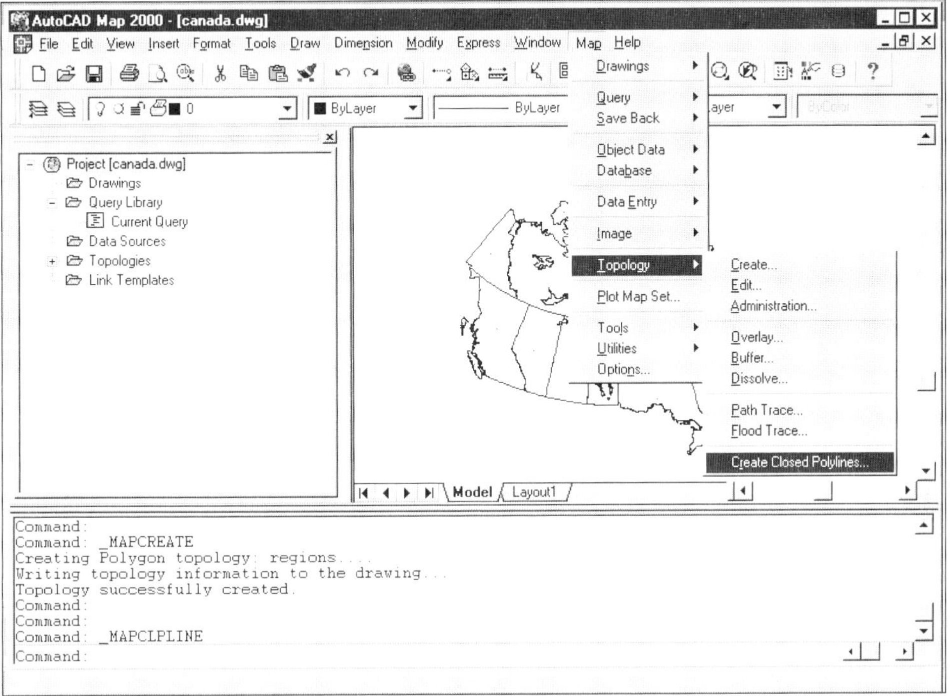

Figure 2–29

AutoCAD Map 2000 utilizes the topology to identify lines that are connected to form polygons. The Create Closed Polylines dialog box will appear. In the upper portion the Topology name **regions** will be shown.

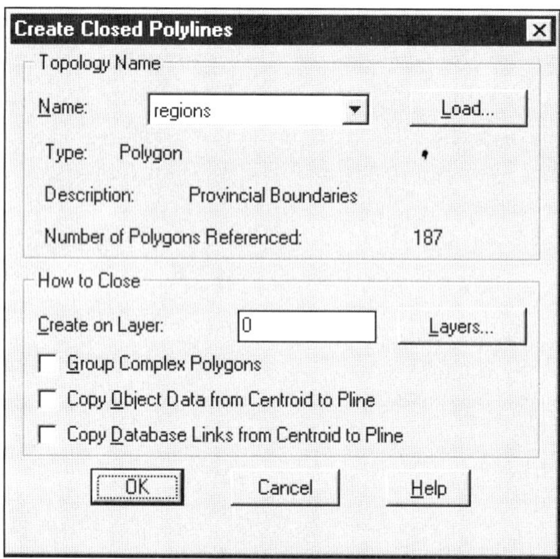

Figure 2–30

Step 10 In the How to Close portion of this dialog box (bottom section) select the **Group Complex Polygons** checkbox. The Copy Object Data from Centroid to Pline and the Copy Database Links from Centroid to Pline checkboxes need not be activated because there is no object data in this base drawing and there are no database links to the centroids. Subsequent chapters deal with establishing database links to drawings.

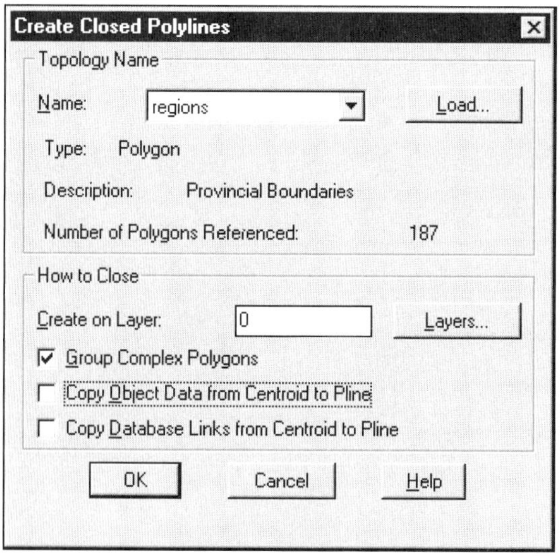

Create Closed Polylines
Topology Name
Name: regions
Load...
Type: Polygon
Description: Provincial Boundaries
Number of Polygons Referenced: 187
How to Close
Create on Layer: 0
Layers...
Group Complex Polygons
Copy Object Data from Centroid to Pline
Copy Database Links from Centroid to Pline
OK
Cancel
Help

Figure 2–31

Click on **OK**.

Complex polygons are those that contain islands shown by other polygons. The second option, copying object data from centroids to plines, is a process for ensuring any data that has been attached to the regional centroids is also attached to the polygon. The ASE Link copying option ensures that any SQL links previously made to the centroid are maintained as links to the polygons.

Step 11 Repeat the hatch procedure from Step 2 and this time the entire province will become selected and will be hatched in upon completion of the command.

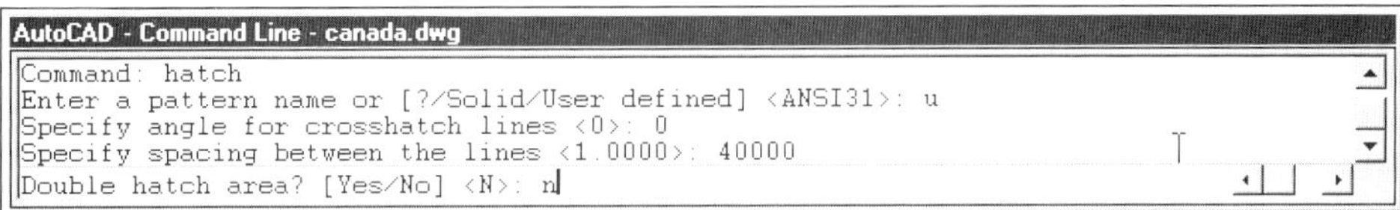

AutoCAD - Command Line - canada.dwg
Command: hatch
Enter a pattern name or [?/Solid/User defined] <ANSI31>: u
Specify angle for crosshatch lines <0>: 0
Specify spacing between the lines <1.0000>: 40000
Double hatch area? [Yes/No] <N>: n

Figure 2–32

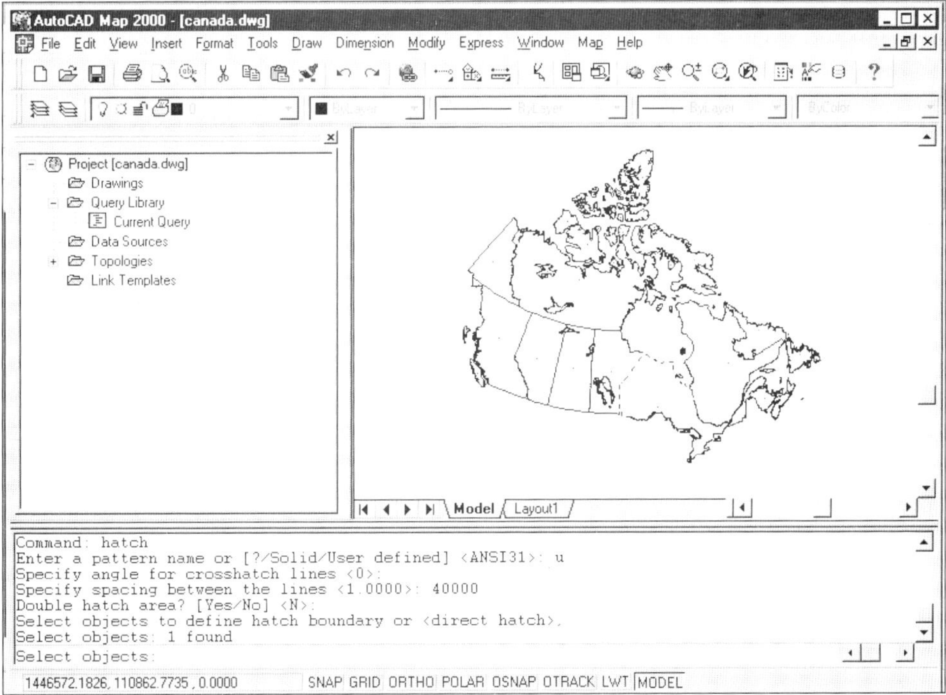

Figure 2–33

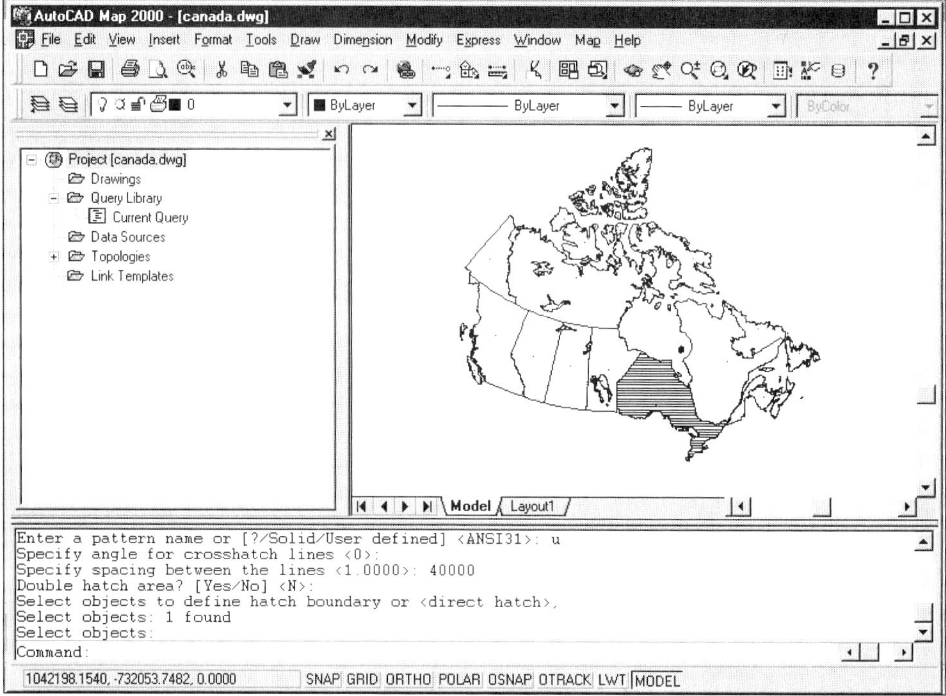

Figure 2–34

This drawing may be saved with a new name or closed without saving the changes.

The capability to create topology will allow you to make use of a wide variety of source drawings that may have been created long before the addition of data or viewing multiple drawings was in fact a possibility. Either continue with the tutorials or exit AutoCAD Map 2000.

TUTORIAL: CREATING NETWORK TOPOLOGY

Network topology is useful whenever you have to work with an interconnected series of lines, such as may be the case with a drainage or road system. In this exercise, you will create topology for a street network drawing accessing the Napanee drawing file provided.

As with the creation of polygon topology, the process begins with a drawing cleanup operation, followed by the creation of topology.

Step 1 Open the **NAPANEE.DWG** file, shown in Figure 2–35.

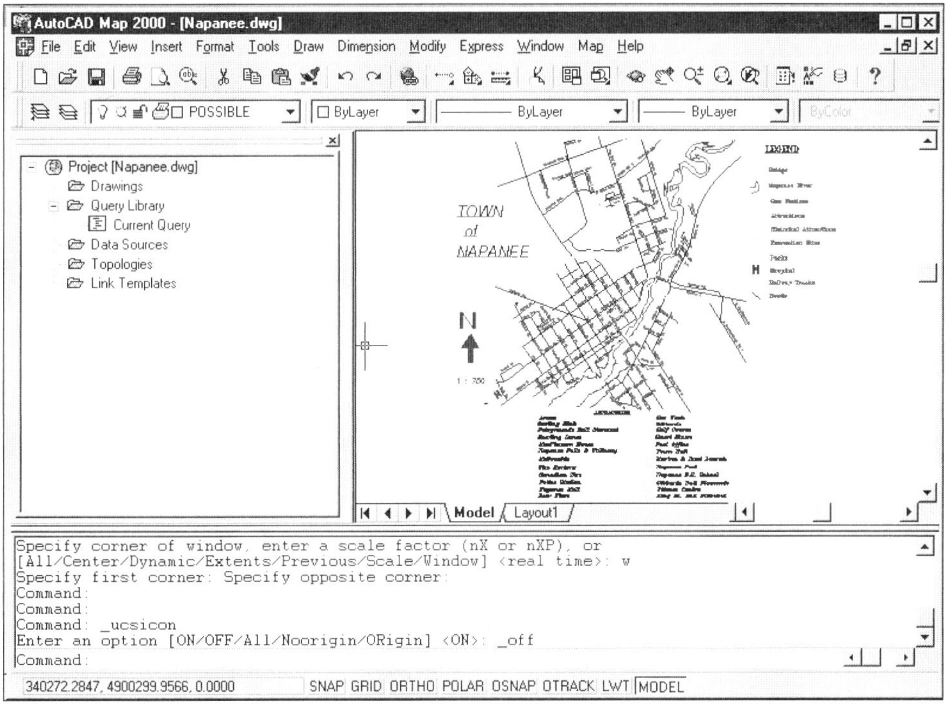

Figure 2–35

Step 2 It is recommended to clean up any drawing prior to creating topology, even though this is not a mandatory step. This procedure will resolve small problems in the drawings that can generate errors when trying to create the topology, such as situations where streets don't quite meet at an intersection or where there are overshoots. Generally it is expedient to have AutoCAD Map 2000 automatically search the drawing for objects to be corrected. Go to the **Map** menu and select **Tools** then **Drawing Cleanup**.

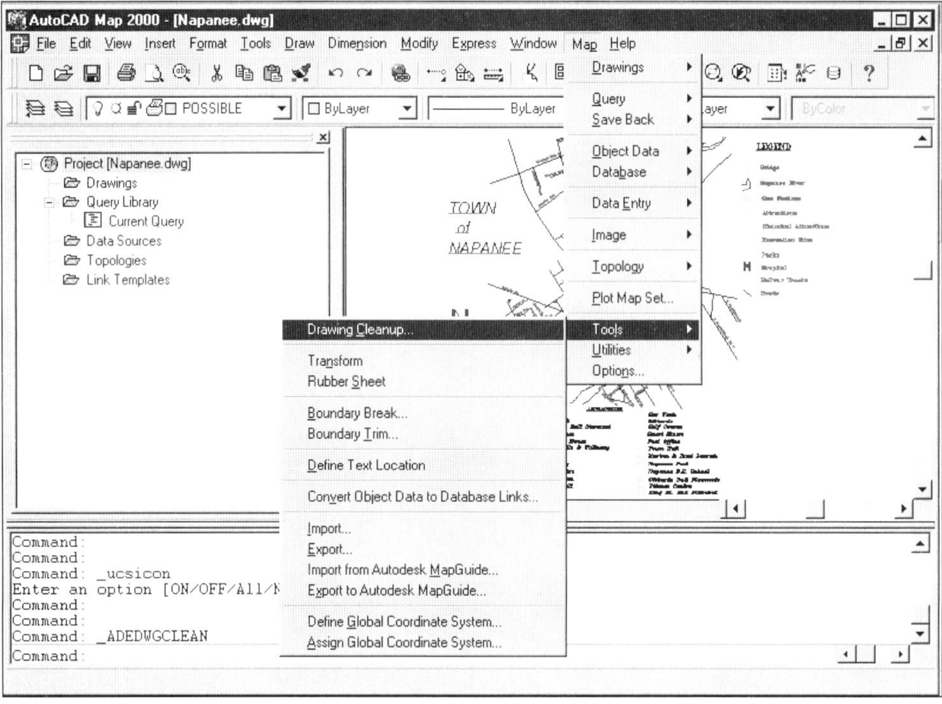

Figure 2–36

Step 3 The Drawing Cleanup dialog box will appear.

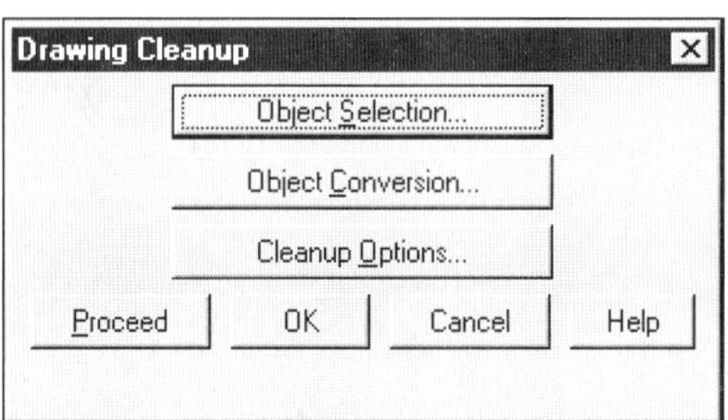

Figure 2–37

Click on the **Object Selection** button. At the Object Selection screen click on **Select Automatically** then click on **OK**.

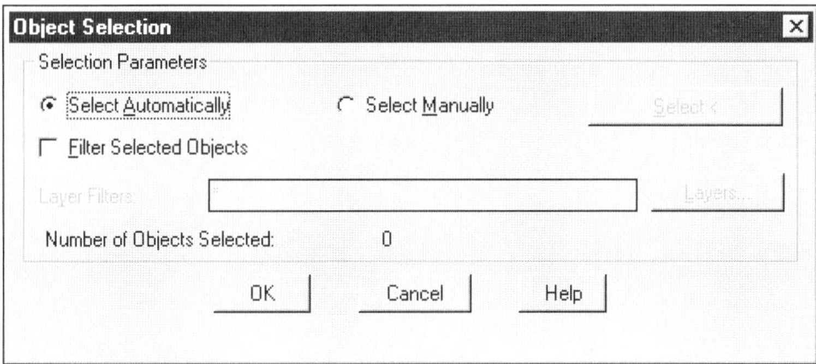

Figure 2–38

Step 4 You will be returned to the Drawing Cleanup dialog box. Click on **Object Conversion**, and in the upper half of the Object Conversion dialog box select **Modify Original Object**. This instructs AutoCAD Map 2000 to make the corrections by altering features already in the drawing. In the lower portion, select **the Line to Polyline** option. This will convert any short line segments to continuous polylines.

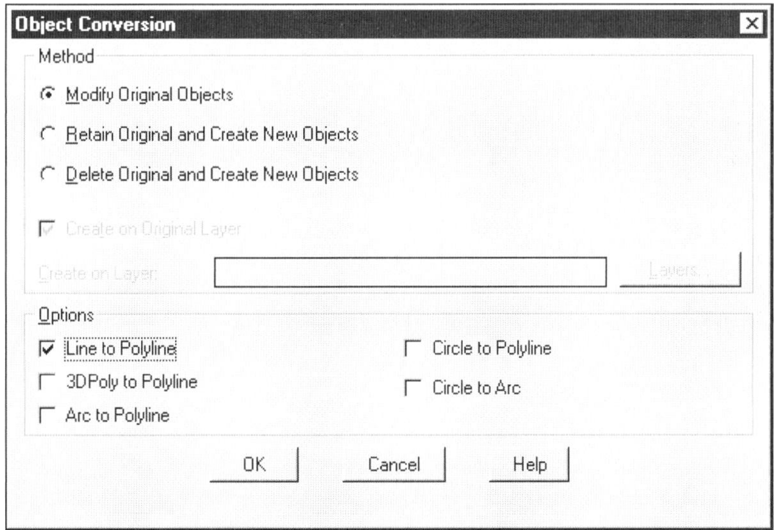

Figure 2–39

Click on **OK**.

Step 5 Click on the **Cleanup Options** button and use the default values.

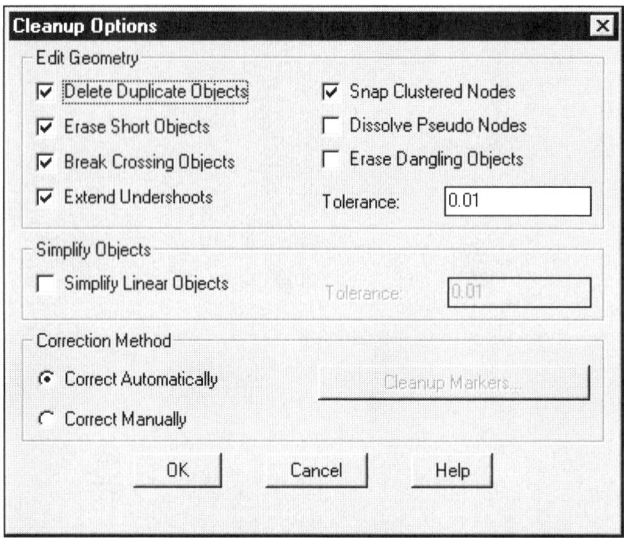

Figure 2–40

Make sure that the **Correct Automatically** option is selected, then click on **OK**. At the Drawing Cleanup screen click on **Proceed**. You will be provided with information in the command line area indicating that objects have been found, modified, deleted, or created.

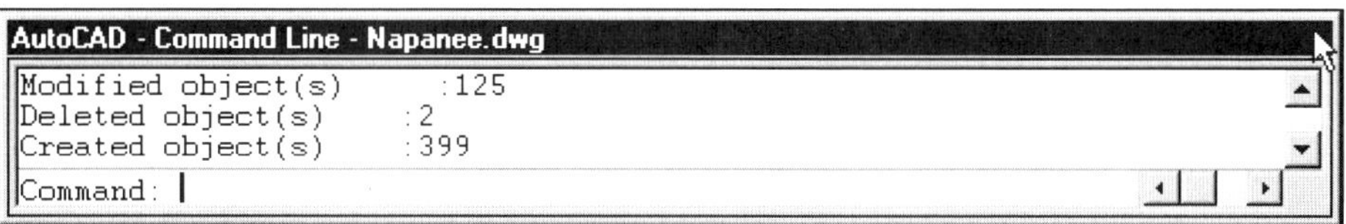

Figure 2–41

Step 6 The next step is to create topology for the network of lines. Go the **Map** menu and select **Topology** then **Create**. The Create Topology dialog box will appear.

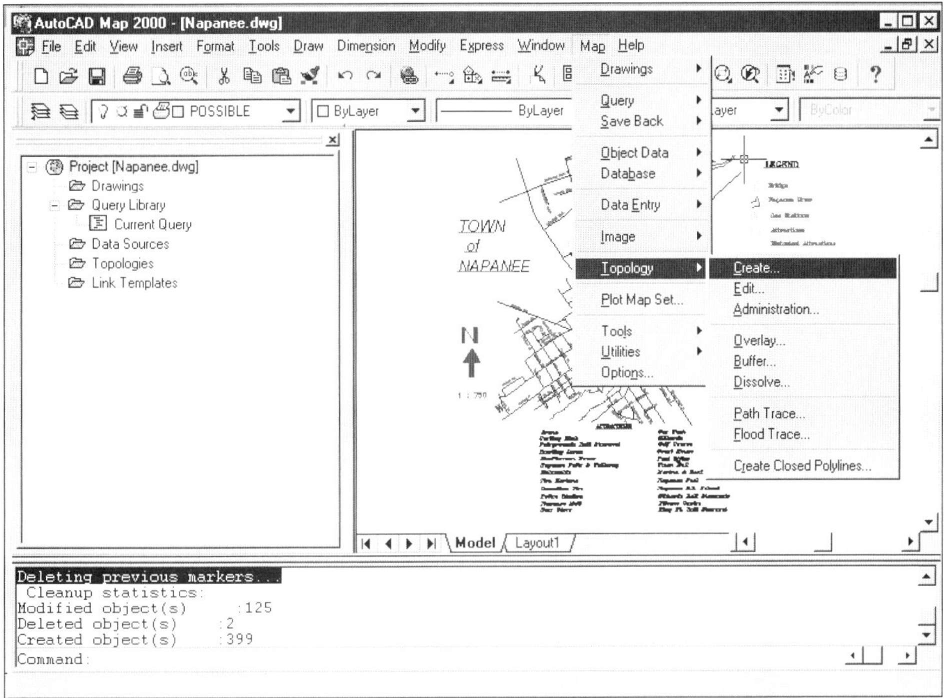

Figure 2–42

Step 7 Set the Topology Type to **Network**, and type in a name, such as **Streets**, and an optional description, such as **Street network**.

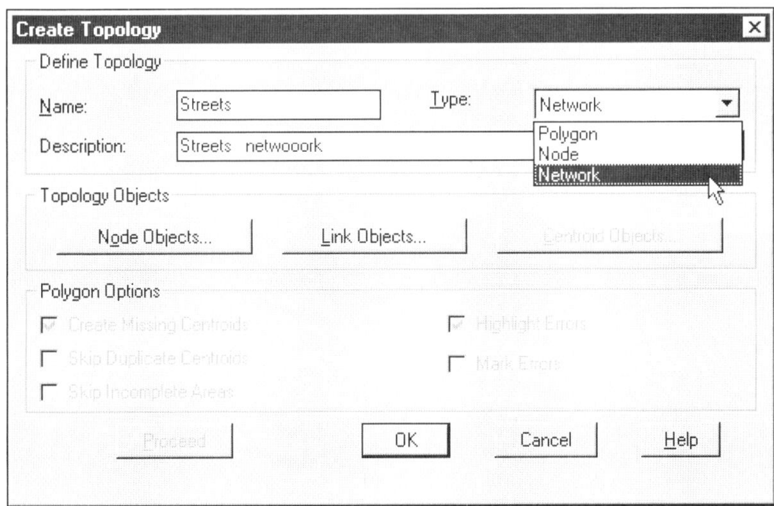

Figure 2–43

Step 8 Click on the **Node Objects** button. In the upper portion of the Node Objects dialog box that deals with Object Selection click on **Select Automatically** and in the lower portion select the **Create Nodes** option.

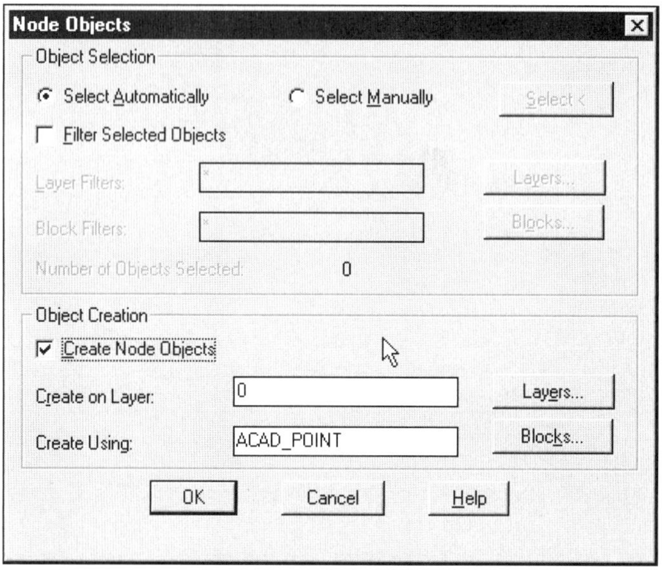

Figure 2–44

This selection has AutoCAD Map 2000 generate Node coordinates for intersections and endpoints of line features. Leave the **layers** and **blocks** at the default setting. Click on **OK** and the **Create Topology** screen will reappear.

Step 9 Click on the **Link Objects** button and select the **Select Automatically** radio button. It is important to note that sometimes one would not use the Select Automatically option. An example of such as case might be when you have a network of roads in the same drawing as a drainage system. Automatic selection would create a network of both sets of lines.

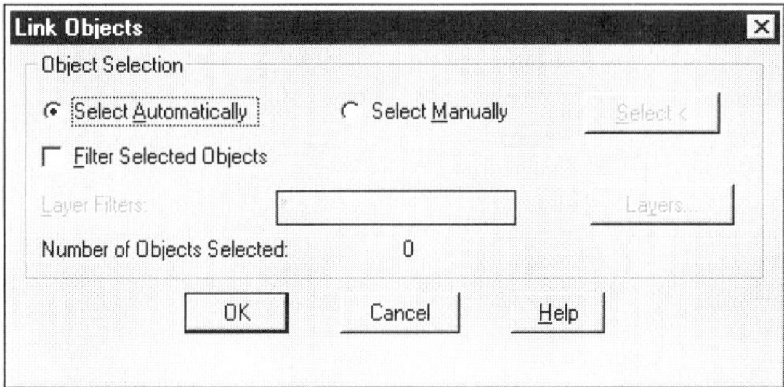

Figure 2–45

This selection has AutoCAD Map 2000 identify links for the topology. One option for future consideration is that of having the topology created on a different layer. A common practice is to create a new layer for the topology, and another is to create the topology on the layer of the feature for which topology is being created. Click on **OK**.

Step 10 The Create Topology dialog box will be presented again. Click on **Proceed**. Information will be provided in the command line area, and you will be informed when the topology has been successfully created.

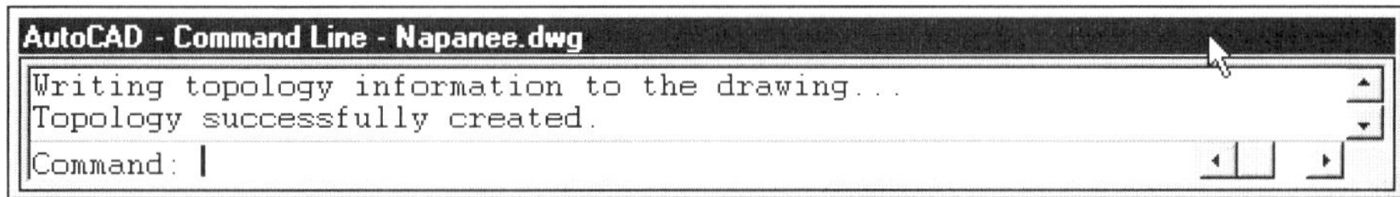

Figure 2–46

Step 11 Go to **File** and **Save As** and save this drawing with the topology added as **NAPTOP.DWG** for future use.

You can now either continue with the tutorials or exit AutoCAD Map 2000.

TUTORIAL: CREATING NODE TOPOLOGY

Node topology can be created from points, blocks, or text. In this exercise however, node topology will be created at the points of intersection of the roads. The provided County drawing will be required.

Step 1 In Map 2000 open the **COUNTY.DWG** drawing file, shown below in Figure 2–47.

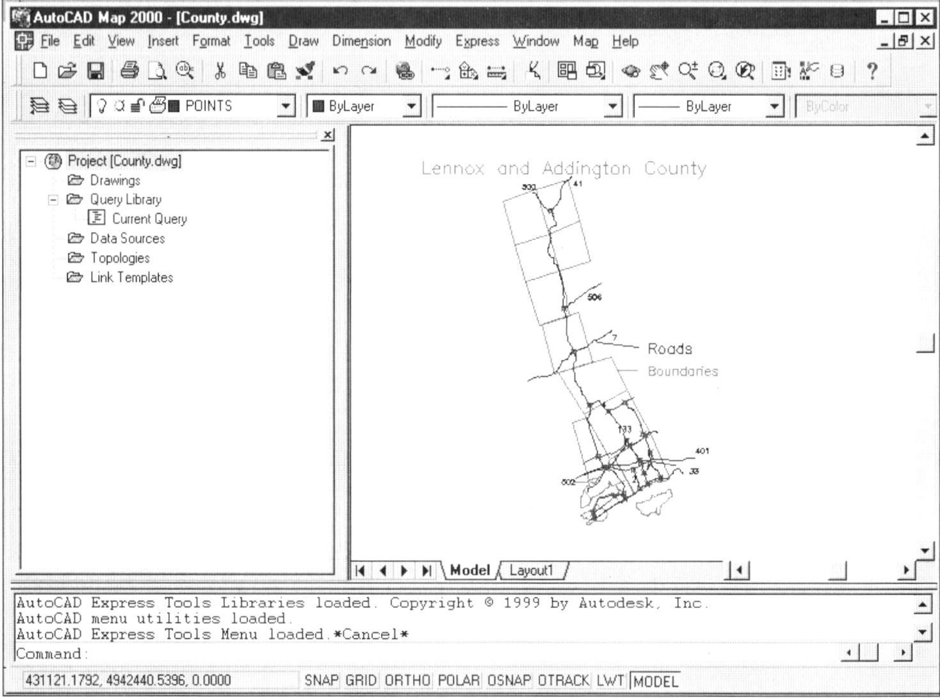

Figure 2–47

Step 2 Go to the **Map** menu and select **Topology** then **Create**. Type in a name and an optional description. The example uses a name of **Crossroads** and a description of **road intersections.**

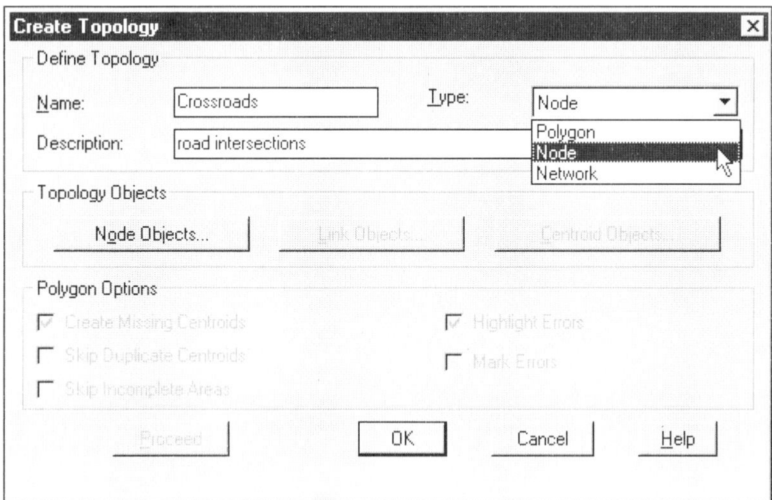

Figure 2–48

Step 3 Set the topology type to **Node**, then click on the **Node Objects** button. At the Node Objects dialog box, click on the **Select Automatically** radio button, then click on **OK.** Information on the progress of the topology creation will be provided in the command prompt area.

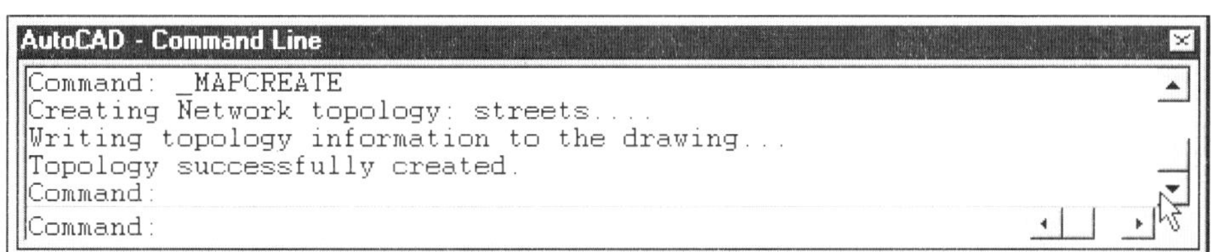

Figure 2–49

Step 4 Save the drawing and either continue with the tutorials or exit AutoCAD Map 2000.

TUTORIAL: NETWORK TRACES USING TOPOLOGY

Network tracing is a technique used, for example, to determine preferred routes across a road network, such as in the case of emergency response vehicles. Similarly, this approach could be applied to a pipeline system or a distribution grid. Network tracing makes use of network topology. For this exercise you will use the NAPTOP drawing that was created in the network topology tutorial.

Step 1 Start AutoCAD Map 2000 and open the **NAPTOP.DWG** drawing. The first step is to load the topology that was created earlier to the drawing file. From the **Map** menu select **Topology** then highlight **Administration**.

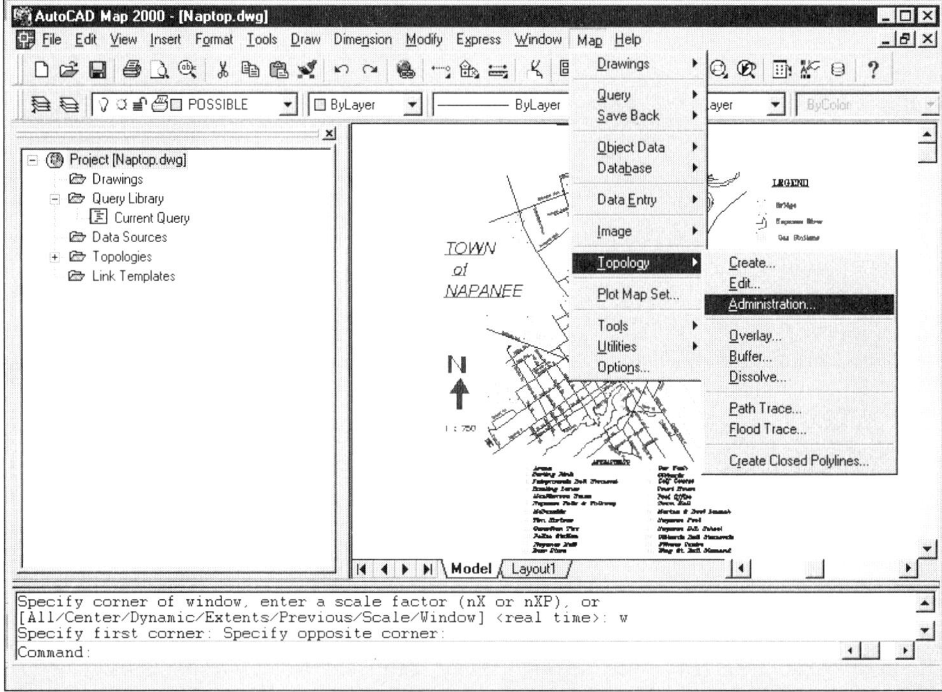

Figure 2–50

Step 2 The Topology Administration dialog box will appear.

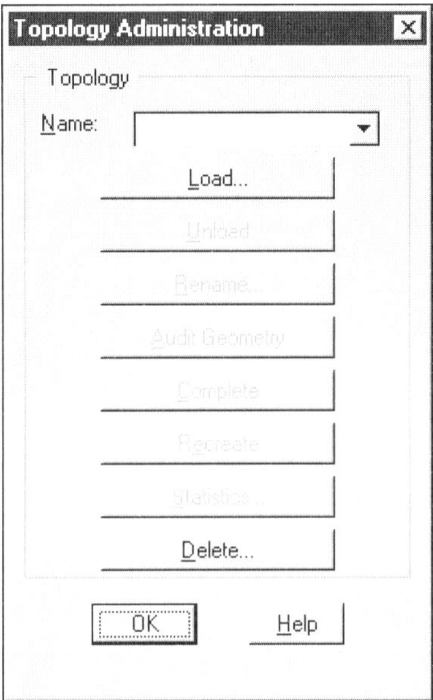

Figure 2–51

Step 3 Click on the **Load** button, and the Load Topology dialog box will be displayed.

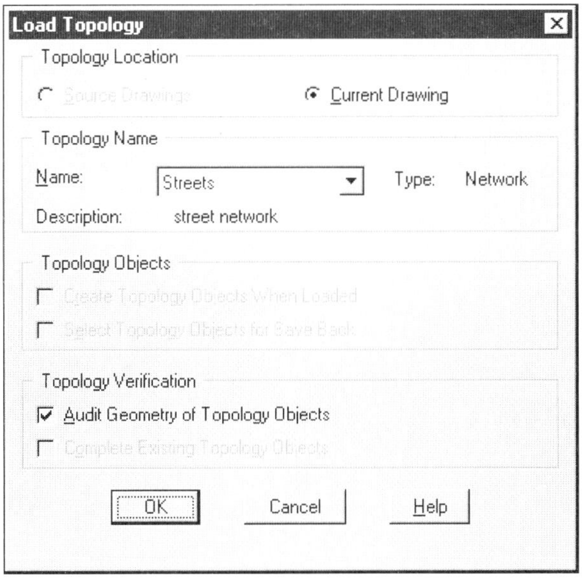

Load Topology
Topology Location
Source Drawings
Current Drawing
Topology Name
Name: Streets
Type: Network
Description: street network
Topology Objects
Create Topology Objects When Loaded
Select Topology Objects for Save Back
Topology Verification
Audit Geometry of Topology Objects
Complete Existing Topology Objects
OK
Cancel
Help

Figure 2–52

The **Streets** topology should be identified in the Name edit box. Click on **OK** and wait briefly while AutoCAD Map 2000 checks the integrity of the topology. A message will inform you of the condition of the topology, if it is complete and intact.

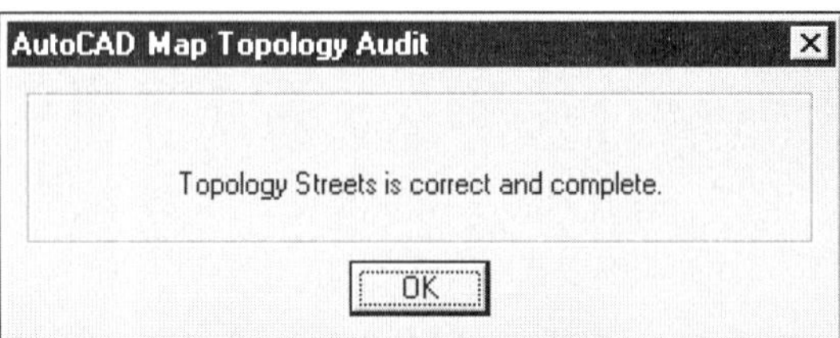

Figure 2–53

Click on **OK** and the Topology Administration box will reappear. Click on **OK** again.

Step 4 You will now use this topology to generate a shortest-distance route map along the street network of Napanee. **Zoom** in on the portion of the map identified in the graphic below (Figure 2–54) to facilitate viewing the streets (see Figure 2–55).

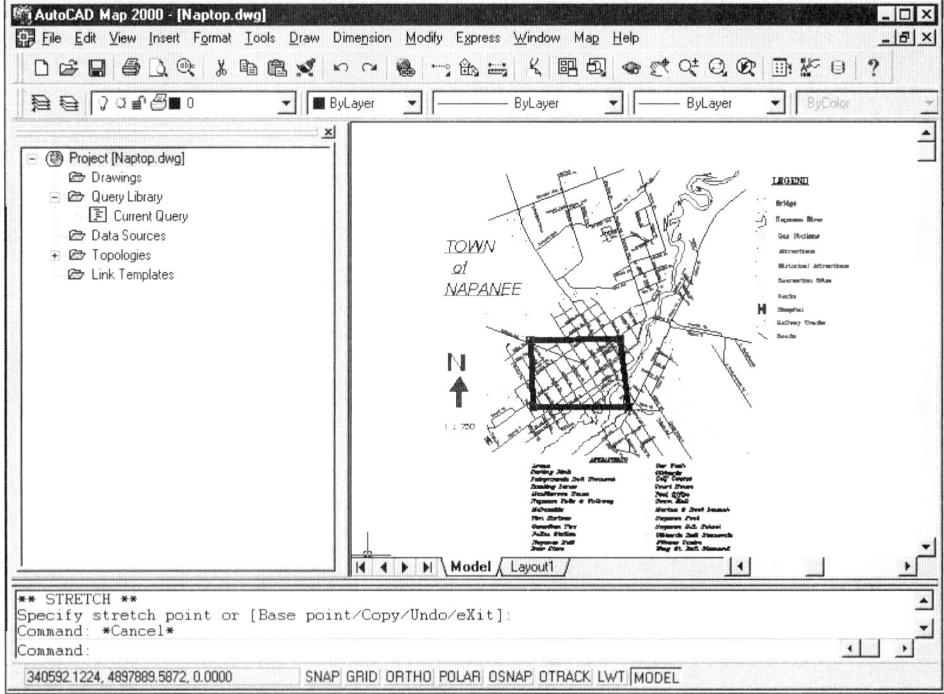

Figure 2–54

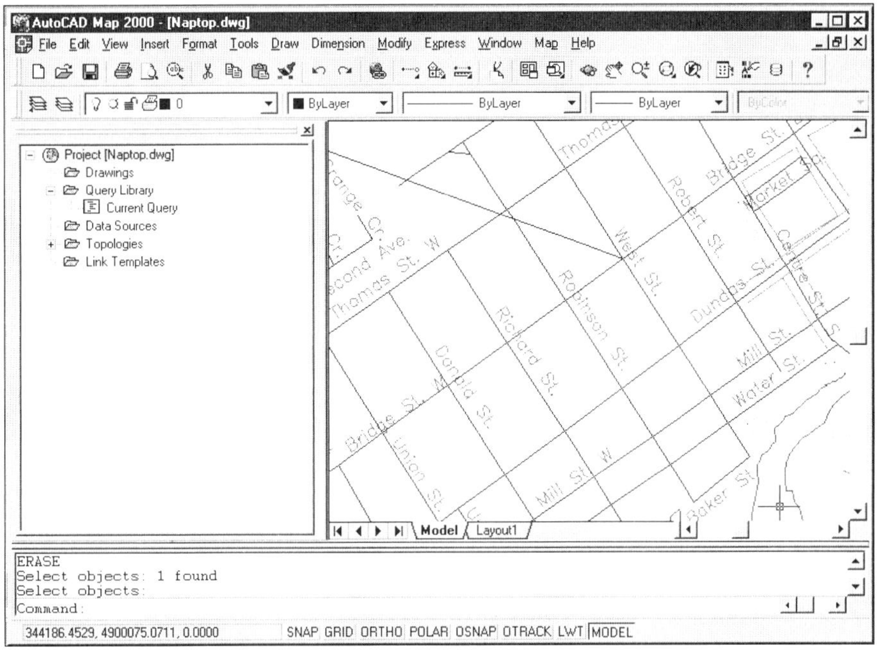

Figure 2–55

Step 5 Go to the **Map** menu and select **Topology** then **Path Trace.**

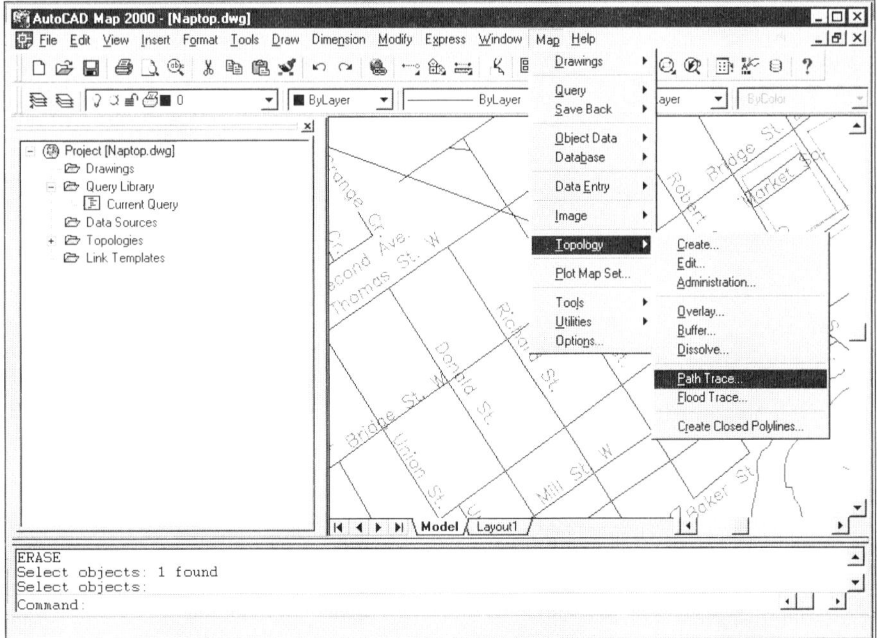

Figure 2–56

The topology file Streets will be listed as the name for the topology in the Shortest Path Trace dialog box.

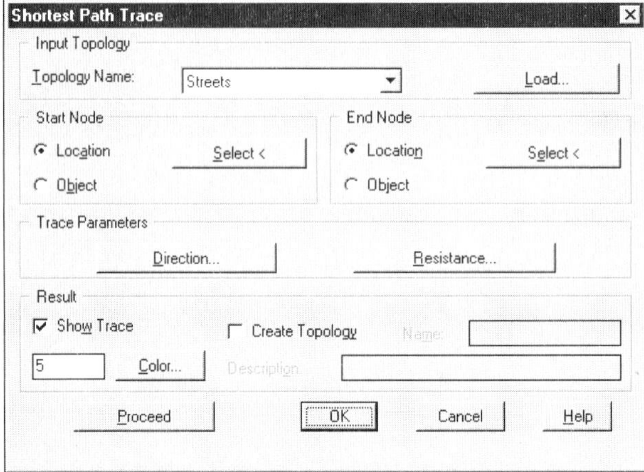

Figure 2–57

Under the **Start Node** make sure that the **Location** radio button is selected and then click on the **Select** button. Place the mouse pointer on an intersection and click.

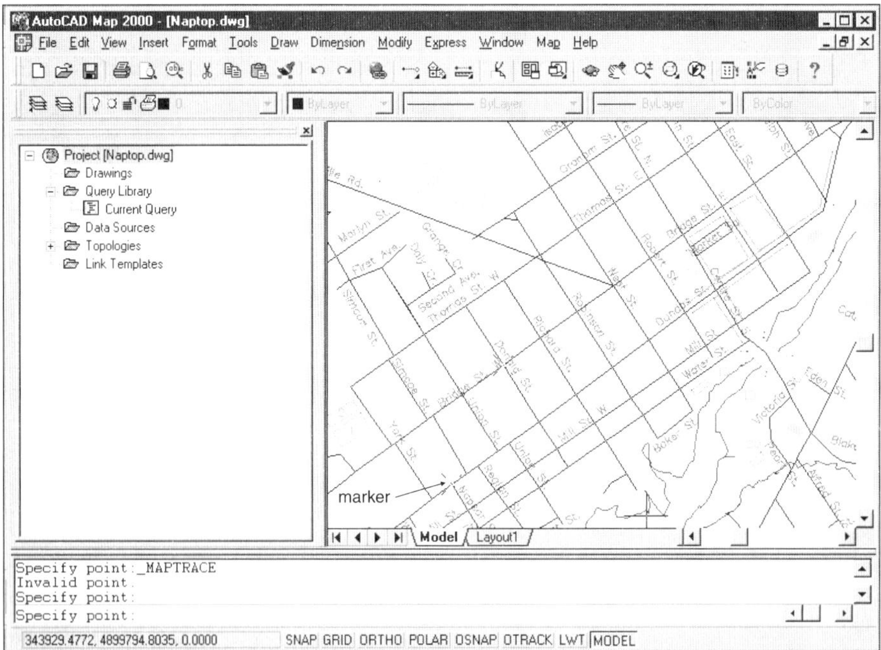

Figure 2–58 *(In this example the intersection selected was the corner of Highway 2 and Napear Street)*

A small marker will appear on the drawing.

Step 6 Click the right mouse button (or press ENTER) and the Shortest Path Trace dialog box will again be presented. Under the **End Node** section choose the **Location** radio button, then click on **Select** and click on another point along the road network.

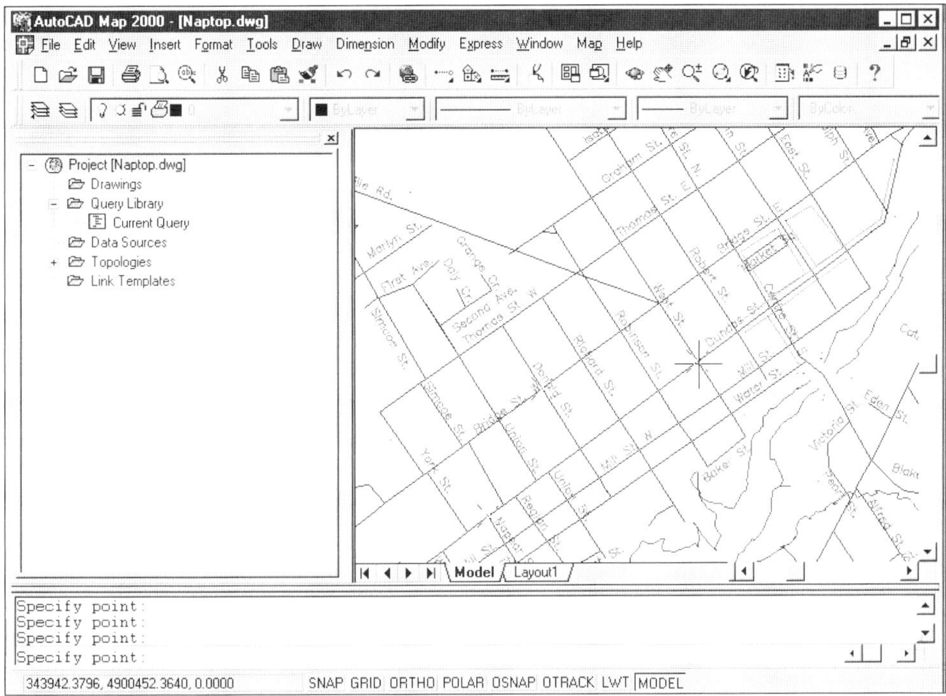

Figure 2–59 *(In this example the second point selected was the intersection of Mill Street and West Street)*

Step 7 Click the right mouse button again to return to the Shortest Path Trace dialog box. Click on the **Color** button to determine the color that will be used to show the shortest route. The Select Color dialog box will appear.

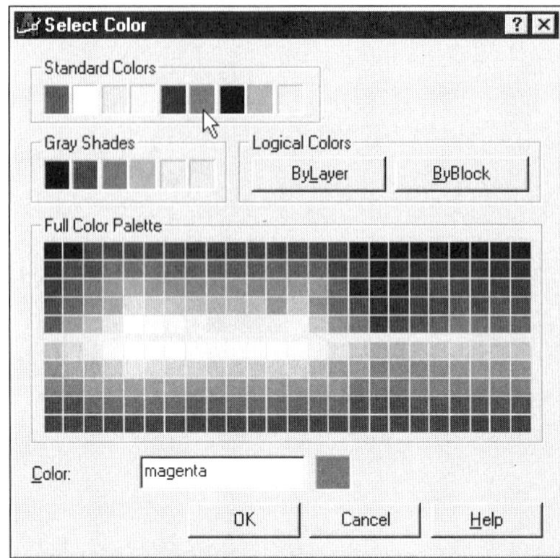

Figure 2–60 *(In this example, the color chosen was magenta)*

Click on **OK**, then click on **Proceed**. Wait a few seconds, and the route will be presented on the screen in the chosen color.

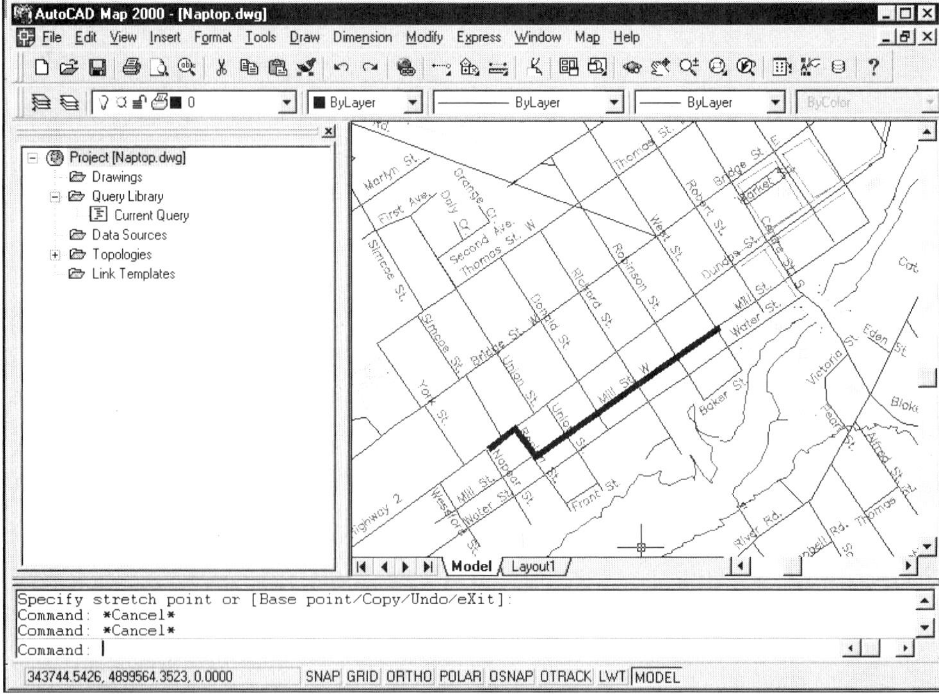

Figure 2–61 *(In the example shown, the route's line width has been enhanced)*

In the process of establishing shortest path traces one can also assign values for *resistance* and *direction* along any route. Resistance refers to the "friction" associated with following a link in the direction that it was created. One might consider the resistance associated with traveling along a road that has numerous traffic lights and intersections compared to a route with fewer such obstacles. *Reverse resistance* refers to the friction associated with traveling in the opposite direction to that which the link was drawn. This allows for varying resistances to be used in the path traces. This enables the user to accommodate, for example, changes in slope in a sewage pipeline. Direction can be established as bi-directional, one-way but the same as the link was created, or one-way and opposite to that in which the link was created. These parameters can be used to emulate real-world conditions to predict travel times across the network.

To start the next part of this tutorial, **zoom** in on a portion of the map similar to that shown below (Figure 2–62).

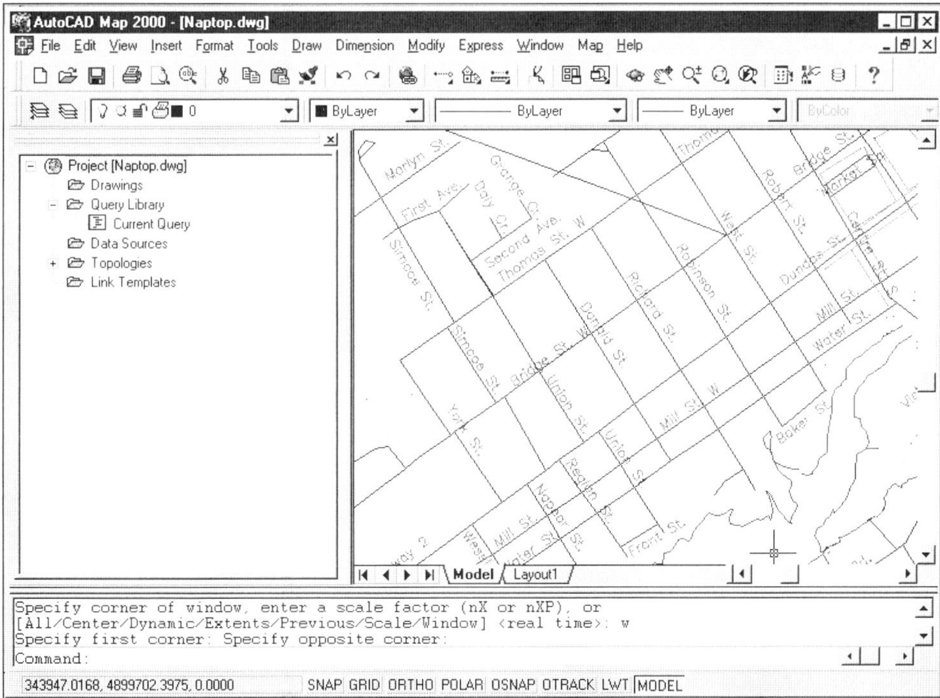

Figure 2–62

Step 8 Go to the **Map** menu and select **Topology** then **Edit**. The Edit Topology dialog box will appear. The topology name should identify the **Streets** topology you created and the object type should be identified as **link**. In the Edit Operation portion of the dialog box select the **Direction** radio button and click on **OK**.

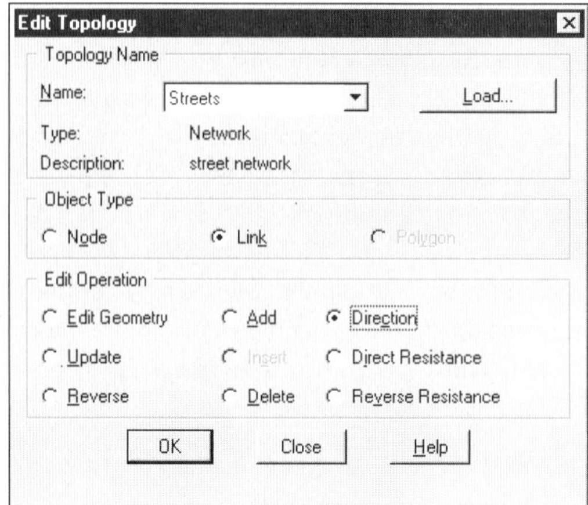

Figure 2–63

Step 9 Refer to the zoomed-in drawing and identify **Dundas Street**. Use the pick box to click on the portions of the street between Donald and Robert Streets. The street sections will be highlighted as they are selected.

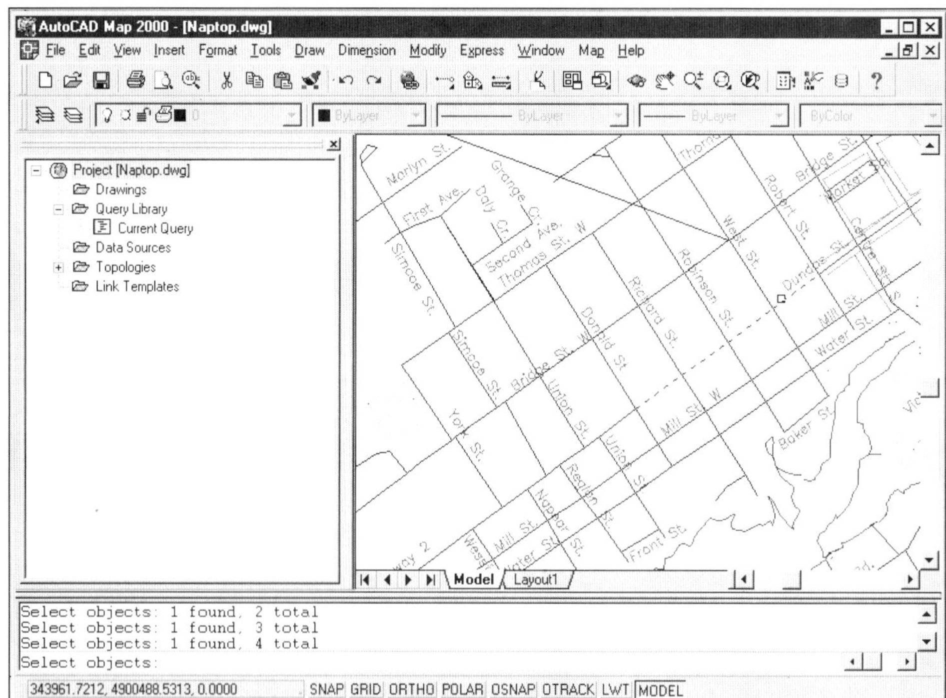

Figure 2–64

Click the right mouse button to signal that you have finished selecting objects. The Edit Direction dialog box will be presented, indicating that you have selected four links.

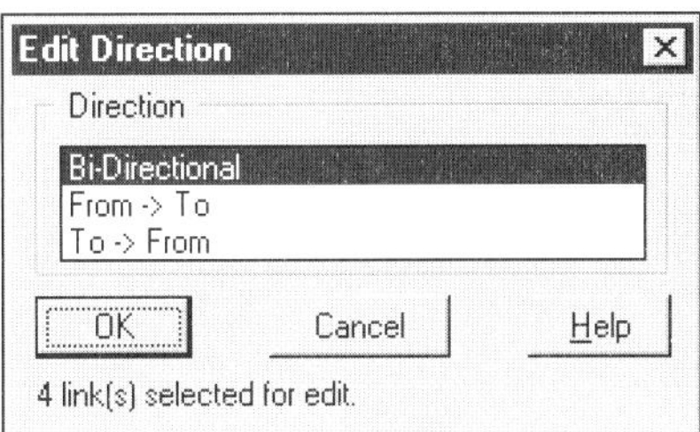

Figure 2–65

The three options allow the directivity to be established. Highlight the **From->To** option in the Direction list. This will assign directivity to these street sections based on the direction in which they were originally drawn. Click on **OK**, and at the Edit Topology screen, click on **OK** again. A message in the command line area will indicate that four links were processed. To end the selection process, click the right mouse button and in the Edit Topology dialog box click on **Close.**

A shortest path trace performed after this data had been added to the drawing would now take into account the one-way directionality of these sections of the street. In the example shown below, a shortest path trace from the corner of Dundas and Robert Streets to the corner of Richard and Dundas would take into account the one way directionality of Dundas Street as it runs from Richard to Robert Streets.

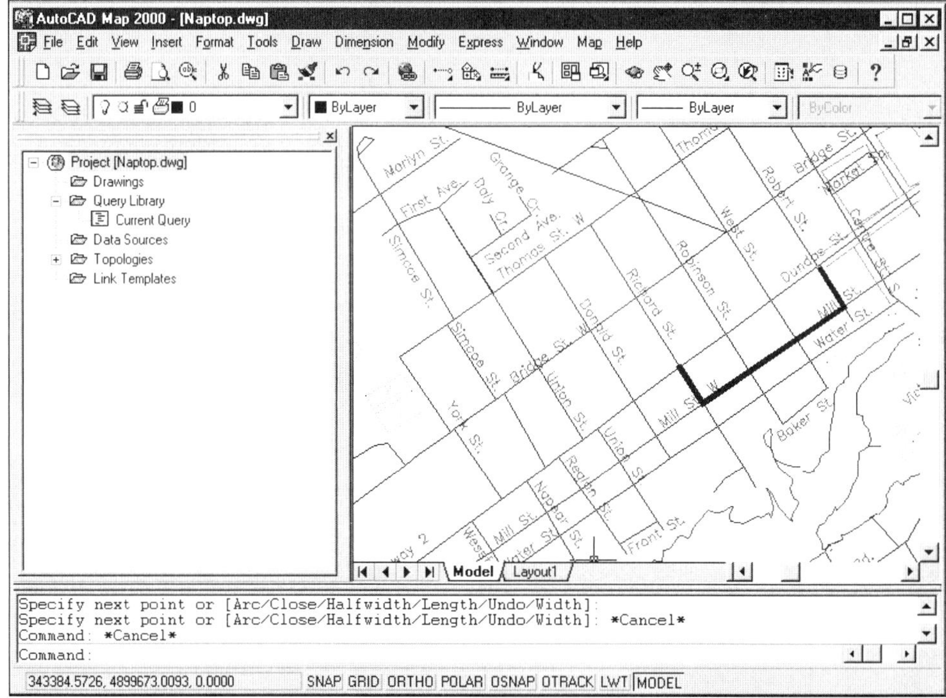

Figure 2–66

Another feature of path traces is the capability of incorporating object data into the trace.

This requires an object data table and a field of data relating to the drawing. For streets, appropriate data might be in the form of speed limits. This will allow the path trace to be based on both distance units from the drawing and speed units from the object data, permitting time calculations yielding paths based on the shortest time required rather than the shortest distance.

Step 10 To complete a time-based path trace, create a new object data table by going to the **Map** menu and selecting **Object Data** then **Define Object Data**. Click on **New Table.**

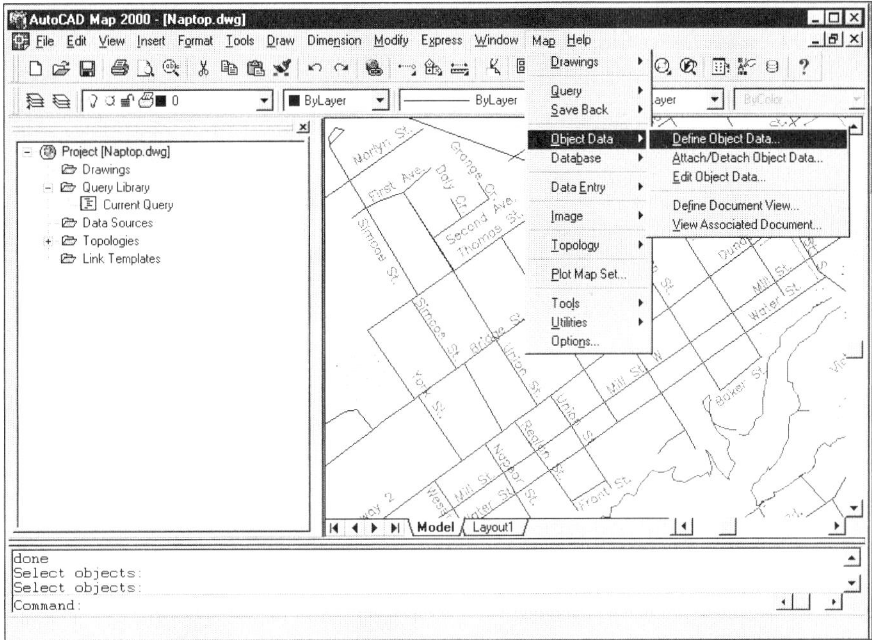

Figure 2–67

In the Define New Object Data Table dialog box enter a table name, **Streetfact,** in this case, ensure that the data type is set to **Integer**, enter a field name of **Speed**, and an optional description of "**average speed in mph**".

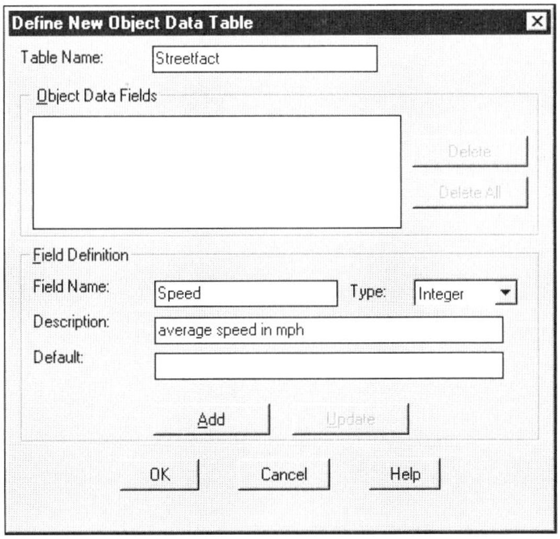

Define New Object Data Table
Table Name: Streetfact
Object Data Fields
Delete
Delete All
Field Definition
Field Name: Speed
Type: Integer
Description: average speed in mph
Default:
Add
Update
OK
Cancel
Help

Figure 2–68

Click on **Add** then **OK**, and at the next dialog box click on **Close**.

Step 11 Go to the **Map** menu and select **Object Data** and then **Attach/Detach Object Data**.

Highlight the Object Data Field described as **average speed in mph**. Enter a value of **20** in the text box

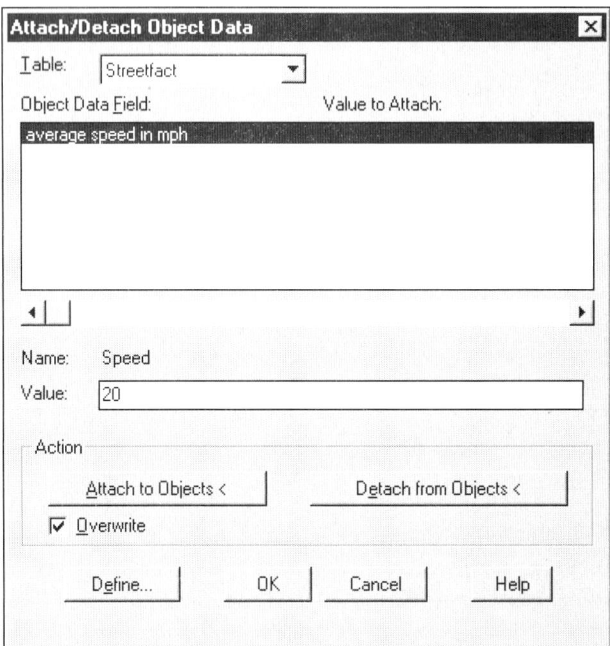

Figure 2–69

Use the mouse pick box to select several sections of Dundas Street. They will become highlighted as the data is attached to them.

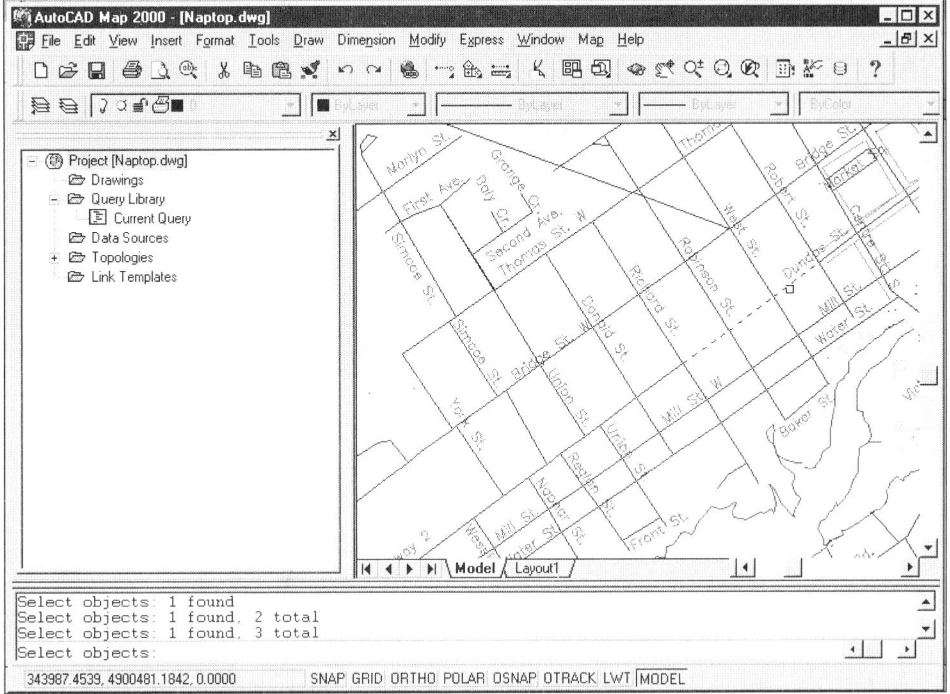

Figure 2–70

Click the right mouse button to signal that you are finished selecting objects. In the command line area of the screen a message will appear indicating the number of objects that had data attached to them. In a typical application, all streets in the network would have object data applied to them.

Step 12 To conduct a shortest path trace utilizing this type of data, go to the **Map** menu and select **Topology** then **Path Trace.** Select a start node and an end node. Click on the **Resistance** button and the Resistance Parameters dialog box will appear.

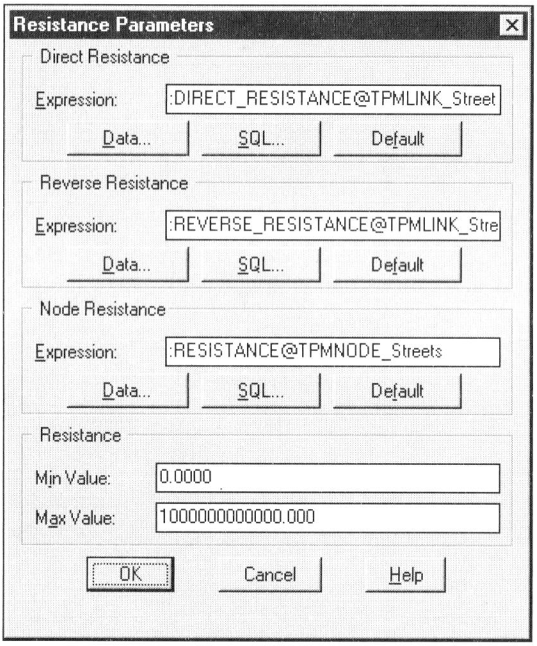

Figure 2–71

In the Expression edit box located in the Direct Resistance portion of the dialog box, enter in the AutoLISP expression **(\ length(\speed@streetfact 5280) 60)**. This dialog box also facilitates setting resistance for the opposite directions and for nodes in the topology. Click on **OK**, then click on **Proceed**. The trace will show a route based on the time taken to get from one node to another based on the directivity in the drawing and based on the speed parameters that were set through the object data. You can save the file with a new name or exit without saving changes.

At this time either continue on with the tutorials or exit AutoCAD Map 2000.

NETWORK FLOOD TRACES

The *networkflood* function is similar to the network path trace except that it traces out all the routes from a particular, selected node. The functionality depends on the data appended to the drawing set. One could, for example, seek out all gas stations within a certain distance from a point if that data had been built into the drawings.

TUTORIAL: BUFFER ZONE ANALYSIS

Buffer zones allow you to identify belts or strips of land that surround a particular topology. The zones can be centered on nodes as exemplified by business catchment areas around a prospective retail site, a containment area for toxic spills, or along links in a network (Figure 2–72), as would be the case in determining properties to be acquired in order to accommodate a major road improvement project.

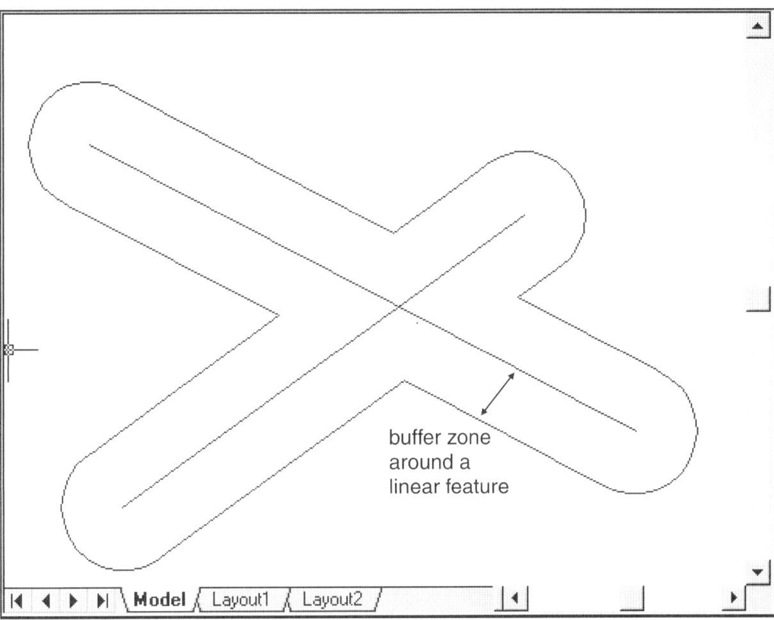

Figure 2–72 *Buffer zone surrounding two linear features*

In this case, the Napanee drawing with the street topology will serve as the base drawing for building a zone surrounding the streets that can be identified as municipal property as opposed to the lot owner's property.

Step 1 Close any drawings that are currently active and open the **NAPTOP.DWG** drawing.

Step 2 Go to the **Map** menu and select **Topology**, then **Administration**.

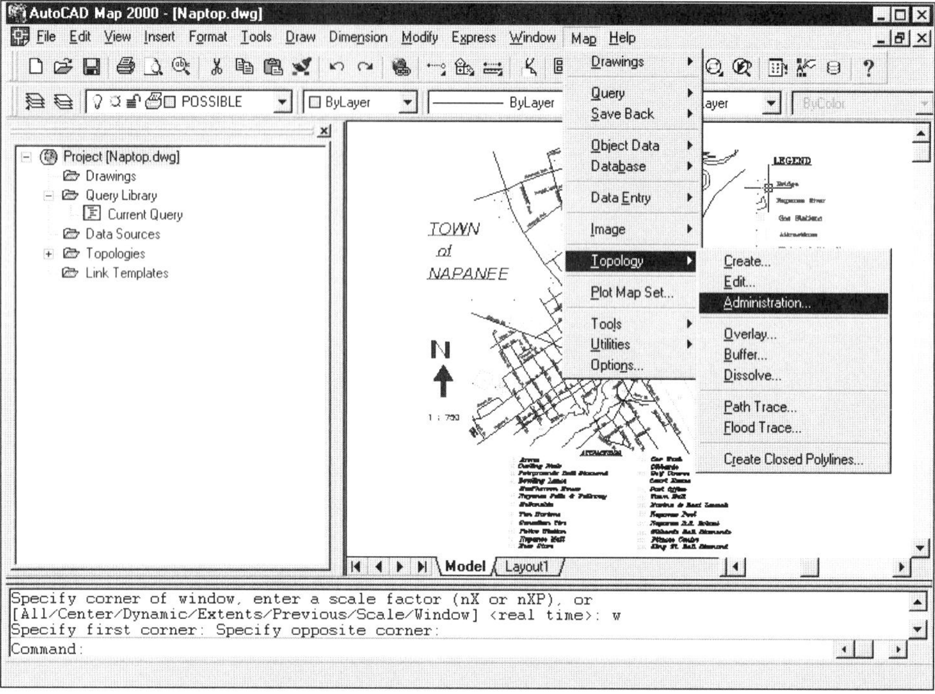

Figure 2–73

In the Topology Administration dialog box click on **Load**. The Load Topology dialog box will appear. The topology file **Street** will appear by default.

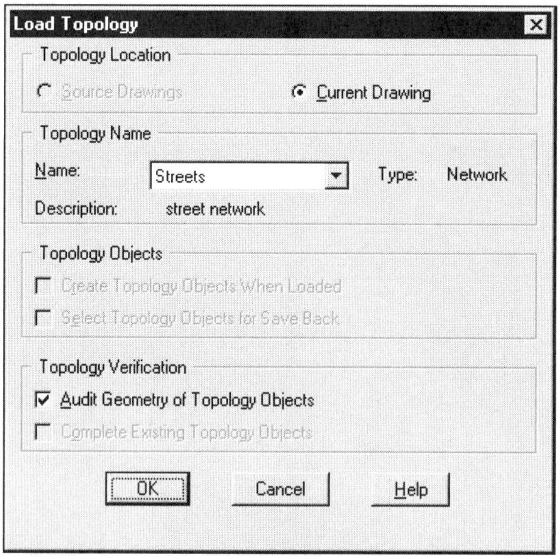

Load Topology

Topology Location

Source Drawings ◉ Current Drawing

Topology Name

Name: Streets Type: Network

Description: street network

Topology Objects

☐ Create Topology Objects When Loaded

☐ Select Topology Objects for Save Back

Topology Verification

☑ Audit Geometry of Topology Objects

☐ Complete Existing Topology Objects

OK Cancel Help

Figure 2–74

Click on **OK**. The AutoCAD Map Topology Audit dialog box will appear, informing you that the topology for Streets is correct and complete. Click on **OK**, and at the Topology Administration box click on **OK** again.

Step 3 Go to the **Map** menu, then select **Topology** and **Buffer**.

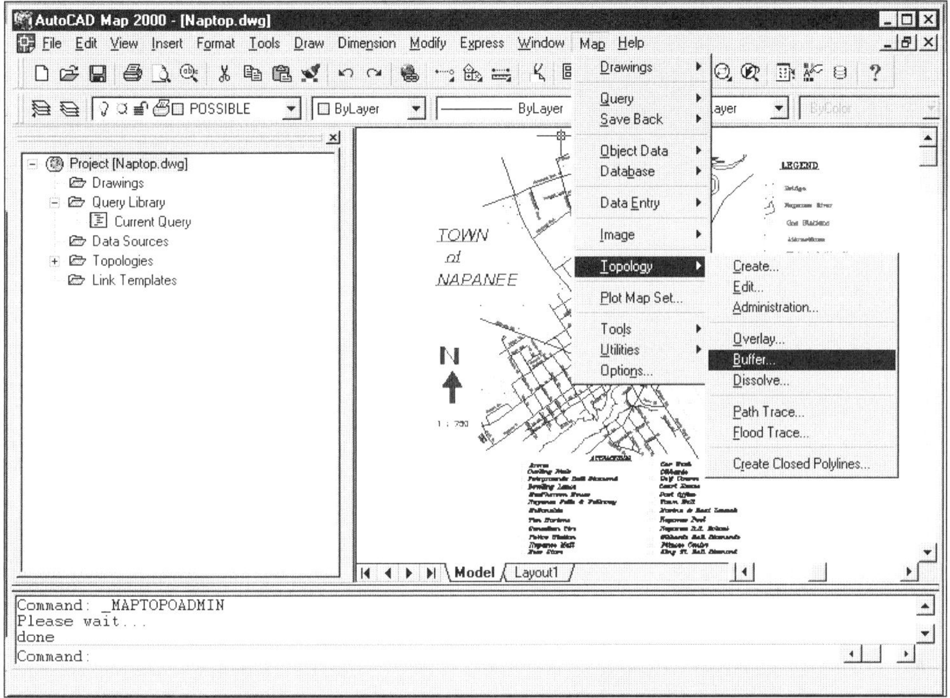

Figure 2–75

The Buffer Topology dialog box will appear.

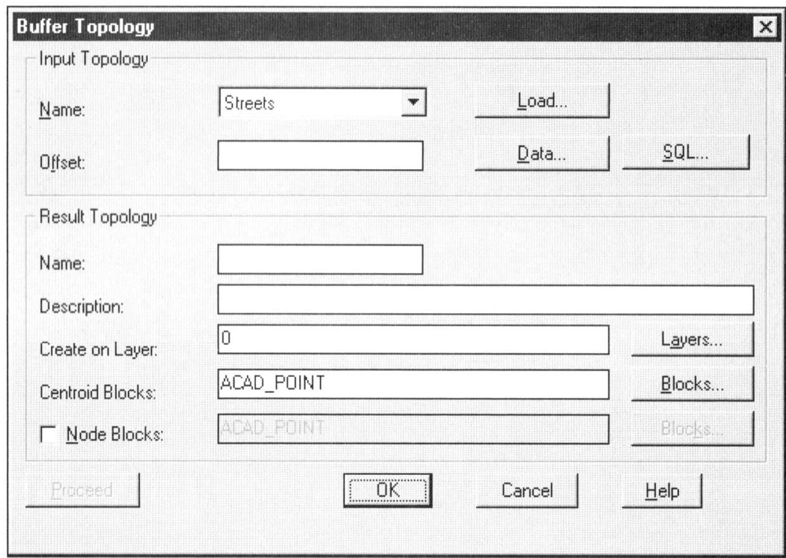

Figure 2–76

In the upper section of this dialog box click on the **Data** button. The Data Expression dialog box will appear.

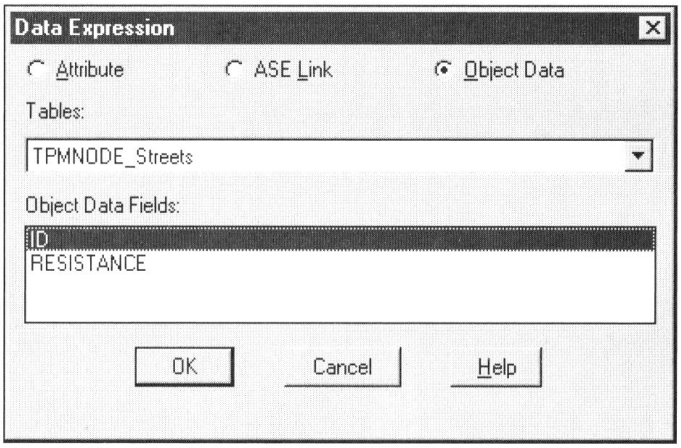

Data Expression
Attribute ASE Link Object Data
Tables:
TPMNODE_Streets
Object Data Fields:
ID
RESISTANCE
OK Cancel Help

Figure 2–77

Select the **Object Data** radio button to indicate the source for data. From the pull-down list of tables, choose **TPMDESC_Streets** and select the **Tolerance** item as the Object Data field then click on **OK**.

In the Offset edit box of the Buffer Topology dialog box replace TPMDESC_Streets with a value of **30** units so the box reads as: **TOLERANCE@30**.

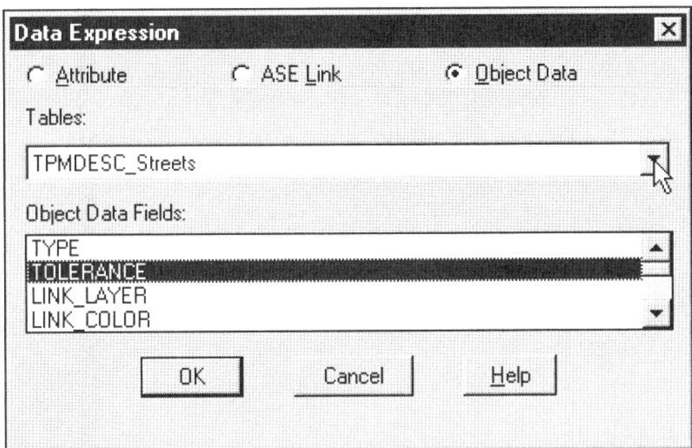

Figure 2–78

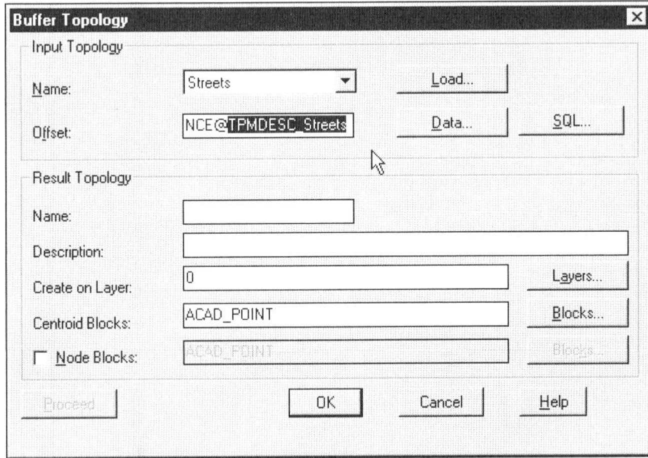

Figure 2–79

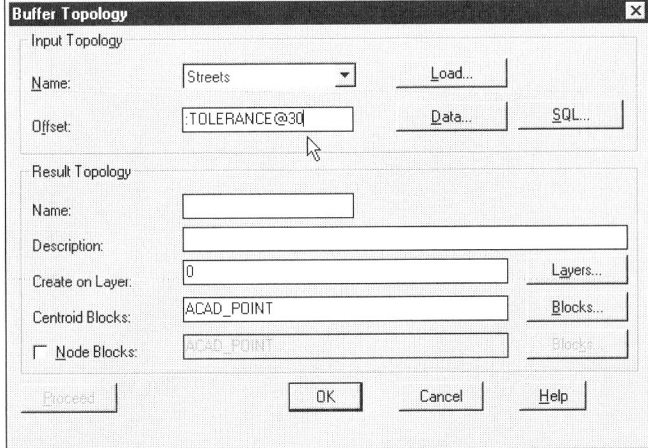

Figure 2–80

Step 4 Enter a Name for the new topology to be created. In this case, the name **Buffer** was used and a description of **Buffer Zone** was applied. Click on **Proceed.**

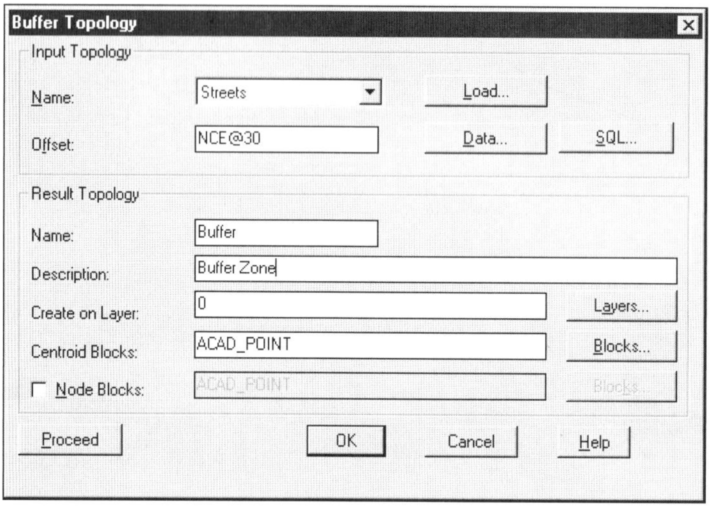

Buffer Topology
Input Topology
Name: Streets | Load...
Offset: NCE@30 | Data... | SQL...
Result Topology
Name: Buffer
Description: Buffer Zone
Create on Layer: 0 | Layers...
Centroid Blocks: ACAD_POINT | Blocks...
Node Blocks: ACAD_POINT | Blocks...
Proceed | OK | Cancel | Help

Figure 2–81

The new topology will be created and buffer zones will be shown on the map.

Step 5 To view the buffer zone, **zoom** in on a section of the drawing.

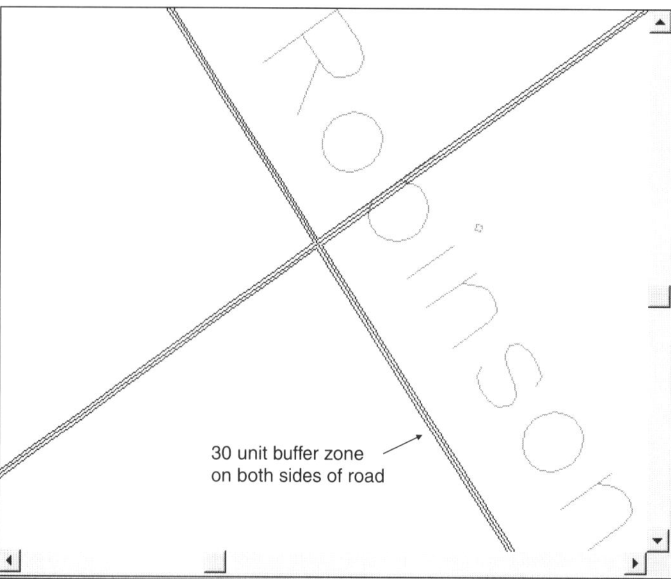

Figure 2–82

Do not save the changes to this drawing as you either continue with the tutorials or exit AutoCAD Map 2000.

TUTORIAL: OVERLAY ANALYSIS

One of the most powerful and commonly used analysis techniques in GIS is the *overlay*. This process positions one layer of data over another using topology, allowing for analysis of the relationship between the features. Topology overlays can be completed using nodes with polygons, networks with polygons, or polygons with polygons. To enhance the overlay process, several options are available, including intersection, union, identity, erase, clip, and paste, depending on the topologies being used.

More information on the specifics of each application is available in the online help guide.

In this tutorial, you will utilize a map of the northern half of Lennox and Addington County that contains hypothetical information showing a nature park area as well as a nearby, overlapping mining claim (Figure 2–83).

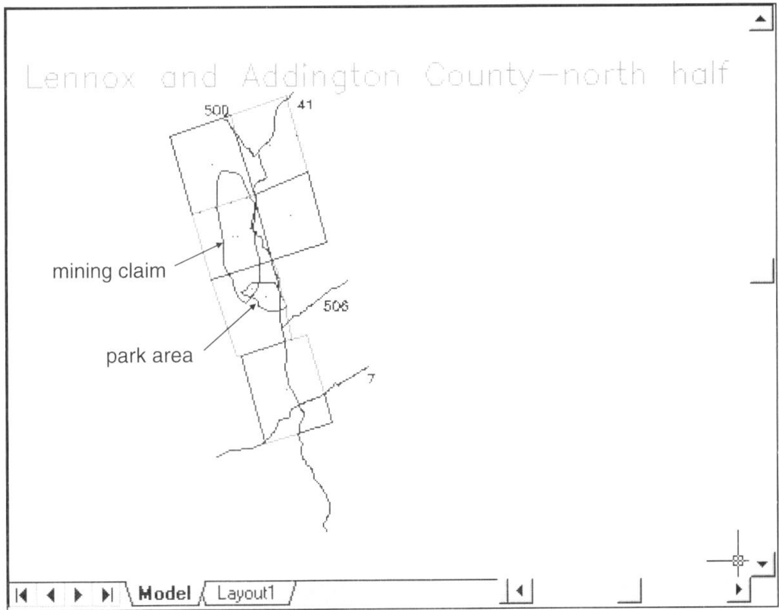

Figure 2–83 *Two regions resulting from overlay analysis*

The exercise will reveal how to use the overlay technique to identify an area of concern or intersection of the two conflicting land uses.

It is important to realize that the base map being used in this exercise had two layers, Mining and Park_area, created with each layer showing only the one feature. In each case, with all other layers frozen and turned off, a polygon topology was created for the feature; once for Park_area and once for Mining, following the example of earlier tutorials.

Step 1 Open the drawing called **LNAOVER**.

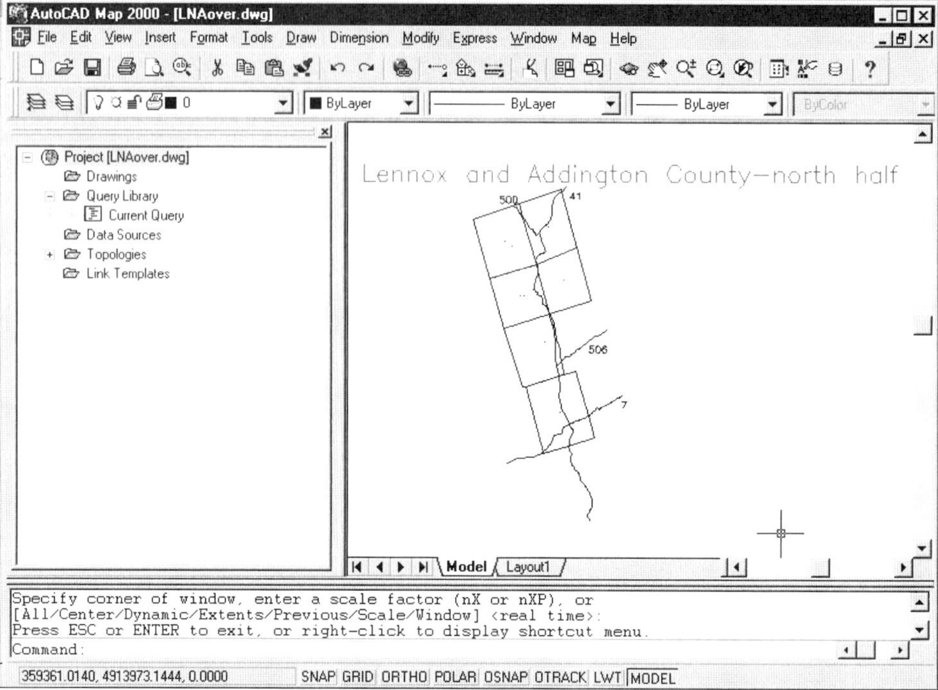

Figure 2–84

Verify the existence of the Park_area and Mining layers. Go to the **Format** menu and then select **Layer** to activate the Layers Properties Manager dialog box. View the list of layers in the drawing, then ensure that the Park_area and Mining layers are both turned **off**.

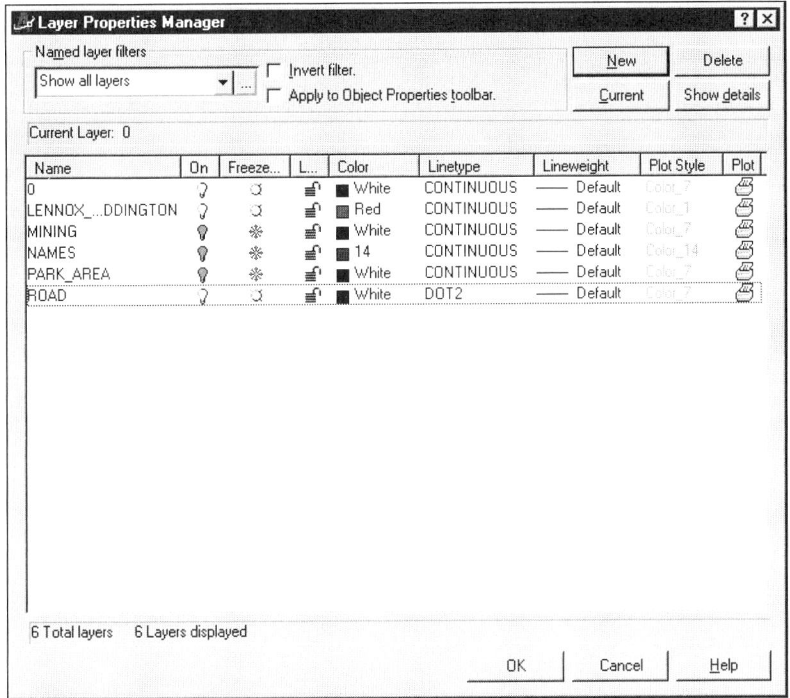

Figure 2–85

Step 2 Go to the **Map** menu, then select **Topology** and highlight **Administration**. In the Topology Administration dialog box click on the **Load** button.

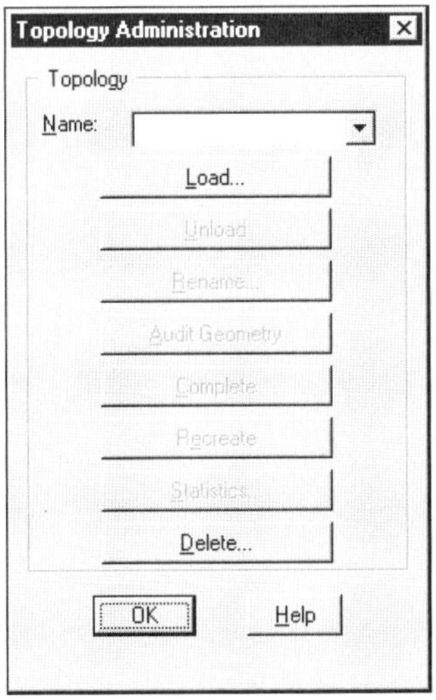

Figure 2–86

From the list available select **Mining** as the topology to load and then click on **OK.** A message will be presented informing you that the topology is **Correct and Complete**. Click on **OK** at this message.

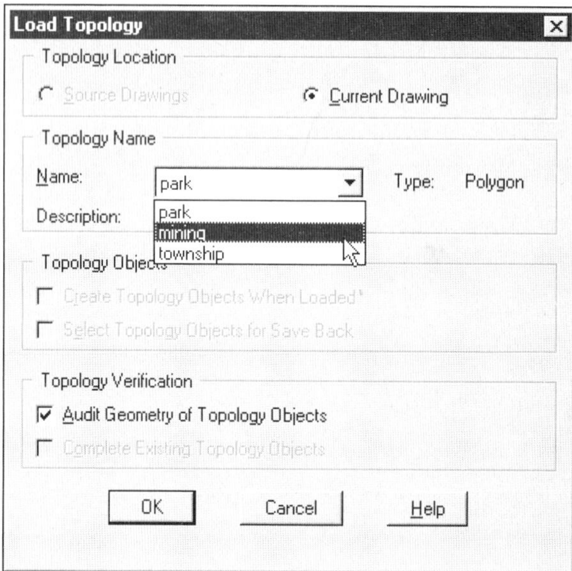

Figure 2–87

Step 3 The Topology Administration dialog box will appear again. Repeat the Load process for the **Park** topology, clicking on **OK** when you are informed that the topology is correct and complete. You will be returned to the Topology Administration dialog box. Click on **OK**.

Step 4 Go to the **Map** menu and select **Topology** then **Overlay**.

The Overlay Topology dialog box will be presented. In the **Source** edit box you should see the Mining topology and in the **Overlay** edit box the Park topology should be listed. The **Operation** should be set to **Intersect**. This setting will find the areas of the two topologies that are shared and reflect an area of concern.

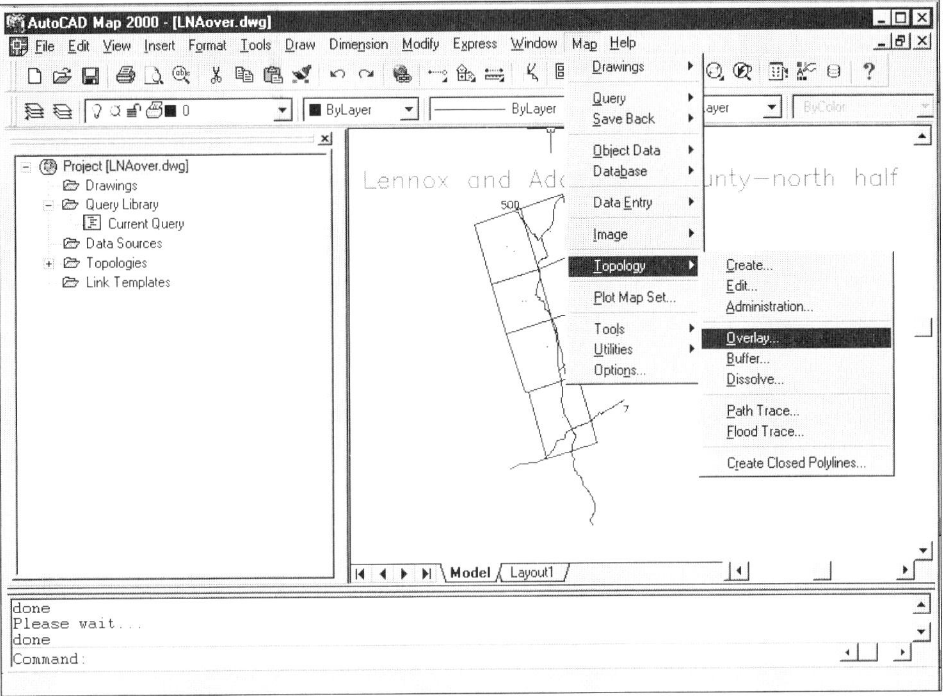

Figure 2–88

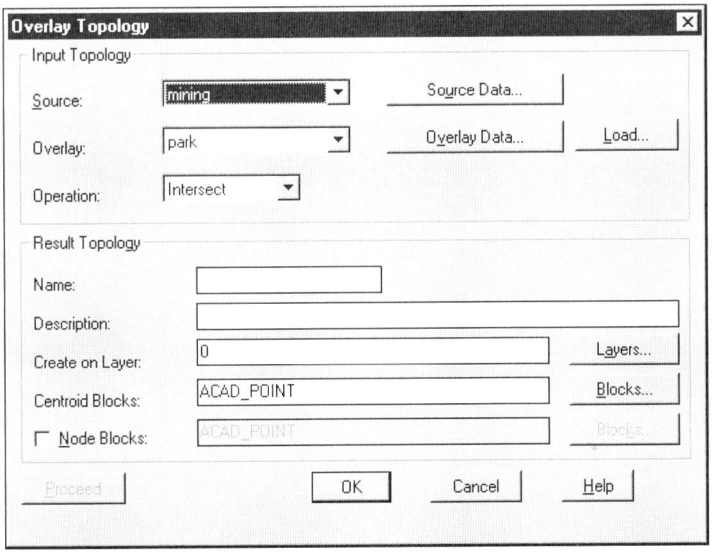

Overlay Topology

Input Topology

Source: mining — Source Data...

Overlay: park — Overlay Data... — Load...

Operation: Intersect

Result Topology

Name:

Description:

Create on Layer: 0 — Layers...

Centroid Blocks: ACAD_POINT — Blocks...

Node Blocks: ACAD_POINT — Blocks...

Proceed — OK — Cancel — Help

Figure 2–89

Step 5 Type in **Conflict** as a name for the resulting topology and enter a description of **Area of overlap**.

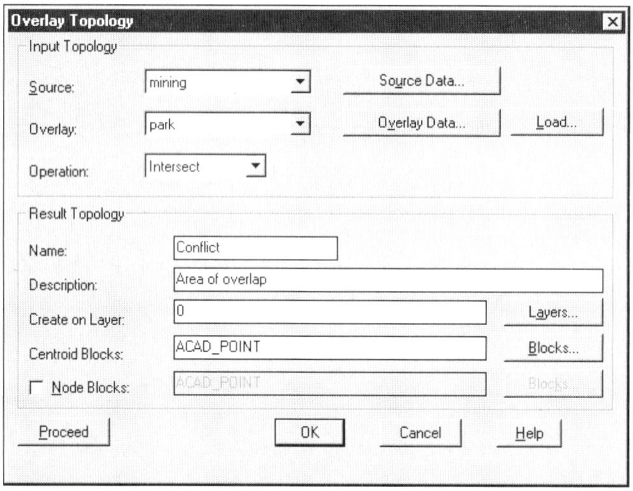

Overlay Topology
Input Topology
Source: mining — Source Data...
Overlay: park — Overlay Data... — Load...
Operation: Intersect
Result Topology
Name: Conflict
Description: Area of overlap
Create on Layer: 0 — Layers...
Centroid Blocks: ACAD_POINT — Blocks...
Node Blocks: ACAD_POINT — Blocks...
Proceed — OK — Cancel — Help

Figure 2–90

Click on **Proceed.**

The resulting intersection of the two regions will be drawn on the screen.

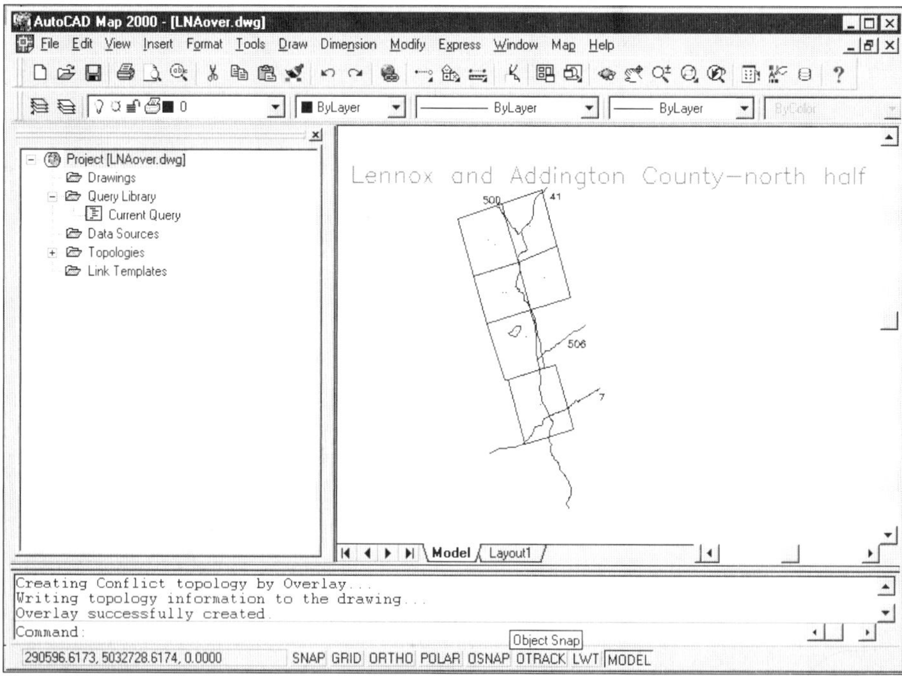

Figure 2–91

The small region shown is the area that is common to both topologies, Mining and Park_area.

Step 6 Either continue on with the tutorials or go to File then Exit.. Do not save changes made to the drawing.

SUMMARY

Topology is what separates a "dumb" drawing from an intelligent map. Different types of features, nodes, networks and polygons can have topology allowing AutoCAD Map 2000 to distinguish things like the direction of a line or which polygon a point is within, facilitating spatial analysis. Whenever topology is to be created for a drawing, a Drawing Cleanup should be performed to correct small errors that may have been produced when the map was originally created. Once the map files have topology, then a variety of analytical techniques can be used. Network traces can be used to determine shortest routes or, if resistance values are added to the topology, fastest routes. This technique could be used for planning for a variety of routing problems ranging from emergency responses to school bus routes. A Network Flood can be used to assess features within a given distance from a selected starting point. This procedure could be employed to locate for example, fast food restaurants, ski resorts, or hotels within specified distances from given starting points. Buffer zone analysis is a commonly used procedure in GIS. AutoCAD Map 2000 can make use of topology to determine zones of a specified size that surround features in the drawing. Overlay analysis allows the user to study the relationship between two or more variables. When combined with the collection of data over a period of time, Overlay analysis can be repeated to perform a temporal analysis of the relationship. This could be utilized to study, for example, the leaching of contaminants from a land fill, the growth pattern of a forest fire, or changes in demographic characteristics of a region.

REVIEW QUESTIONS

1. What is topology?
2. What are three types of topology?
3. Where do nodes occur?
4. What could you do after opening a drawing to determine easily whether or not the file contained closed polygons?
5. What are centroids?
6. What is the difference between a shortest path trace and a network flood?
7. What is the general purpose behind mapping a buffer zone?
8. How does an overlay differ from buffering?
9. Describe the function of the Topology Administration dialog box.
10. What are two overlay operations?

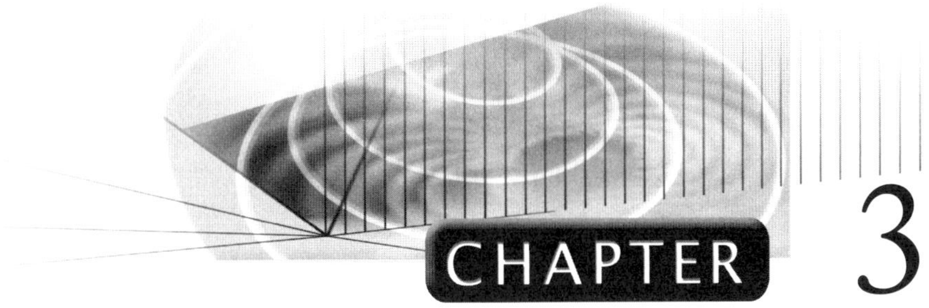

3

Raster Images

OBJECTIVES

After completing this chapter you will know how to

- import raster images
- use raster images as drawing backdrops
- how to georeference raster images

KEY TERMS

Raster
Pixel
Tif
Georeferenced

In GIS, *raster images* are generally associated with photographic images taken of the earth from a source like a satellite. These kinds of images can be of varying precision levels, and the information received can range from radar-based to infrared to conventional photographs. Other commonly used sources of raster images include scanned images and orthophoto images. The following exercise introduces raster images using a sample image downloaded from the Internet site **www.ngdc.noaa.gov.8080/production/html/BIOMASS/night.html,** which has been named USALIGHT.TIF. Prior to beginning this exercise, ensure that this TIFF file has been copied to your computer.

Although this tutorial utilizes a TIFF, a variety of other raster image file formats can be used. AutoCAD Map 2000 identifies 14 different types of raster images that it can directly import.

TUTORIAL: VIEWING RASTER IMAGES

Step 1 Start AutoCAD Map 2000 with a **new** drawing as the project drawing. Go to the **Insert** menu item and select **Raster Image**. The Select Image File dialog box will appear.

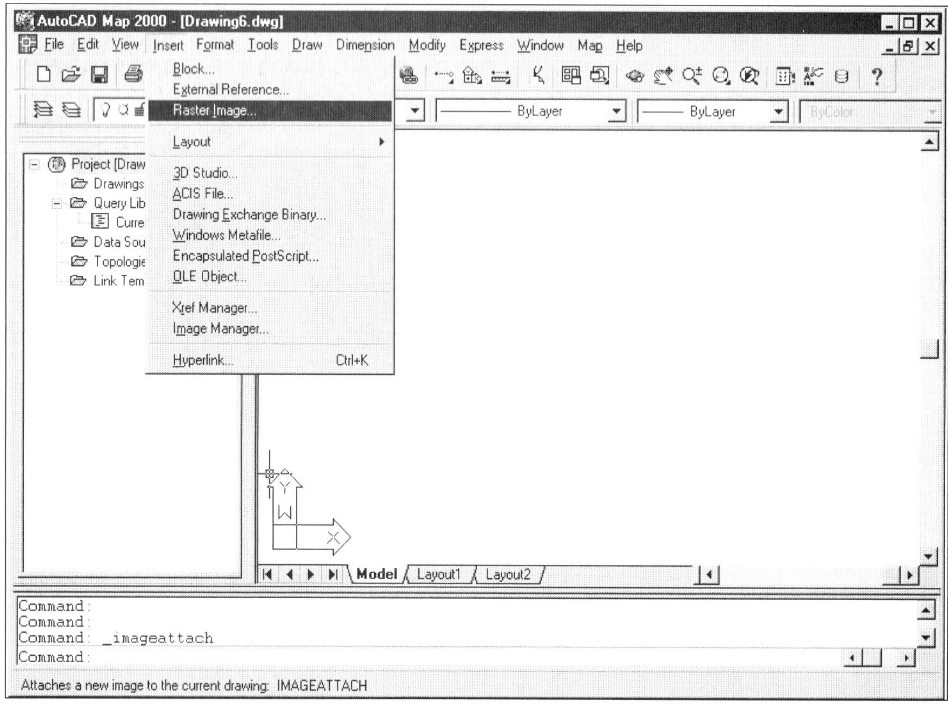

Figure 3–1

Step 2 Select the directory where you installed the **USALIGHT.TIF** and highlight it. The image will be presented as a thumbnail on the right side of the screen.

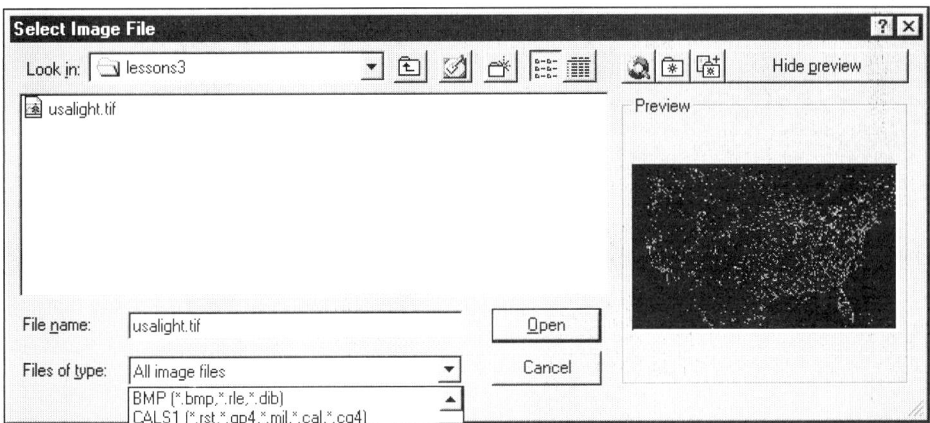

Figure 3–2

To view a list of the different file formats that can be imported directly into AutoCAD Map 2000 you can click on the **File of Type** option.

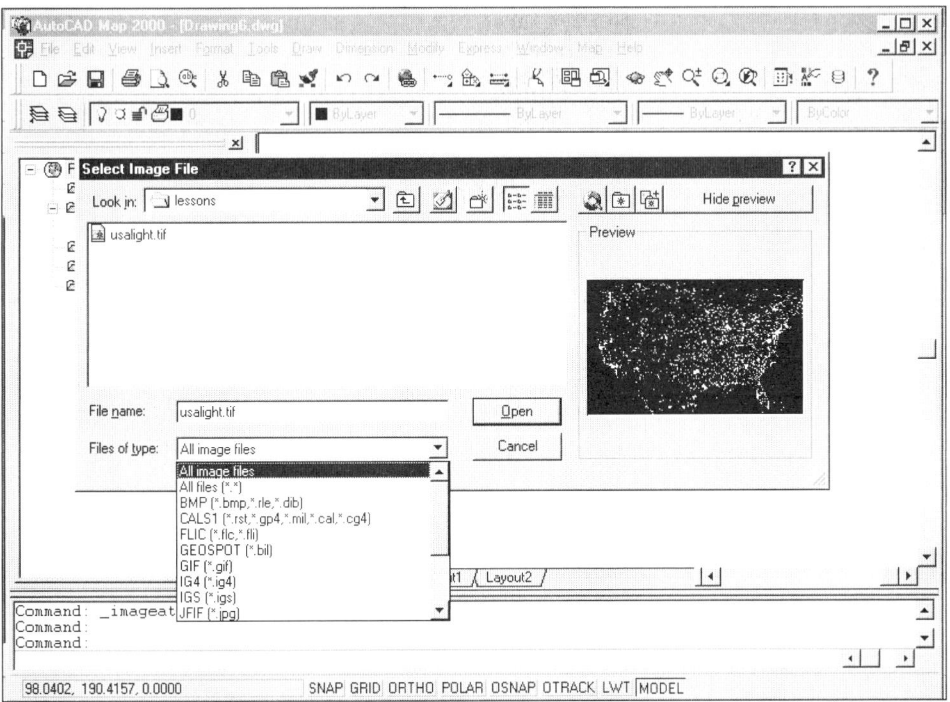

Figure 3–3

Click on **Open** and the Image dialog box is presented. Click on the **Details** button.

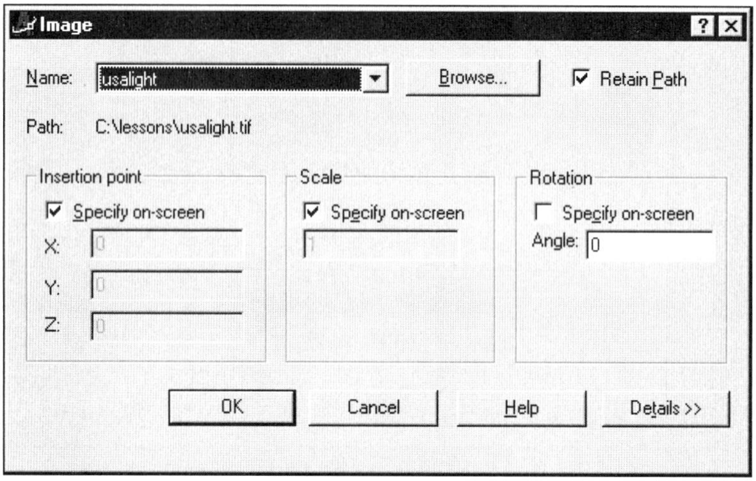

Figure 3–4

This dialog box provides additional information about the image, including its resolution, in this case, 300 by 300 dots per inch. Also shown are the width and height of the image in pixels. This image is 800 pixels wide by 480 pixels tall, which leads to the rectangular shape of the image when it is presented on the screen.

Step 3 In the Image dialog box Scale area, click on the **Specify on Screen** checkbox to remove the checkmark and then select the **edit** box below, replace the **1** with a value of **10.**

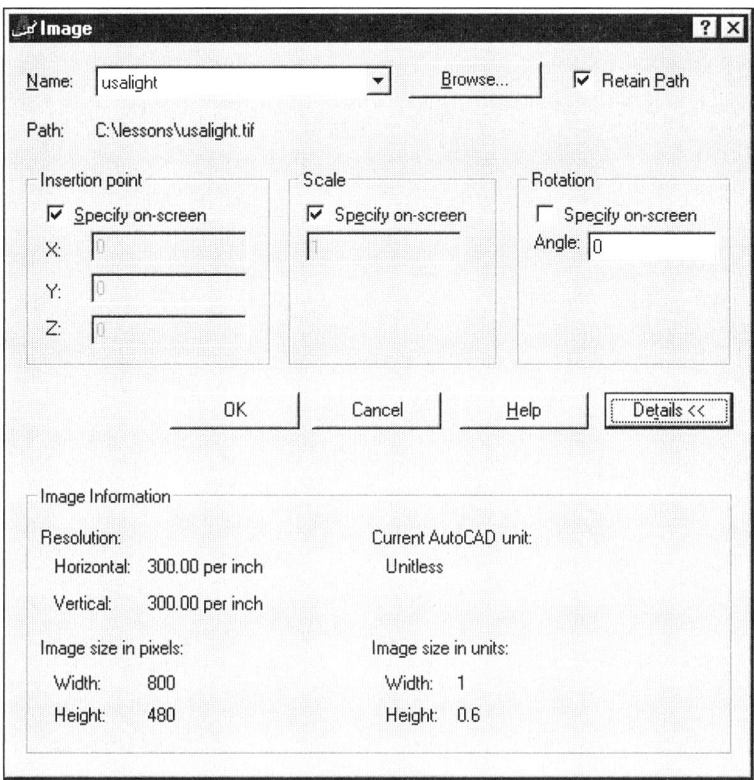
Image
Name: usalight
Browse...
Retain Path
Path: C:\lessons\usalight.tif
Insertion point
Specify on-screen
X: 0
Y: 0
Z: 0
Scale
Specify on-screen
1
Rotation
Specify on-screen
Angle: 0
OK
Cancel
Help
Details <<
Image Information
Resolution:
Horizontal: 300.00 per inch
Vertical: 300.00 per inch
Current AutoCAD unit:
Unitless
Image size in pixels:
Width: 800
Height: 480
Image size in units:
Width: 1
Height: 0.6

Figure 3–5

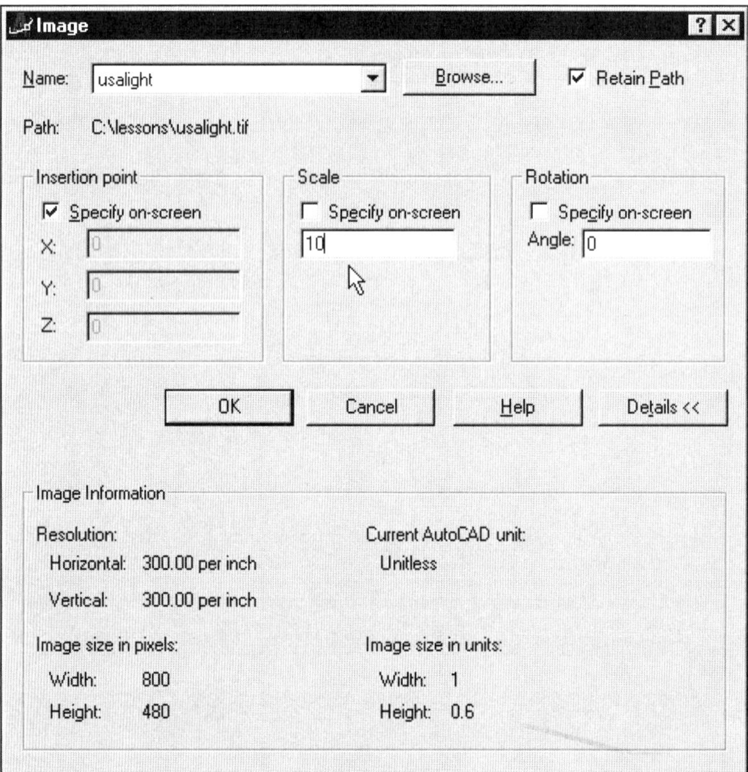

Figure 3–6

Click on **OK.** The mouse pointer will have a "frame" attached to it to indicate the dimensions of the image you are loading. Click the left mouse button and the image will be dropped on the screen.

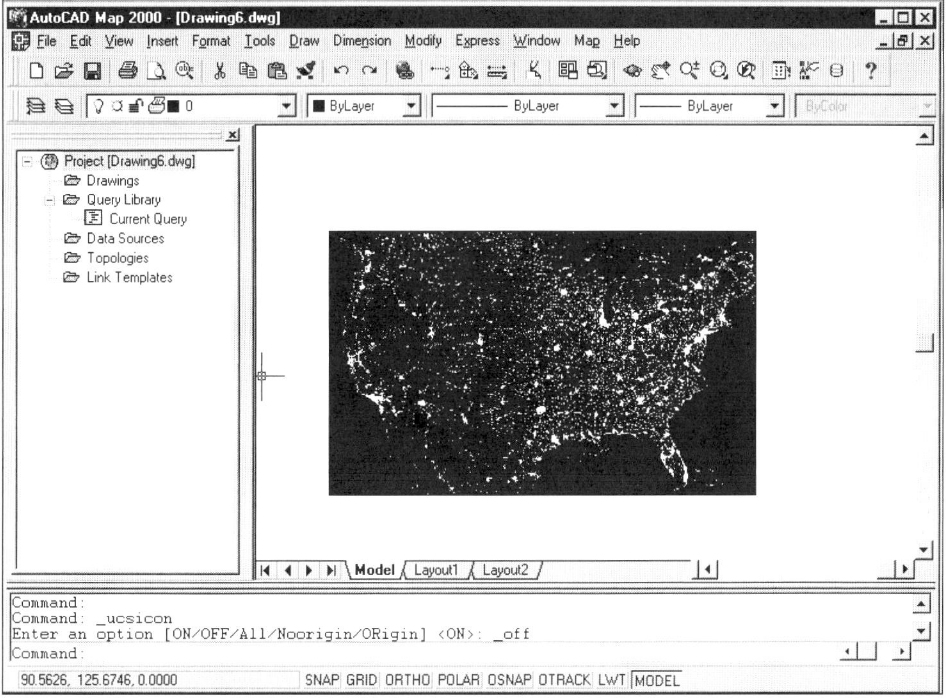

Figure 3–7

This image, which shows the lights of the United States by night, is now available for use as a drawing backdrop, for example.

Step 4 Save this file as **TIF.DWG**.

Step 5 Create a new layer in this drawing called **urbanarea**, make it **current**, and assign it the color **cyan.**

Step 6 Use the large areas of light on the raster image to help you identify urban areas. Draw a closed polyline around three such areas on your image, as shown in the example below.

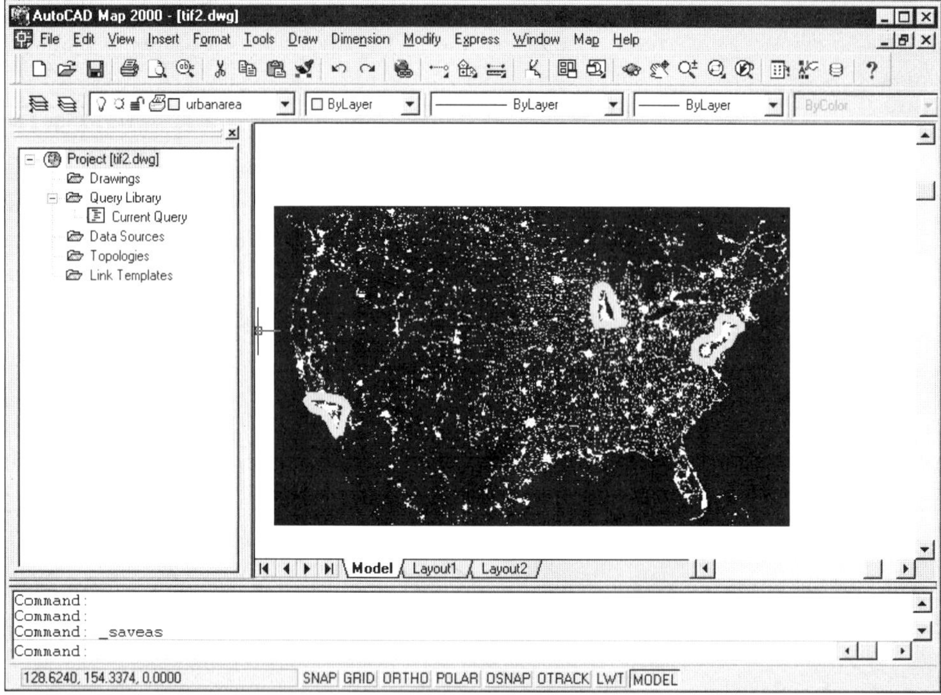

Figure 3–8

Step 7 Save this drawing as, for example, **TIF2.DWG**.

A common application is to utilize a georeferenced raster image in conjunction with a line drawing. *Georeferencing* of a raster image is a two-stage process. The first step is to assign size units to the pixels in the raster in order to establish an accurate scale for use in distance, area, and direction determinations. The second stage is to anchor the raster accurately to the surface of the earth by establishing the coordinates for a corner of the raster.

Step 8 Open the **USA.DWG** drawing. Move the crosshairs across the map and notice the changing coordinate values in the lower left portion of the screen.

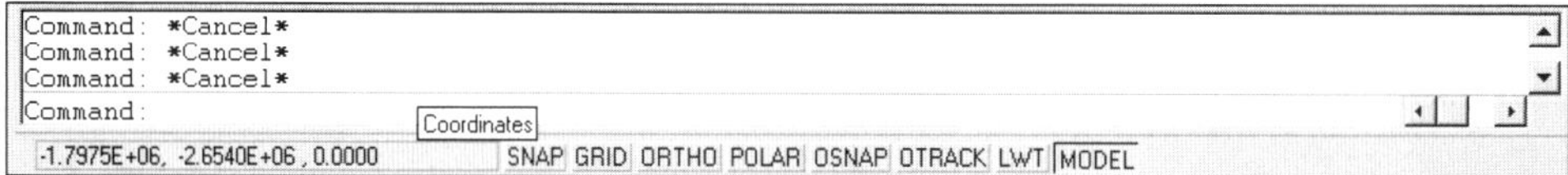

Figure 3–9

These coordinates can be used to identify an insertion point for a raster image.

With the map of the United States on the screen go to the **Insert** menu and select **Raster Image**. Open the **USALIGHT.TIF** file. In the Image dialog box remove all checkmarks from the Image

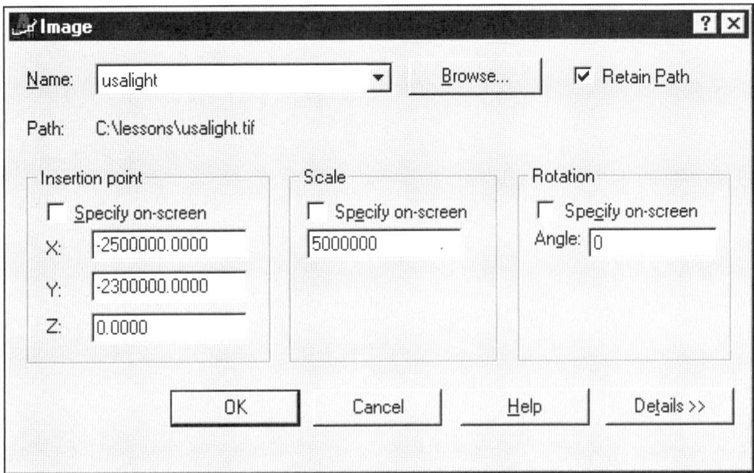

Figure 3–10

Parameters options **Insertion Point, Scale** and **Rotation**. In the **Insertion** section, enter **–2.5E+006** for the x value and **-2.3E+006** for the y value (-2500000 and –2300000 respectively) and in the **Scale** text box enter a value of **5000000.** Leave the Rotation angle at **0**. Click on **OK** and the raster will be placed on top of the vector file. (In this case this will not be a perfect alignment between the two maps since they are of different projections.)

Step 9 To better see both maps go the **Layer Properties Manager** control and change the color of the United States drawing to **red**. (Select all layers and change the color of them all simultaneously.) Click on **OK** to display the drawing in red.

Step 10 To place the vector map on top of the raster map go to the **Tools** menu, select **Display Order** then **Send to Back.**

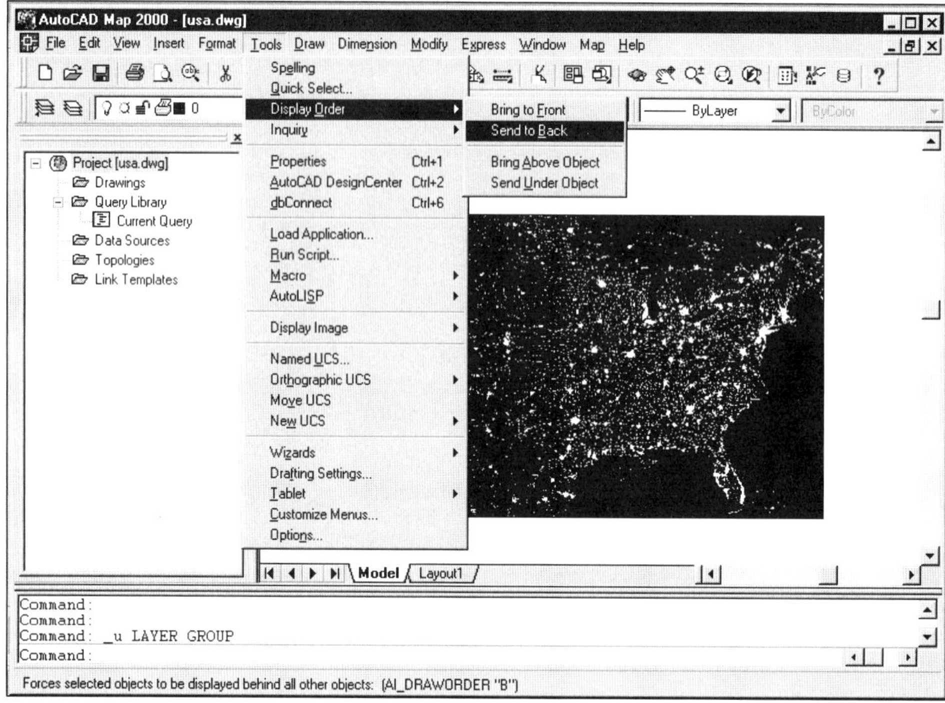

Figure 3–11

Pick the frame around the raster image then click the right mouse button to end the selection process and the combined images will become visible.

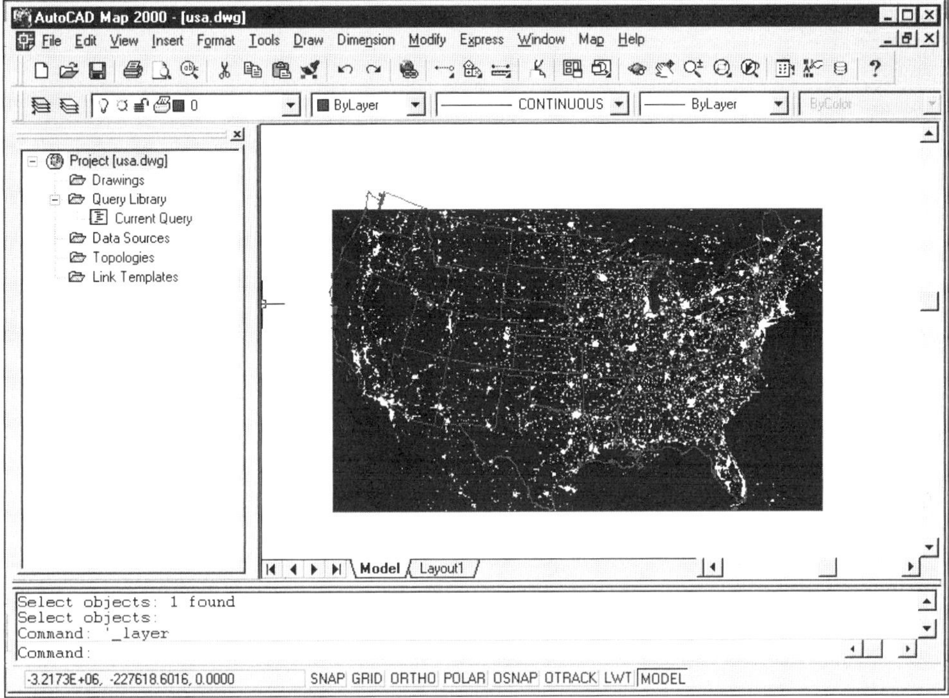

Figure 3–12

Either continue with the tutorials or exit from AutoCAD Map 2000. Do not save the changes to the drawings.

SUMMARY

The availability of raster images is increasing and the number of applications for their use is also. In this chapter, an image captured from the internet was used as a base map for "heads-up-digitizing" which is in itself, a powerful GIS technique. If features are drawn on top of a raster image file they will be geographically correct with respect to the image itself and with respect to any other features that are digitized or drawn. If the image is to be used in conjunction with other map files for GIS applications then it should be georeferenced. Georeferencing the raster image establishes a location on the earth's surface for the lower left hand corner of the image and assigns a distance unit for each pixel. The location for the lower left corner must be in terms of the same projection and coordinate system that will be used by any attached drawings. The numbers used in Step 8 were derived from the coordinates shown in the bottom left corner of the AutoCAD Map 2000 screen when the USA.DWG file is open. In some cases, procedures like Rubber Sheeting may be required to align a vector file to a raster file. This process is covered later in the text.

REVIEW QUESTIONS

1. What is the difference between a raster file and a vector file?

2. Can you identify major urban centers on the satellite image, and if so, what is the population threshold before a center becomes recognizable? (Check this threshold value in an atlas.)

3. What is georeferencing?

4. What are 3 types of image files that can be accessed by AutoCAD Map 2000?

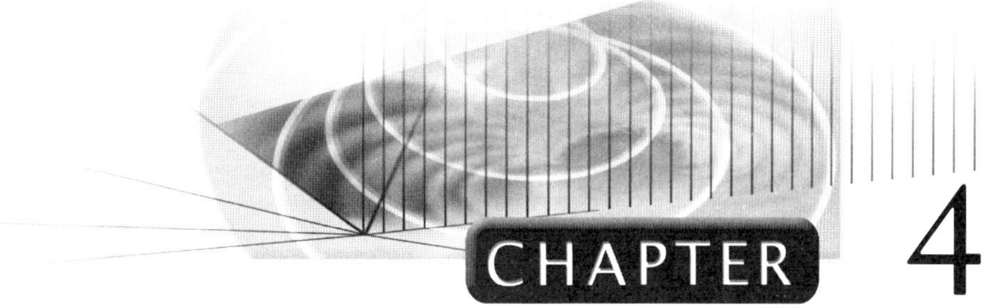

4

GIS Techniques

OBJECTIVES

After completing this chapter you will be able to utilize saving techniques employed in the process of creating and editing data in GIS identified as:

- Edge Matching
- Rubber Sheeting
- Mapbook printing
- Importing non-DWG drawings

KEY TERMS

Edge Matching
Property Conditions
Save Set
Source drawings
Rubber Sheeting
Mapbook
Extents
Duplicate objects
Dangling objects
DXF
Buffer Fence

In GIS, one often encounters drawings that require major modifications to be usable. One may find that available drawings must be combined to provide coverage of an area. This can occur whenever municipal boundaries change and maps from different sources have to be utilized. *Edge matching* can often resolve this problem.

In some cases where maps were generated utilizing slightly different coordinate systems or base points, one map may have to be stretched, or *rubber sheeted*, to align with an adjacent map. When such multiple drawings are combined, the resulting detail often dictates that a very large paper map be printed to allow the user to access the details. Mapbook printing can alleviate this problem.

In many cases the drawing files are available, but in non–AutoCAD Map 2000 DWG format. In this case, a conversion to DXF format will allow those files to be used. The knowledge of how to use these GIS procedures can save a great deal of time when working with files provided by other sources.

TUTORIAL: EDGE MATCHING

Edge matching is a technique that is used to align two geographically adjacent drawings that may have been generated by different people or at different times and as a result do not line up correctly. This is shown in the road network map of Napanee below, which is made up of a upper, new portion called Napnew and an older lower half identified as Napold (Figure 4–1).

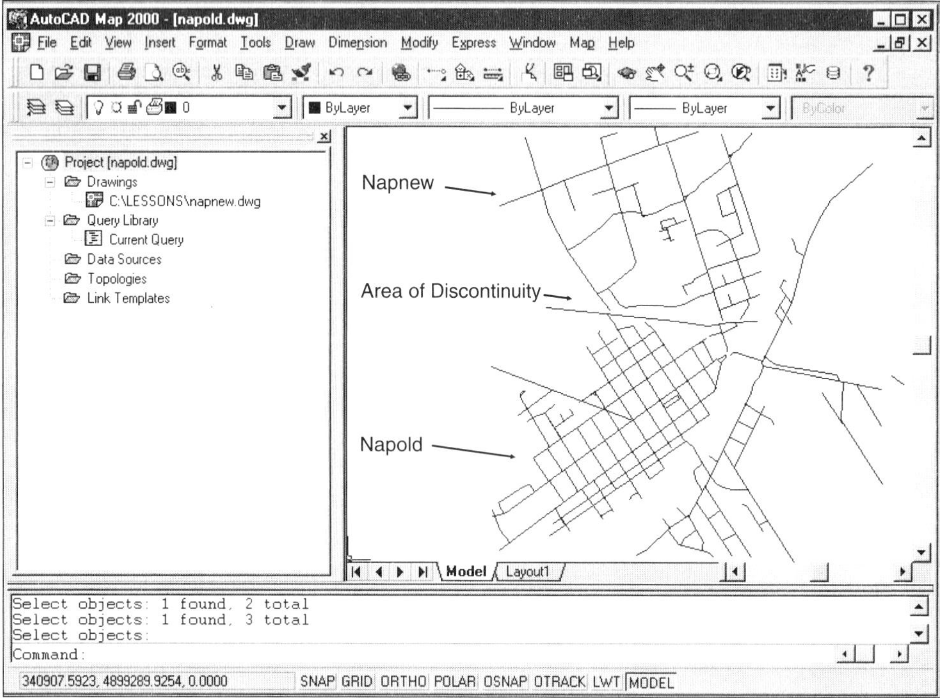

Figure 4–1 *Both halves of Napanee*

To align the two halves, you conduct a query of the area in question to combine the two drawings and then clean up the region where the two drawings should interface.

Step 1 Start AutoCAD Map 2000 and open **NAPOLD.DWG**. This file shows the lower half of the road network of Napanee.

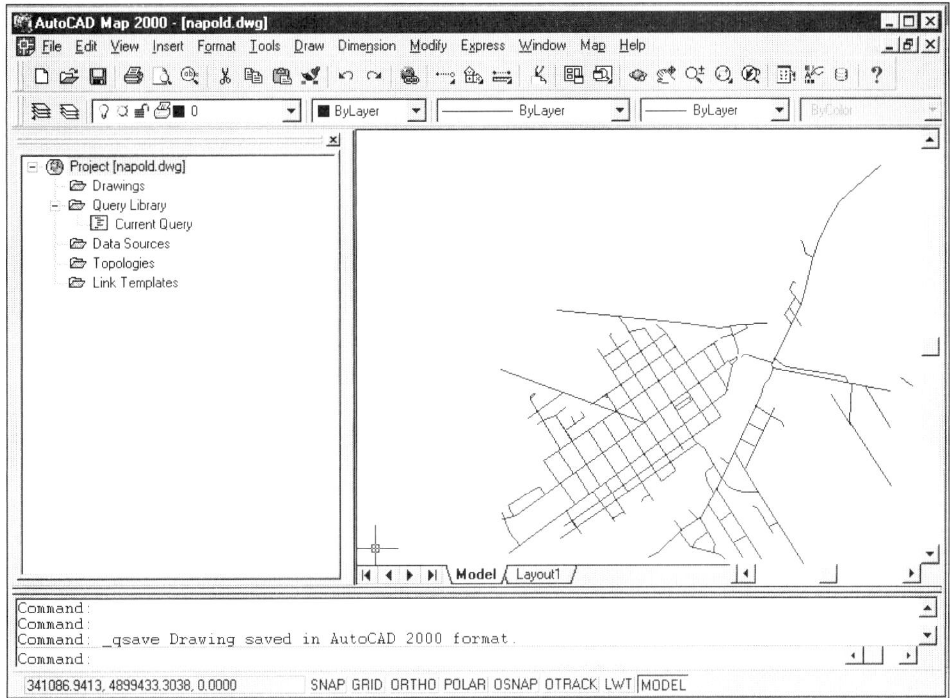

Figure 4–2 *Town of Napanee, NAPOLD.DWG*

Step 2 Go to the **Map** menu, then select **Drawings** and then choose **Define/ Modify Drawing Set**. Click on **Attach**, then at the **Select Drawings to Attach** dialog box, highlight the **NAPNEW** file and click on **Add.**

Click on **OK**. The NAPNEW drawing will be attached and made active (Yes will appear beside it in the Define/Modify Drawing Set). Click on **OK** at the Define/Modify Drawing Set dialog box.

Step 3 To view both drawings, go to the **Map** menu and select **Drawings** then **Quick View Drawings**. At the Quick View Drawings dialog box click on **OK** to see both halves of the road network. The roads will not be joined.

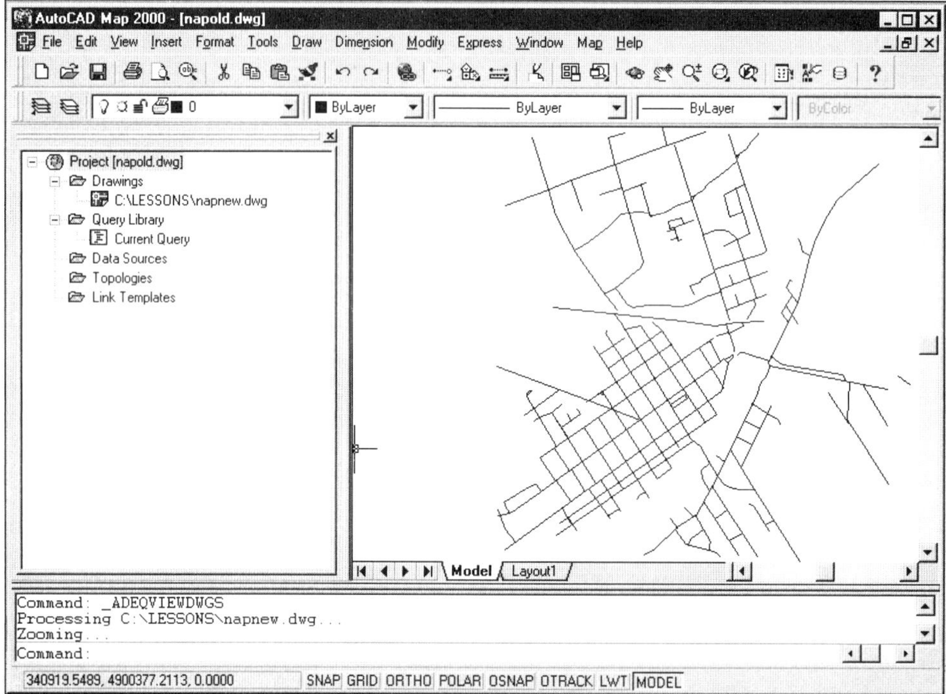

Figure 4–3

Step 4 Go to **Map** then **Query** and select **Define Query**.

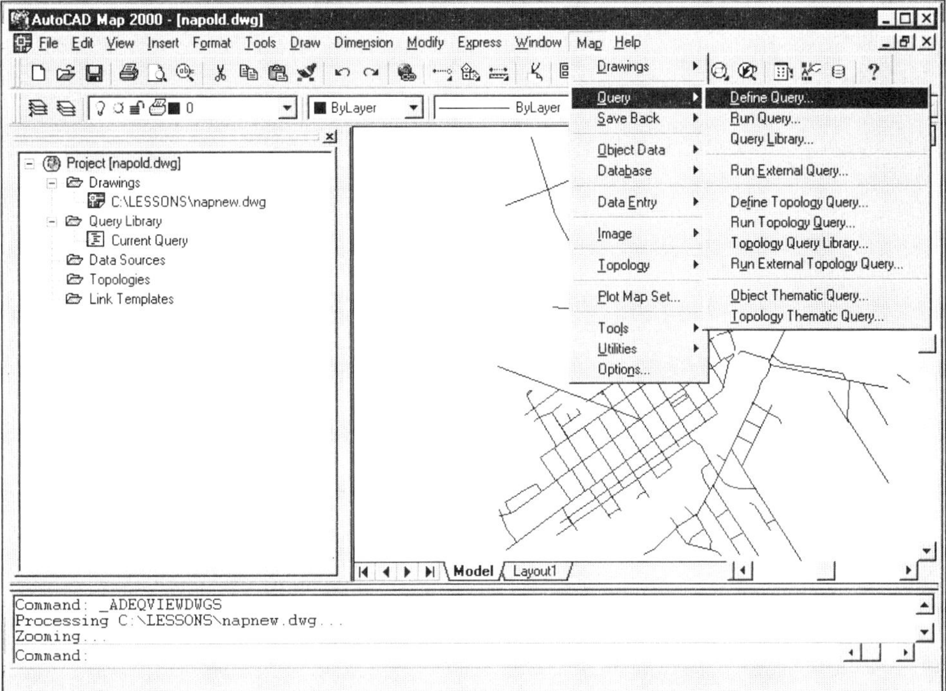

Figure 4–4

Select **Location**. Although in earlier queries a circle was used to select the area, this time select **Buffer Fence** as the defining type. Set the Selection Type to **Crossing**. Click on **Define**, and draw a boundary line that identifies the general zone along which the two drawings should be

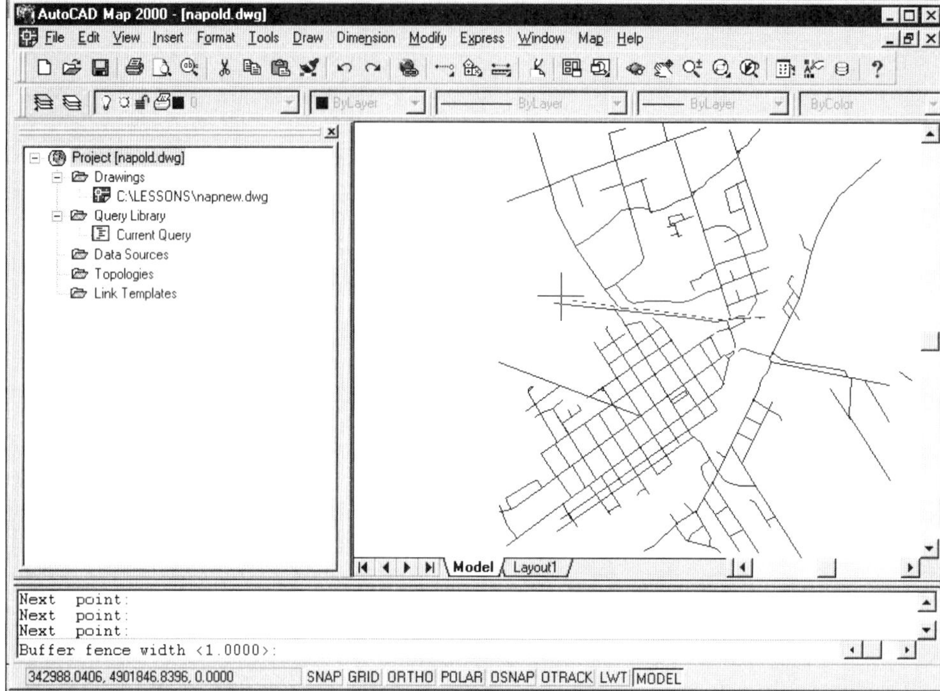

Figure 4–5

Edge Matched but are not. In the Command Prompt text box set the **Buffer Fence** to a width of **100.**

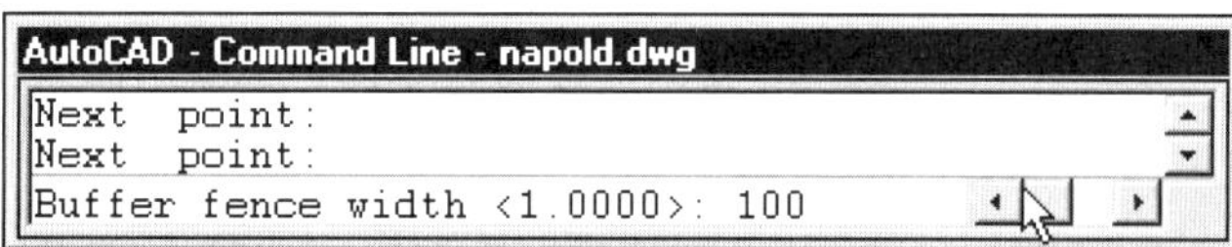

Figure 4–6

A zone will appear identifying a region 100 units on either side of the buffer fence.

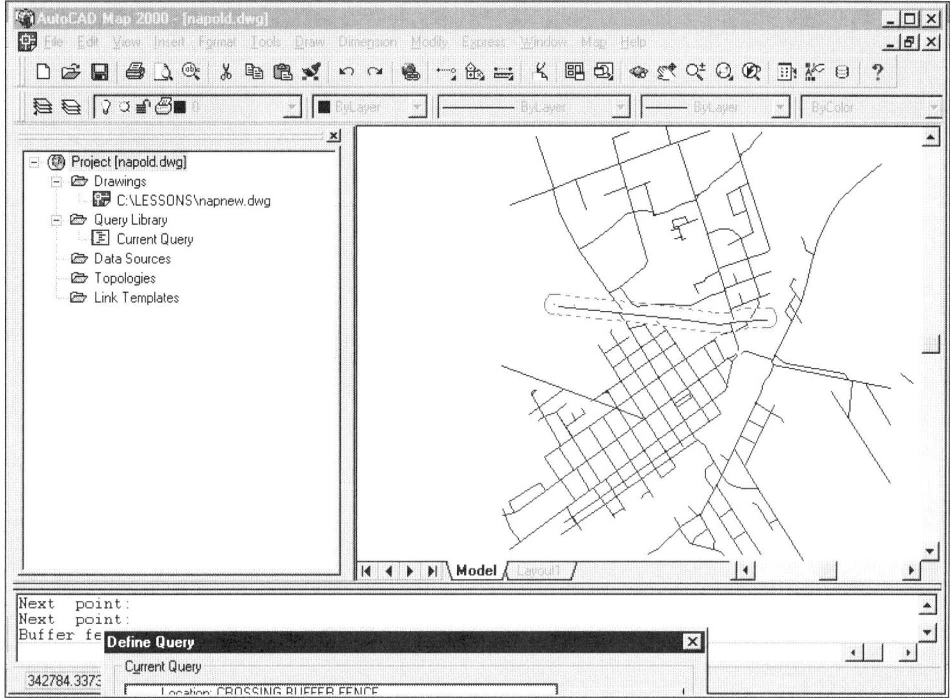

Figure 4–7

Step 5 At the Define Query dialog box click on the **Property** button. The Property Condition dialog box will appear.

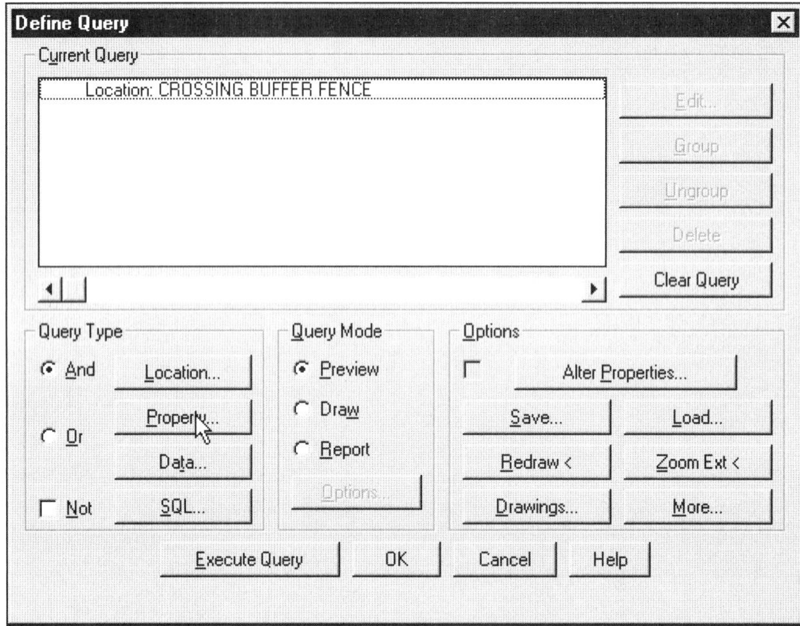

Figure 4–8

At the Property Condition dialog box select **Layer**, then click on the **Values** button. The Select dialog box will pop up.

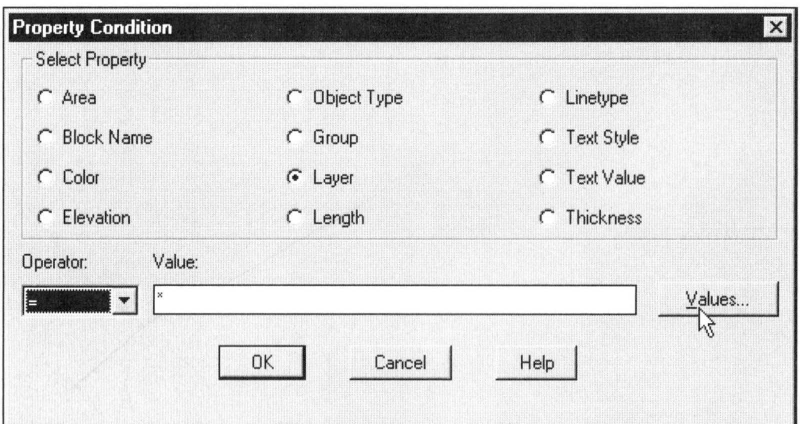

Figure 4–9

Pick layer **0**, then **OK** (The layer 0 is chosen for tutorial purposes, but any layer may be selected).

e-presented. Click on **OK**.

S

nature of the query is shown, set
cute Query.

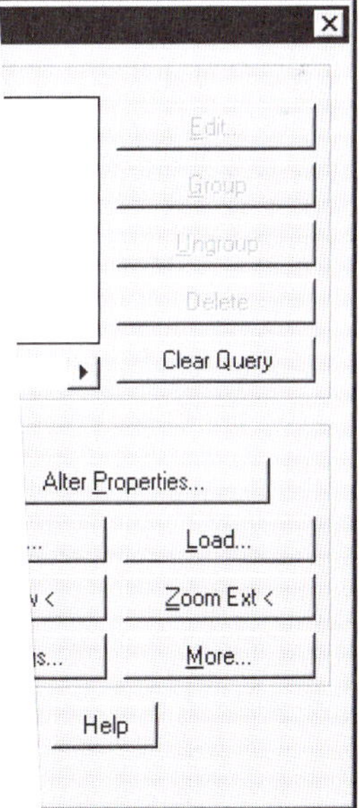

You will be provided information in the command prompt area of the screen that indicates the number of objects being queried. This process has identified the objects that are not aligned within the **Buffer Zone.**

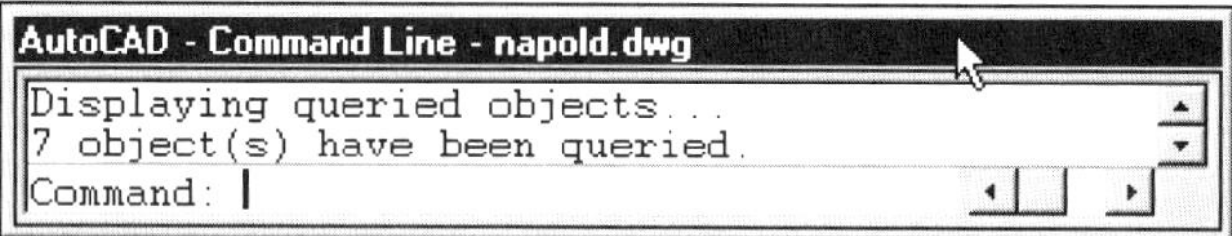

Figure 4–12

Step 7 Go to the **Map** menu and select **Tools** then **Drawing Cleanup**. The Drawing Cleanup dialog box will appear.

Step 8 Click on **Object Selection** and in the Object Selection dialog box select the **Select Automatically** option.

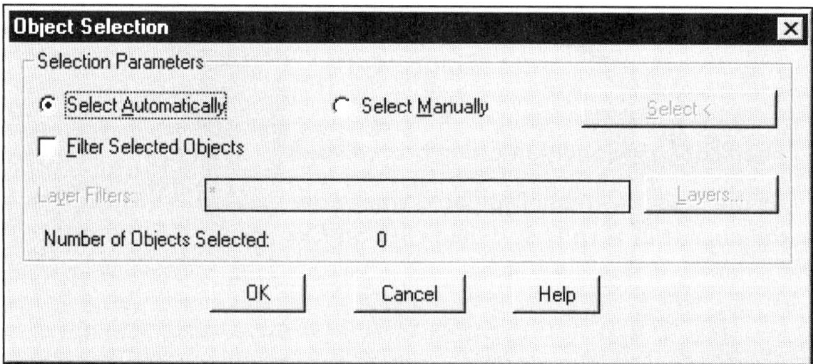

Figure 4–13

Click on **OK.**

Step 9 You will be returned to the Drawing Cleanup dialog box. Select **Object Conversion**, highlight the **Modify Original Objects** option, and click on **OK**. Again, you will be presented with the Drawing Cleanup dialog box. Click on **Cleanup Options**.

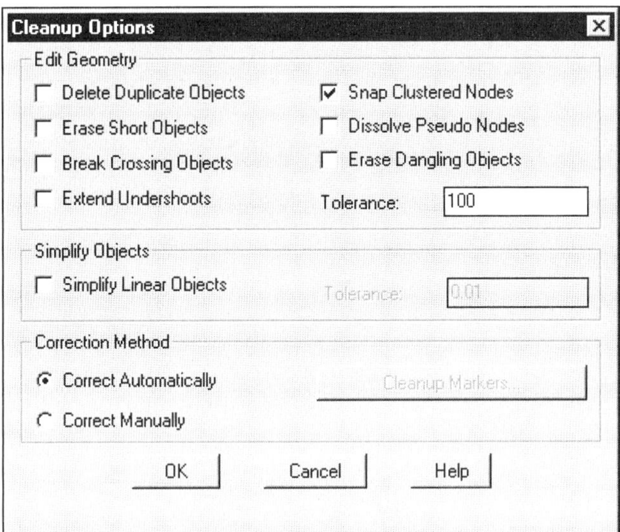

Figure 4–14

Select the **Snap Clustered Nodes** checkbox and deactivate all other options. Set the tolerance to **100.** This value was chosen because it is just larger than the distance by which the maps don't align. Have the correction method set to **Automatically** and click on **OK.** At the **Drawing Cleanup** screen click on **Proceed.**

Step 10 When prompted to Add the objects to the Save Set select **Yes**. The broken boundaries will now be joined. The last step is to save the modifications back to the original source drawings. Go to the **Map** menu and select **Save Back** and then **Save to Source Drawings**. Save the queried objects in this case.

It is important to note that occasionally a map will cause error messages to appear on the screen. You can always click on the Close option, and if you have difficulty with this preventing you from completing Step 10 you can just go to save new joined drawing with a new name.

Either continue with the tutorials or exit AutoCAD Map 2000. You may close the drawing without saving the changes.

TUTORIAL: RUBBER SHEETING

Rubber sheeting is a technique that can be used to make features in a drawing conform to specified locations. Rubber sheeting might need to be used in the case where an old paper map digitized for use in AutoCAD Map 2000 is shown by a new survey to contain discrepancies. In this process, identifiable locations from the drawing, such as benchmarks, are moved to the new location on the drawing. In this technique, all points in the drawing are not moved uniformly and as a result, the size, shape, and locations of some features may become distorted. This implies that the integrity of work done using "rubber-sheeted" drawings may be questionable.

Step 1 Open the **ROAD** drawing file. Select the **Map** menu, then **Drawings** and the **Define/Modify Drawing Set** item. Attach the **PARCEL** drawing. Then select **Map** and **Quick View Drawings.**

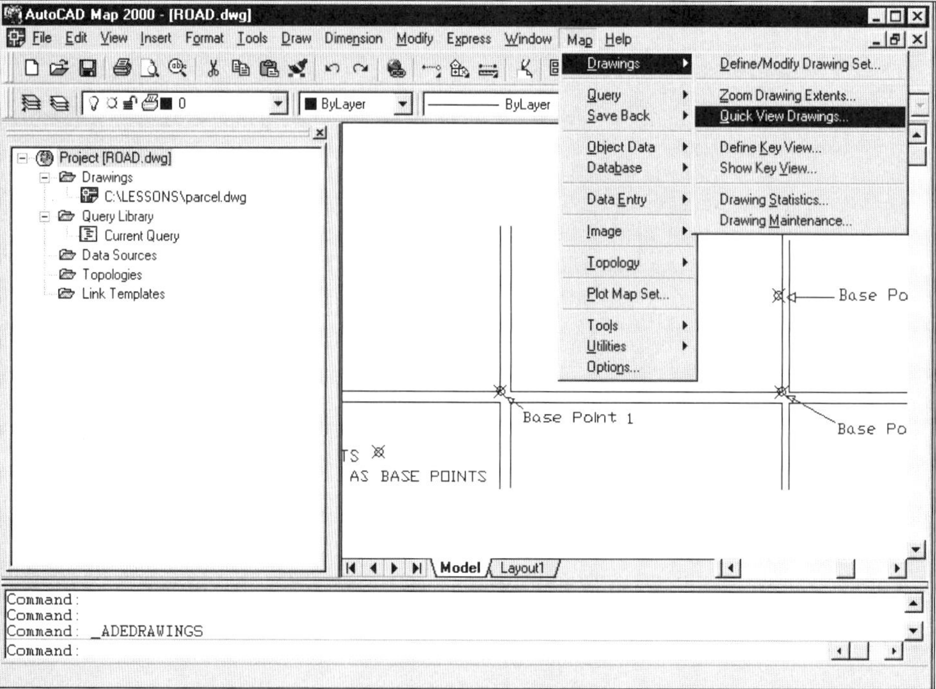

Figure 4–15

Select the **PARCEL** drawing and click on **OK.** The two drawings will be presented.

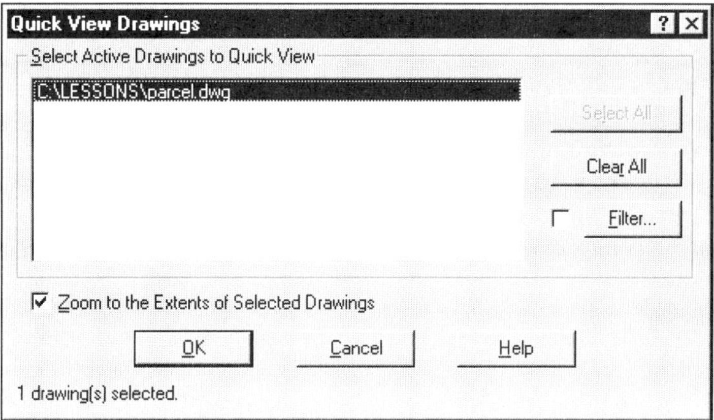

Figure 4–16

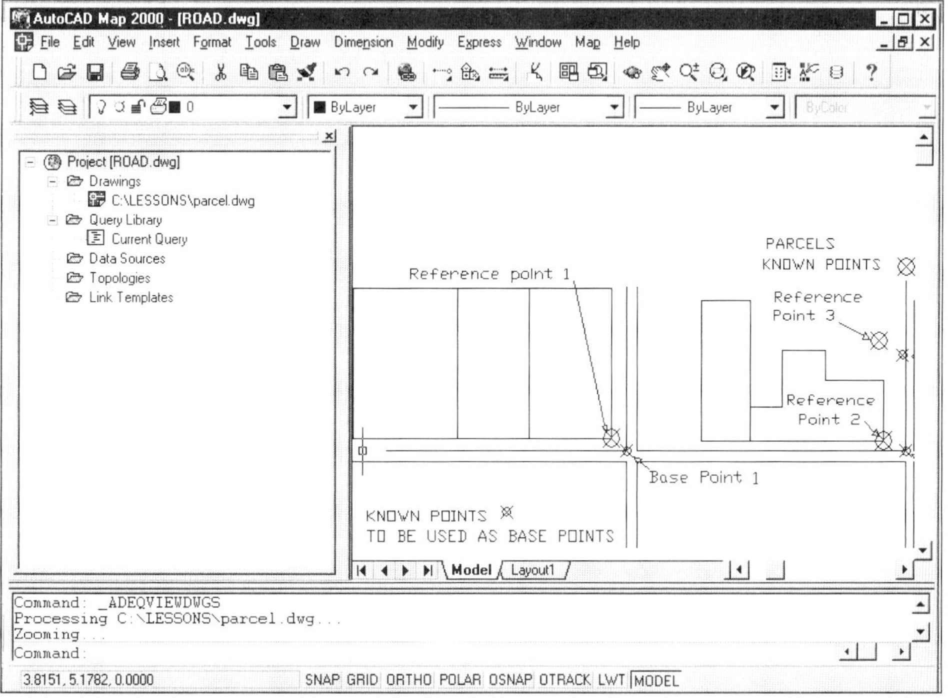

Figure 4–17

Step 2 Two sets of identified points are shown, one for each of the drawings. The ROAD.DWG file contains three points identified as numbered Base Points and the PARCEL.DWG file contains three numbered points identified as Reference Points. These are points that are known locations; for the purpose of this exercise, they are known to be coinciding in the real world. This procedure will align the coincident points; Reference Point 1 with Base Point 1, Reference Point 2 with Base Point 2, and Reference Point 3 with Base Point 3. It is the Reference Points that are assumed to be correct. Go to the **Map** menu item, and select the **Map Tools** option and highlight **Rubber Sheet**.

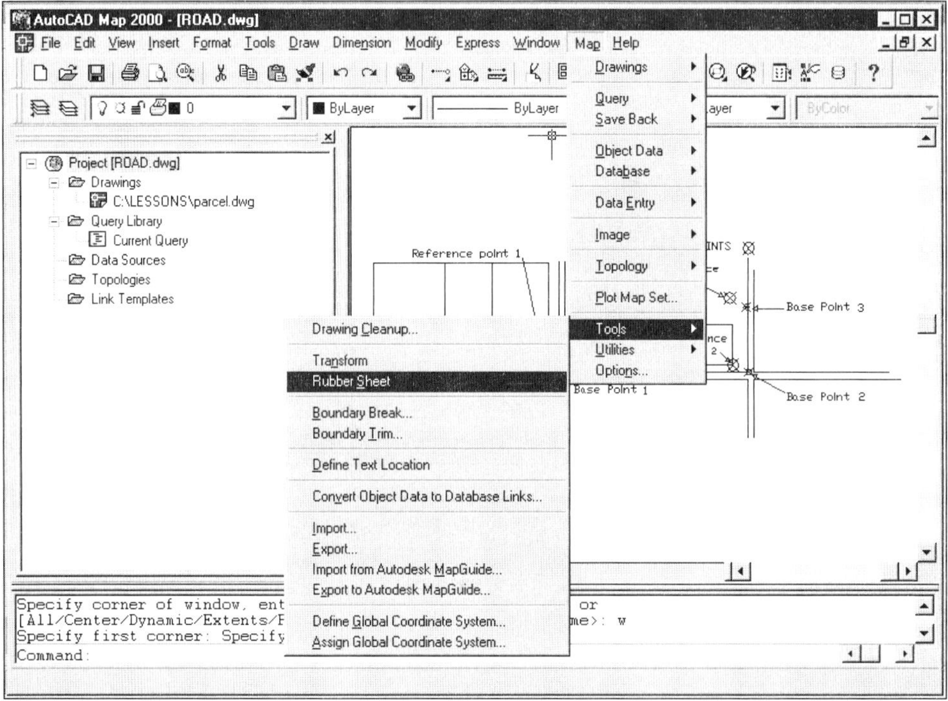

Figure 4–18

Step 3 In the command prompt text box you will be prompted for Base Point 1. Click on the known Base Point 1 on the ROAD map and then, when prompted for a Reference Point, click on the corresponding Reference Point 1 of the PARCEL drawing. In this example, the PARCEL map is taken as the correct drawing and the ROAD drawing is being stretched to match.

Step 4 Repeat the point selection process for Base Point 2 and Reference Point 2. Continue for the third point, and then when prompted for a fourth point press ENTER. Generally, the more points used, the more accurate the new drawing will be in its representation of the real world.

Step 5 You will be prompted to select the objects to be transformed. This can be done either by identifying an area or some objects. Type the letter **s** to select objects and press enter.

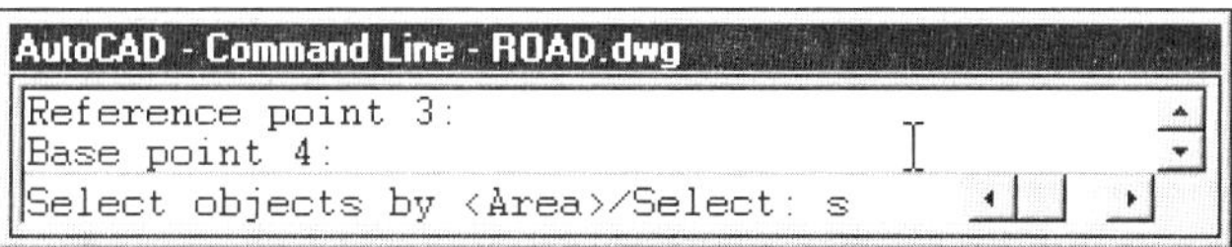

Figure 4–19

Click on each of the components of the **ROAD** drawing. When all the sections are highlighted press **enter**. The **ROAD** drawing will be shifted to align with the **LOTS** drawing.

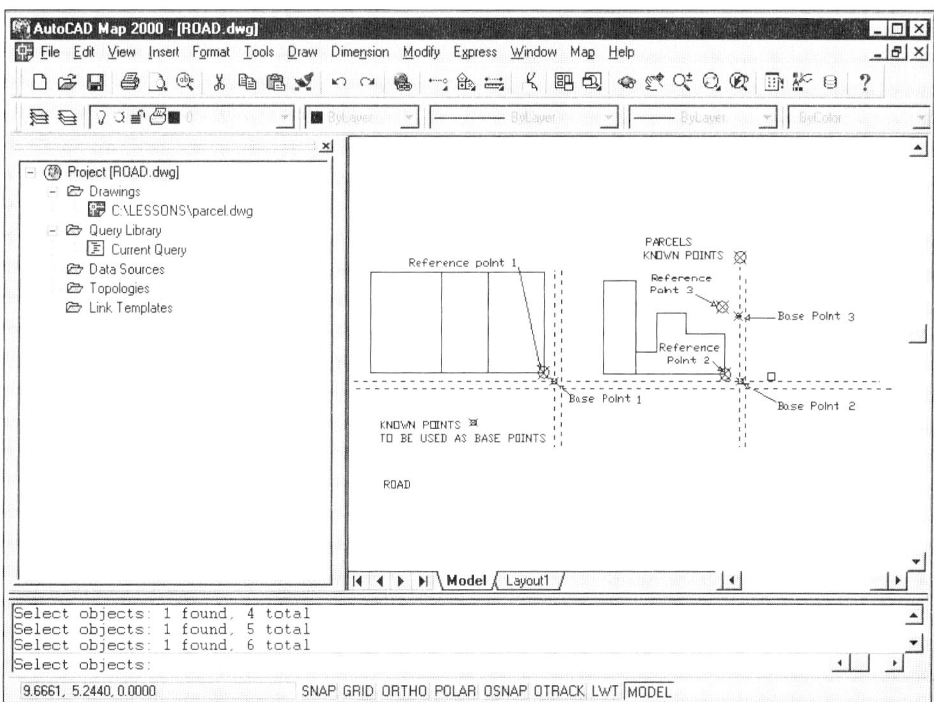

Figure 4–20

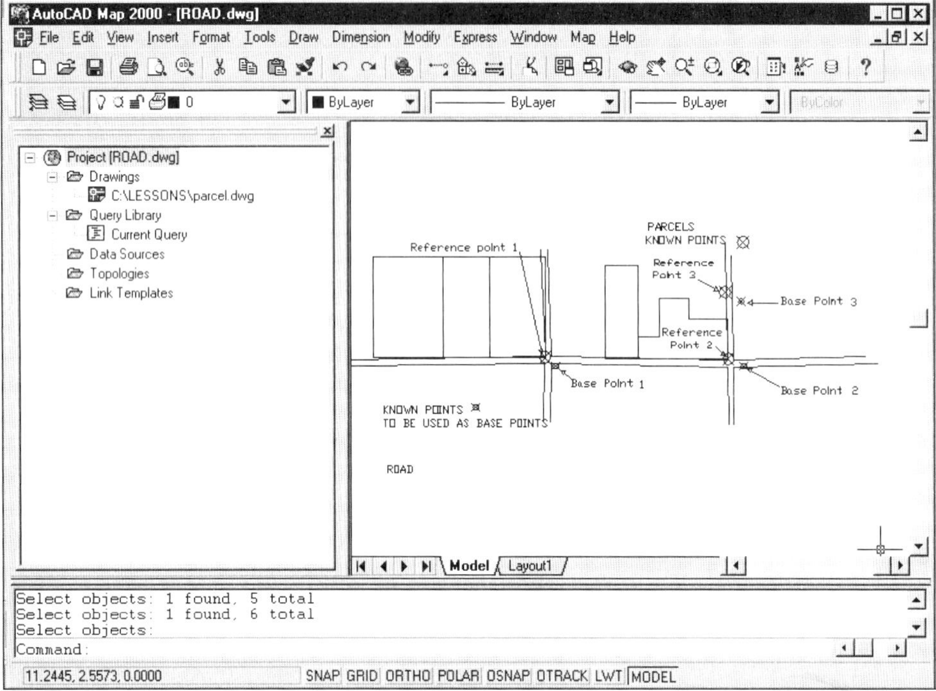

Figure 4–21

Step 6 The modified drawing can be saved by selecting **Map** and then **Save Back** and **Save to Source Drawings**. Either continue with the tutorials or exit AutoCAD Map 2000.

TUTORIAL: MAPBOOK PRINTING

Mapbook printing allows you to take what would be a large, plotted paper copy of a map and break it down into a series of smaller maps that essentially form an atlas from the large original.

In this exercise, drawings from earlier exercises will be utilized in conjunction with additional drawings to illustrate how one can generate a mapbook.

Step 1 Open the **ROAD** drawing, then use the **Map** menu item and the **Drawings, Define/Modify Drawing Set** to attach the **PARCEL, HOUSES**, and **PLOTFRAME** drawings.

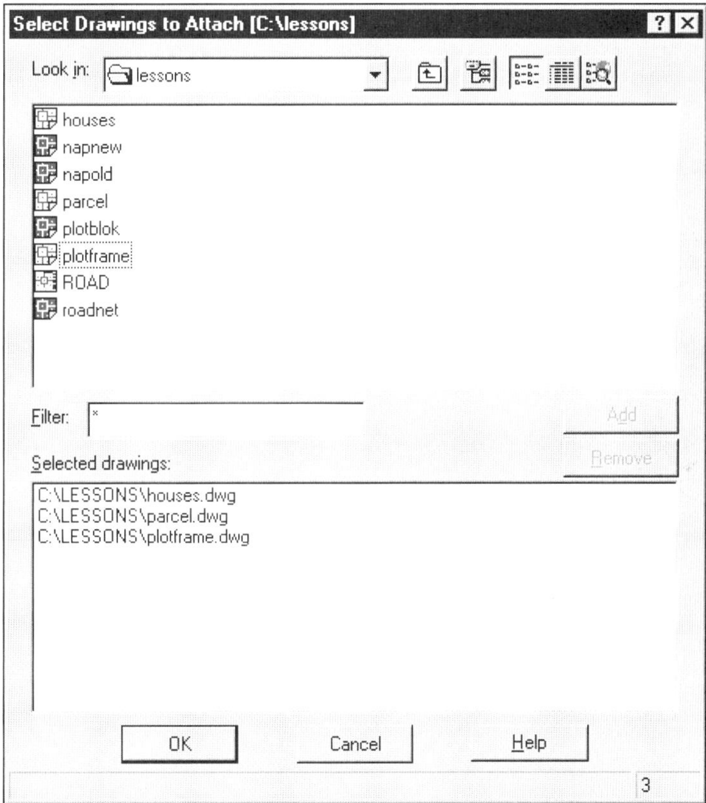

Figure 4–22

The PLOTFRAME drawing consists of closed polygons with object data attached to the polygons. Use the **Map** menu item in conjunction with, **Drawings**, and the **Quick View Drawings** option, to view all the attached drawings.

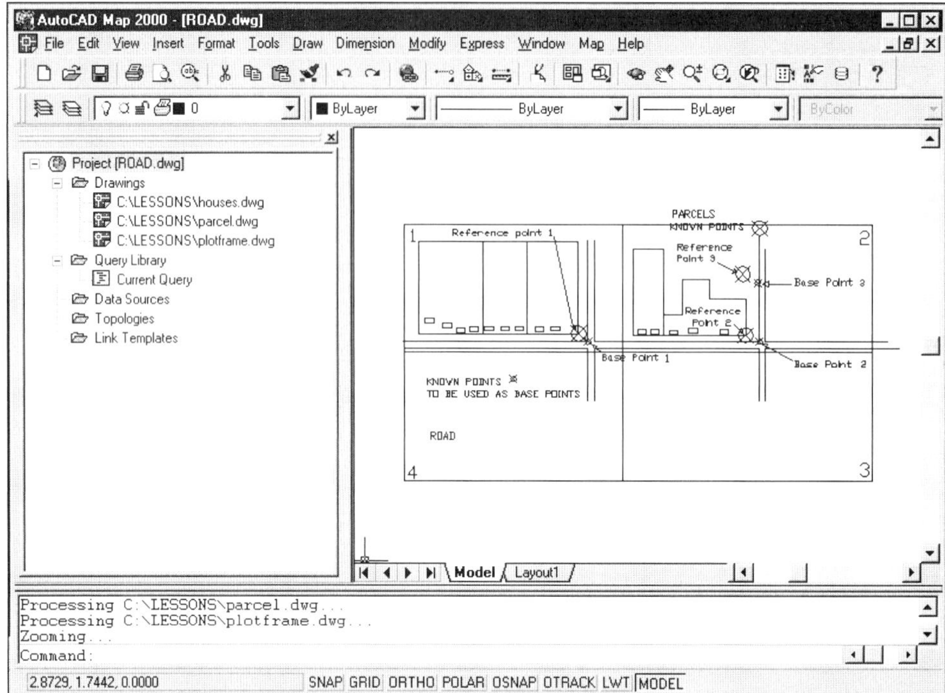

Figure 4–23

You may wish to resize the drawings on the screen through the **zoom** functions.

Step 2 The next step requires that a query be established to identify all the features in all of the drawings. Go to **Map**, then select **Query** and **Define Query**. For the Location option, select **All**, then click on **OK**. Instead of executing the query, click on the **Save** button.

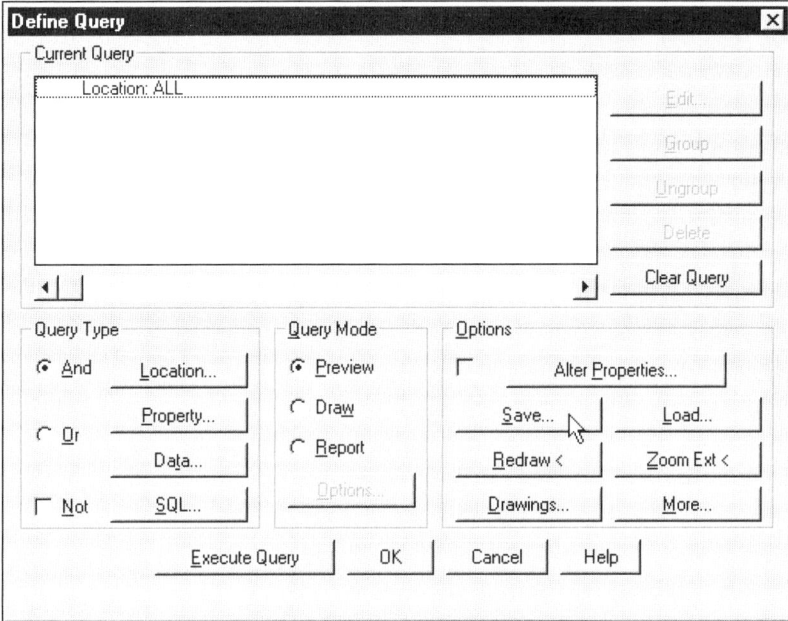

Figure 4–24

Step 3 At the Save Current Query dialog box, select the **New Category** option. At the Define New Category dialog box type in **mapbook**. Click on **OK**. The Save Current Query dialog box reappears.

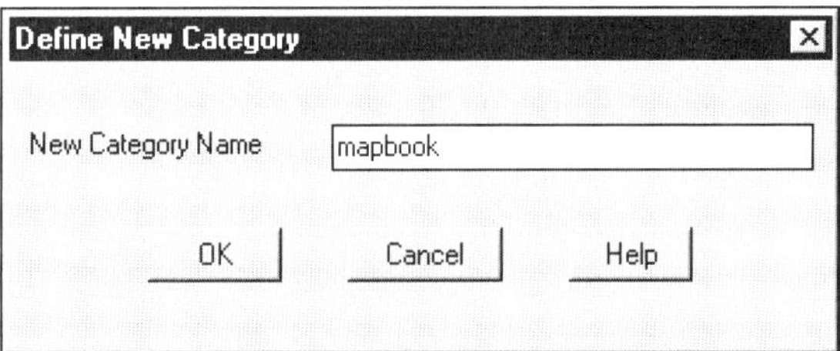

Figure 4–25

For a Name enter **Parcels_Roads_Houses** and for a Description type in **mapbook printing**.

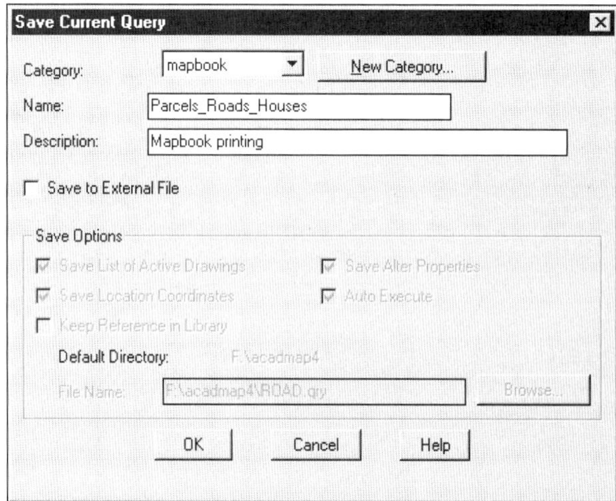

Figure 4–26

Click on **OK**. You will be returned to the Define Query dialog box. Click on **OK** again.

Step 4 Go to the **Insert** menu and select the **Block** item.

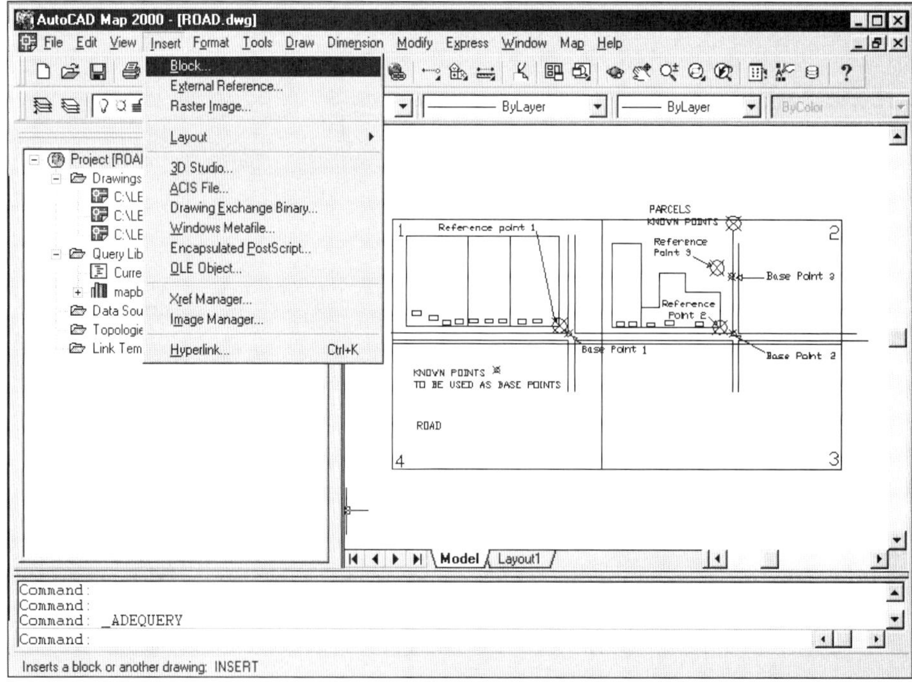

Figure 4–27

At the **Insert** dialog box, click on the **Browse** button and load the

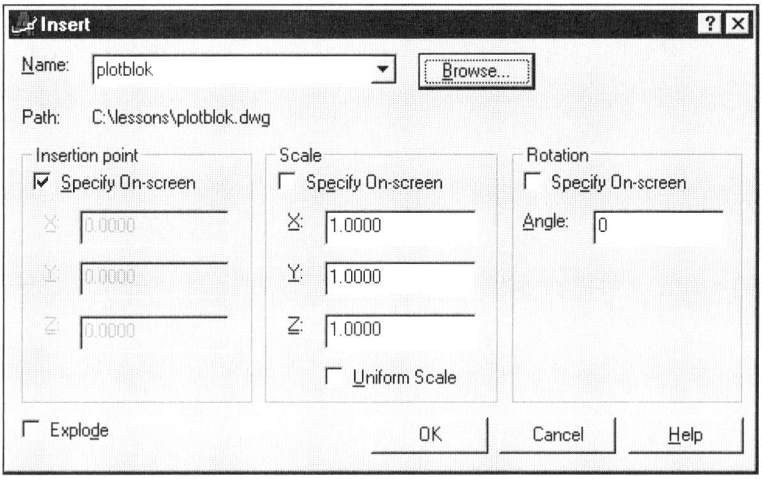

Insert
Name: plotblok Browse...
Path: C:\lessons\plotblok.dwg
Insertion point: Specify On-screen; X: 0.0000; Y: 0.0000; Z: 0.0000
Scale: Specify On-screen; X: 1.0000; Y: 1.0000; Z: 1.0000; Uniform Scale
Rotation: Specify On-screen; Angle: 0
Explode OK Cancel Help

Figure 4–28

PLOTBLOK.DWG file then click on **Open** . The **Insert** dialogue box is presented. Set the **Insertion Point** to **Specify on Screen** and **deselect** the **Scale – Specify on Screen** and click on **OK** . You can place the plot outline on the screen so that the map files are in the main viewport.

Click the left mouse button to place the PLOTBLOK drawing on the screen. You may wish to adjust the view of the drawings by using the zoom or pan features.

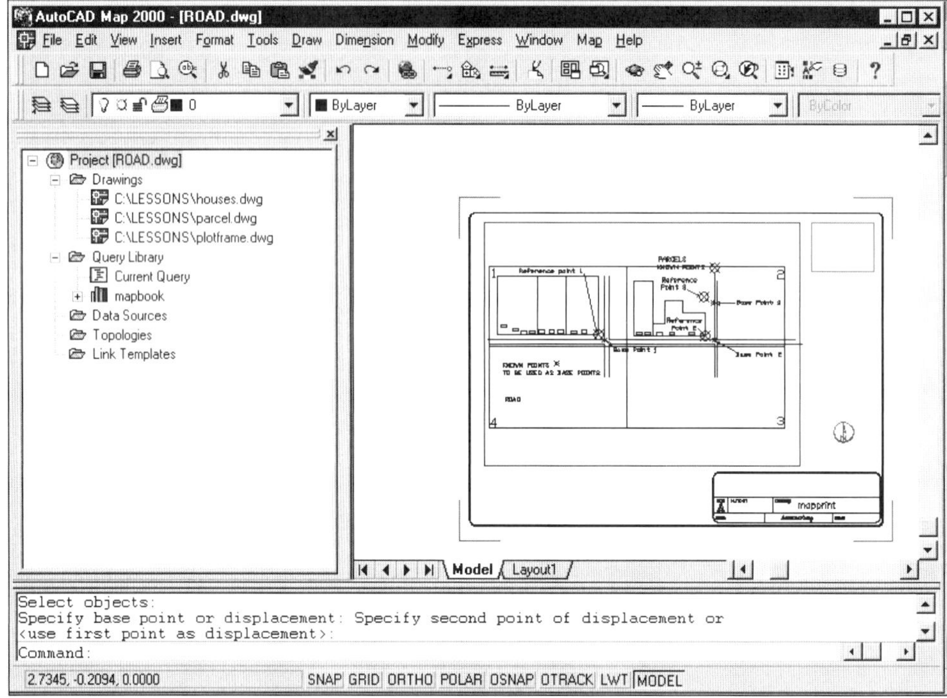

Figure 4–29

When prompted for a title type, in **mapprint** and press ENTER or click the right mouse button.

To plot mapbooks, a Layout must be established. This is most conveniently accomplished by using the wizard.

Step 5 Go to the **Tools** menu and select **Wizards**, then **Create Layout**.

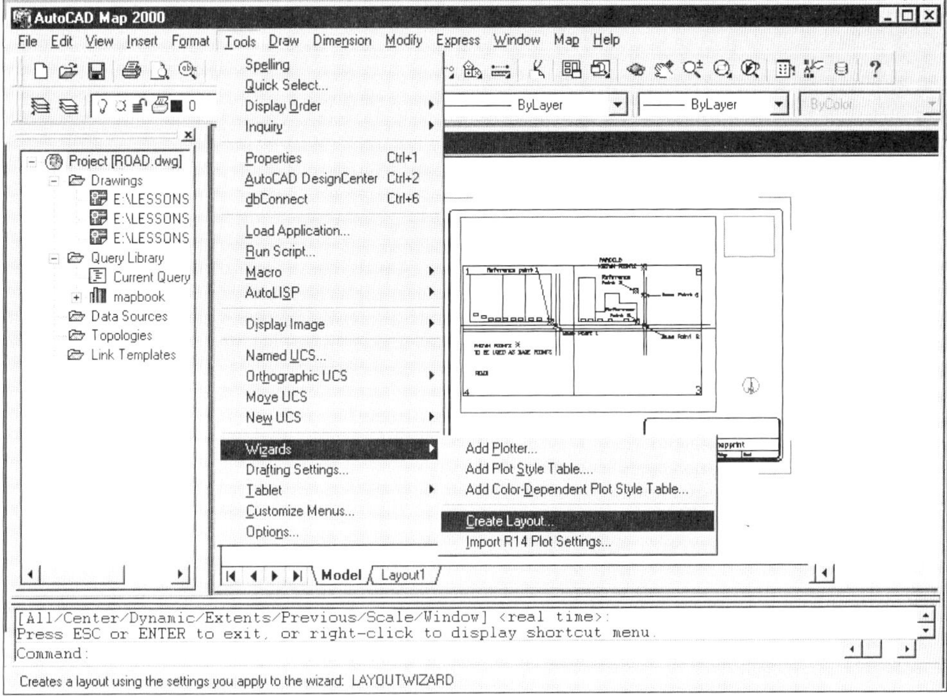

Figure 4–30

The first screen of the wizard, the **Create Layout - Begin** dialog box, is presented. Assign a name to the layout. In this example, the name used was **Map Book.**

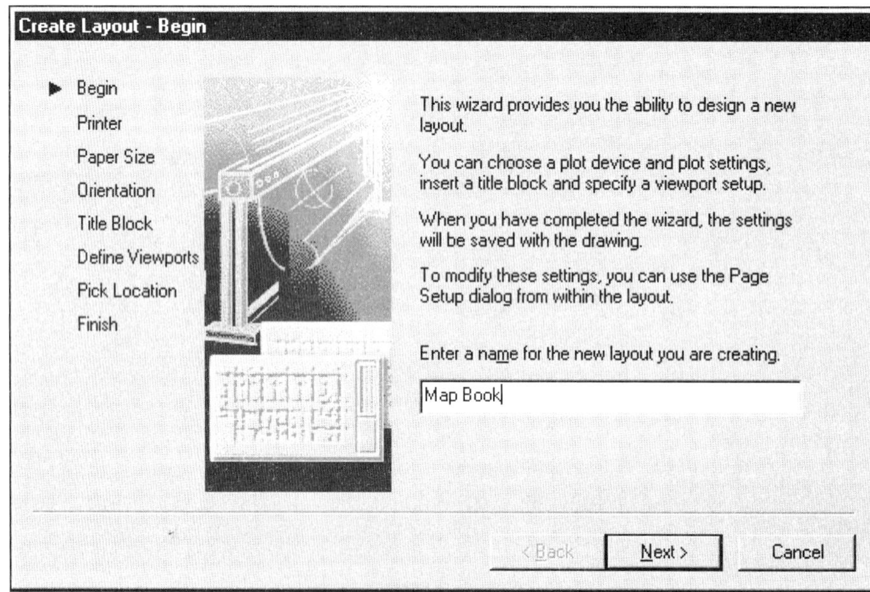

Figure 4–31

Click on **Next.**

The Create Layout - Printer dialog box is presented next. Select your output device. In this case, an HP DeskJet printer is the selection.

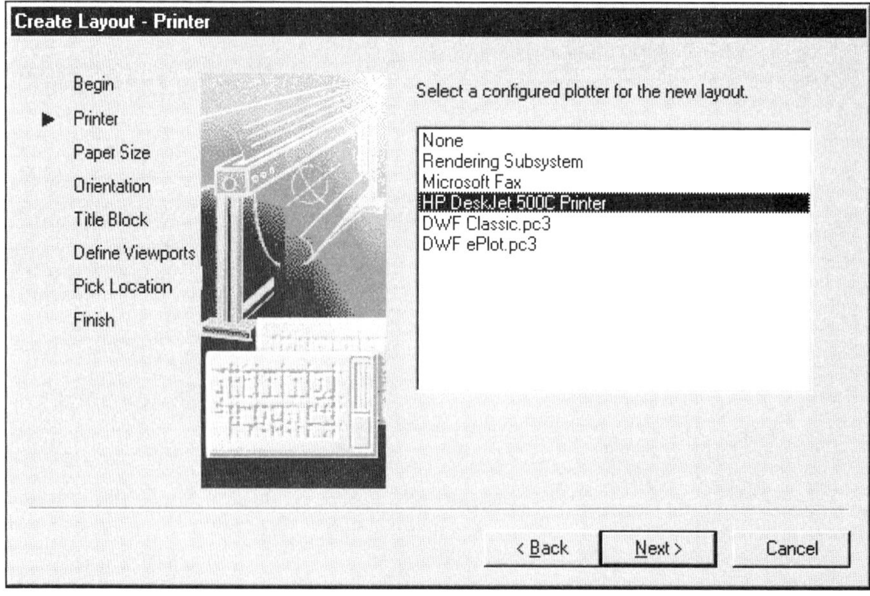

Figure 4–32

Select **Next.**

The Create Layout - Paper Size dialog box is presented.

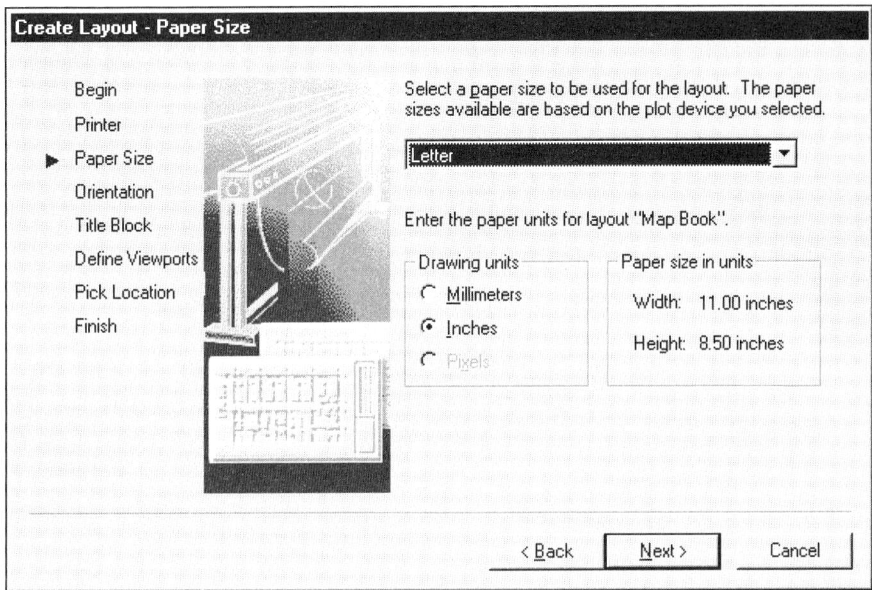

Figure 4–33

Make the appropriate selection for your output devise and select **Next**. The **Create Layout-Orientation** dialogue box is presented to facilitate the selection of Portrait or Landscape options. Select **Landscape** for this example and click on **Next.**

The **Create Layout - Title Block** dialog box is presented next. This screen provides the user to select from a wide variety of Title Blocks. For this example, select **None** then select **Next**.

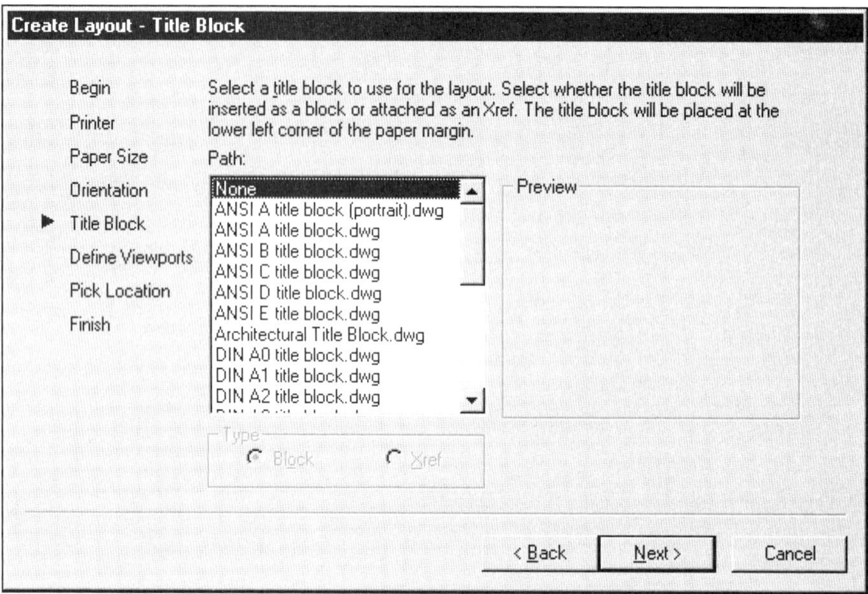

Figure 4–34

The **Create Layout - Define Viewports** dialog box is presented. For this example, select **None** as the Viewport Setup.

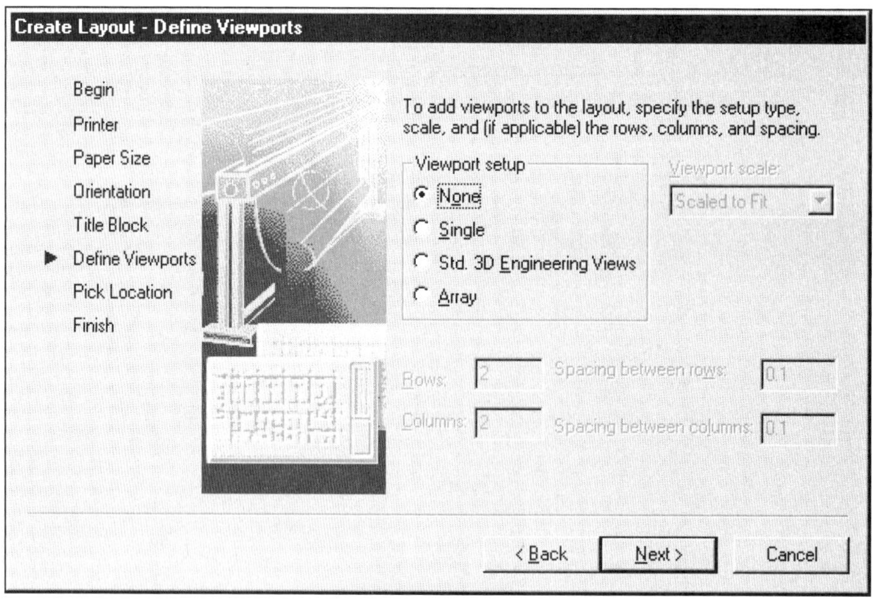

Figure 4–35

Click on **Next**, then Click on **Finish** to finish the wizard. The new Layout will have been prepared and is presented on the screen, overlaying the drawings. To view the drawings you can click on the **Model** tab from the bottom of the drawing screen.

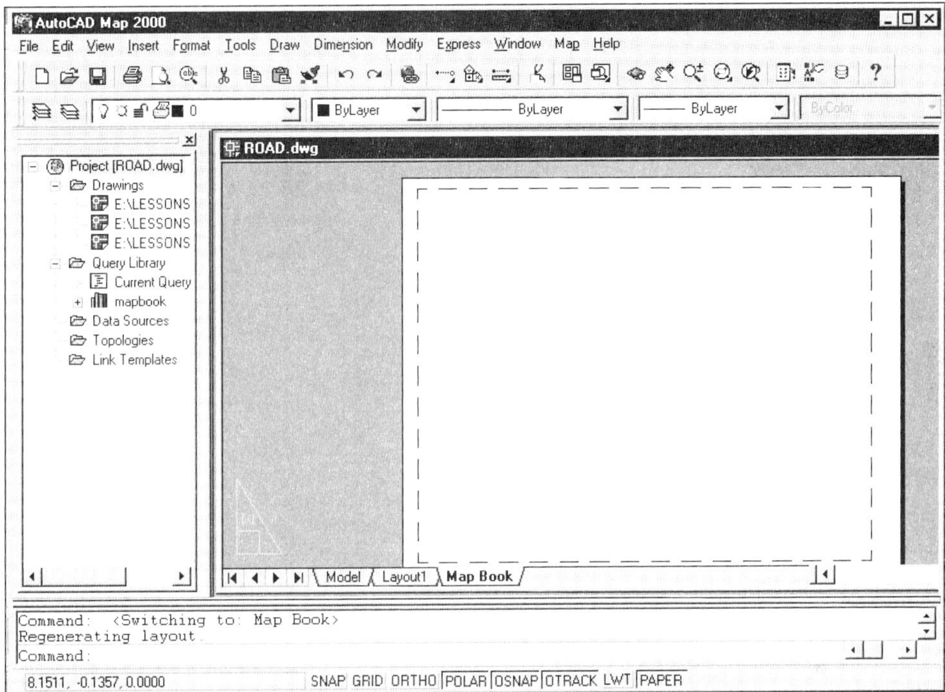

Figure 4–36 *Layout view with drawings not showing.*

Step 6 Go to **Map** then select **Plot Map Set.**

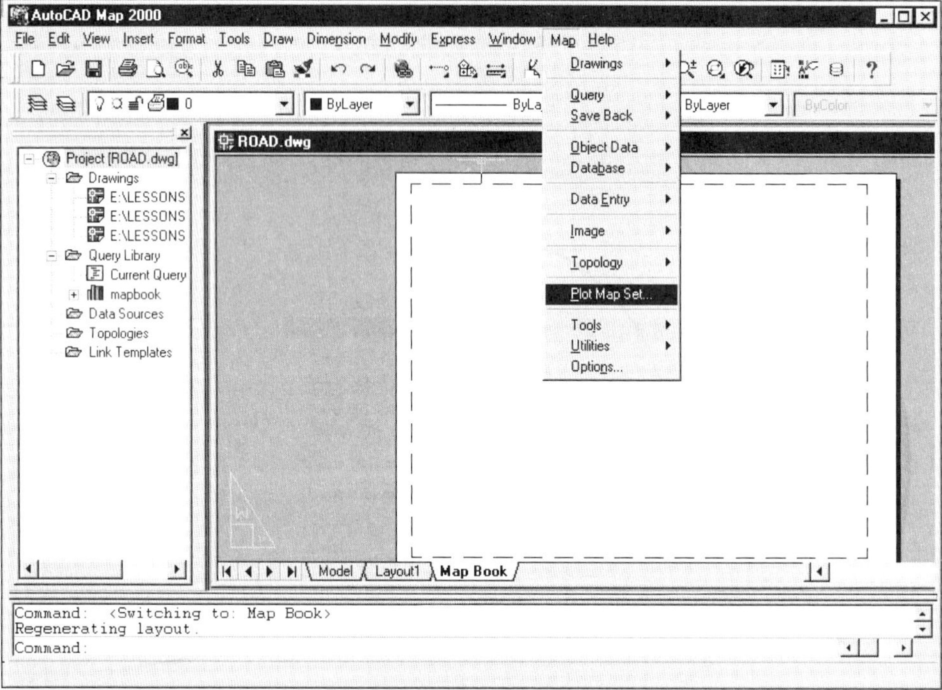

Figure 4–37

The Plot Map Set screen will appear. Click on **New.** In the **Plot Set Definition** dialog box, enter a name (in this case "**test**") and a description (in this instance, "**4 part mapbook**"), then click on **Plot Template Block**. The Plot Template Block dialog box will appear.

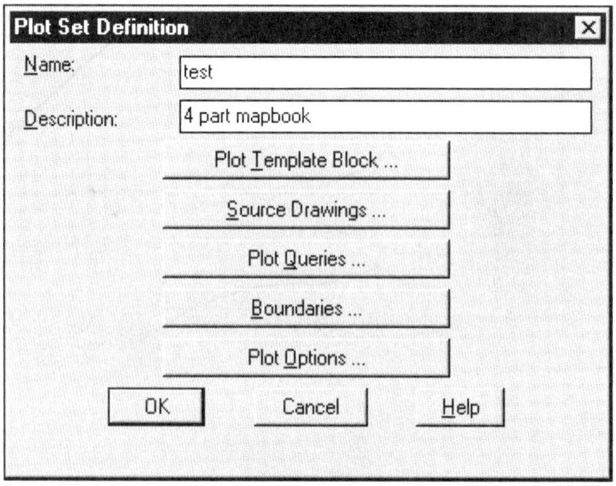

Plot Set Definition

Name: test

Description: 4 part mapbook

Plot Template Block ...

Source Drawings ...

Plot Queries ...

Boundaries ...

Plot Options ...

OK | Cancel | Help

Figure 4–38

The Block name will show as PLOTBLOK. Set the main viewport as **REF_VPORT**. The Reference Viewport Layer checkbox should be activated, showing a checkmark beside it and the edit box located to the right should show **MAIN_VPORT**.

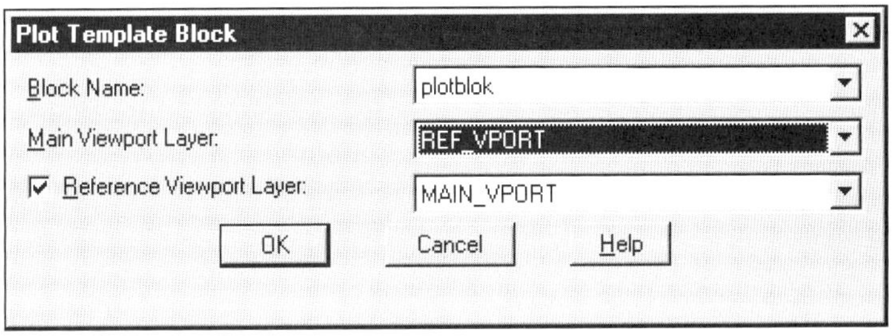

Figure 4–39

Click on **OK**.

Step 7 The Plot Set Definitions dialog box reappears. Click on the **Source Drawings** button. The attached drawings will be listed in the Source Drawing Selection dialog box, in the list of Attached Drawings. Highlight each and then click on the **>>** button to add the files to the list on the right side of the screen, the Selected Drawings list.

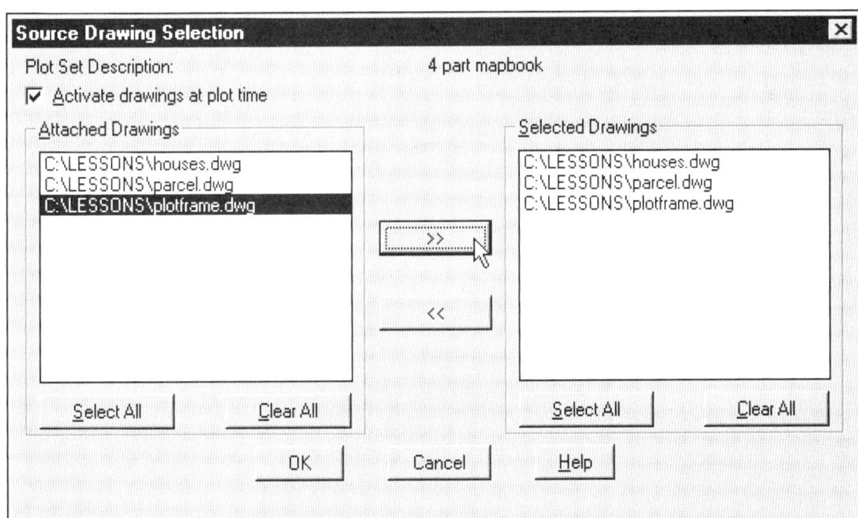

Figure 4–40

Click on **Select All** for the Selected Drawings, then click on **OK**.

Step 8 The Plot Set Definition dialog box reappears. Click on the **Plot Queries** button. The Plot Query Selection dialog box will appear. The category you created in Step 3, mapbook, will be listed, as will the query name, mapbook printing. Select this item from the left-hand column, Queries, and click on the **>>** button to add this item to the right-hand column, Selected Queries. Highlight it in the right-hand column then click on **OK**.

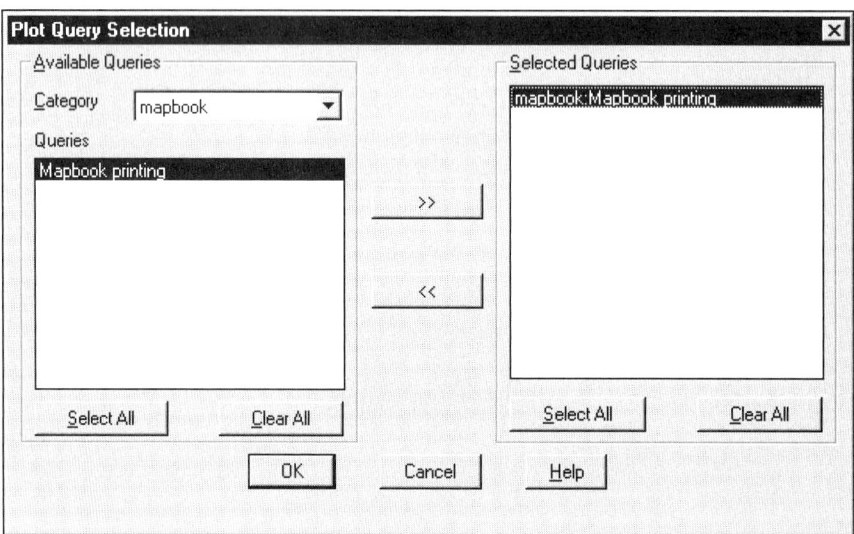

Figure 4–41

Step 9 Click on the **Boundaries** button from the Plot Set Definition dialog box. In the Plot Boundary Definition dialog box, select the file named **PLOTFRAME.DWG**. This drawing previously had object data attached to each polygon in the Boundary layer under the table name Printorder and under a field name of printorder.

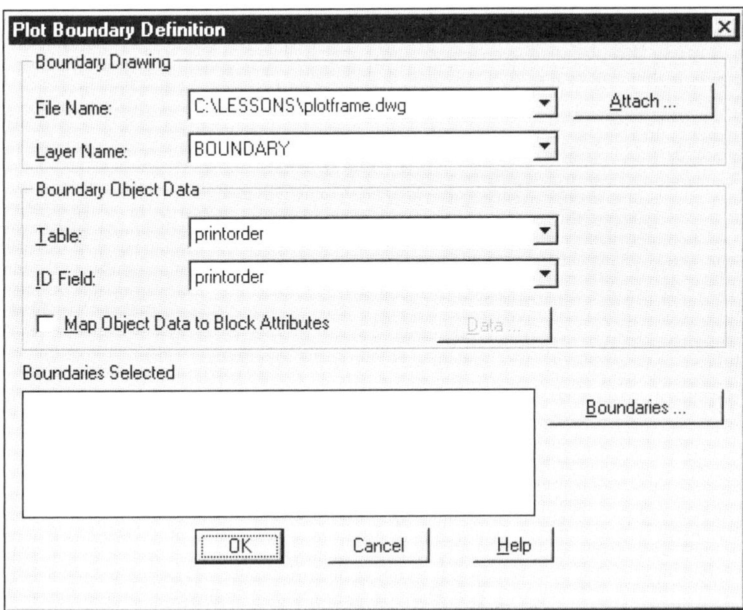

Figure 4–42

Click on the **Boundaries** button.

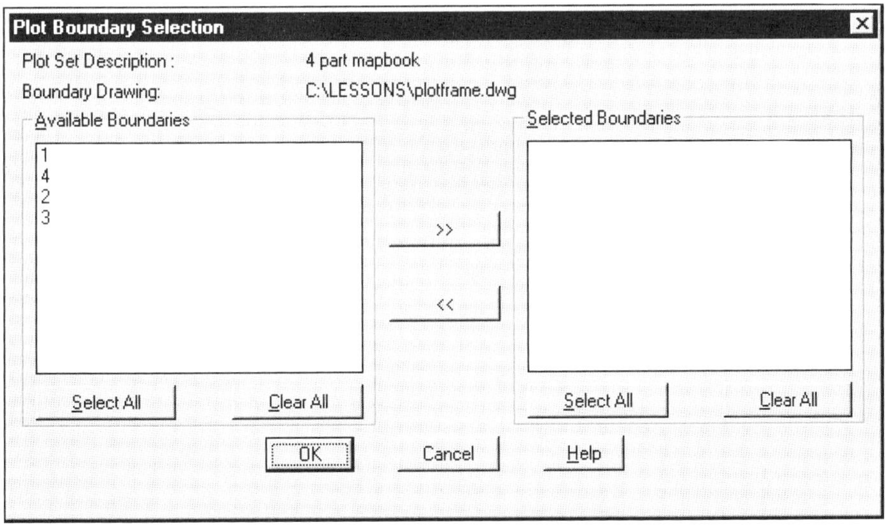

Figure 4–43

The Plot Boundary Selection dialog box appears. From the Available Boundaries click on the **Select All** button, then click on the **>>** button. Use the **Select All** option under the Selected Boundaries to choose all **four** boundaries, then click on **OK** to return to the **Plot Set Definition** screen.

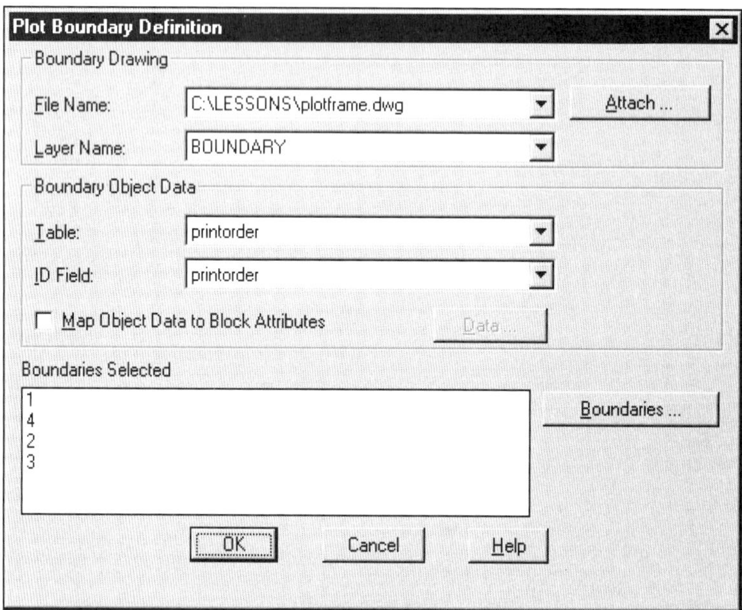

Figure 4–44

Step 10 Click on the **Plot Options** button. The Plot Set Options dialog box will appear. From the pull-down Layout list, select **Map book**. In the Main View Scale select the **Plot to Scale** radio button and enter a scale of **1:100** in the edit box. Enter a **Reference Scale** of **1**. Select the **Trim Objects at Boundaries** radio button then click on **OK.**

Plot Set Options
Layout: Map book
Main View Scale
Plot to Fit Plot to Scale Main Scale: 1:100
Reference View Scale
Reference View: Enabled Reference Scale: 1.0000
Boundary Edge Options
Buffer Boundary Distance 0.1000 Select <
True Rectilinear
Trim Objects at Boundaries
Template Block Settings
Select Layer ...
OK Cancel Help

Figure 4–45

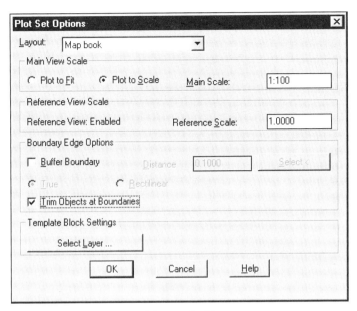

The Plot Set Definition dialog box is presented again. Click on **OK**, and the Plot Map Set dialog box appears.

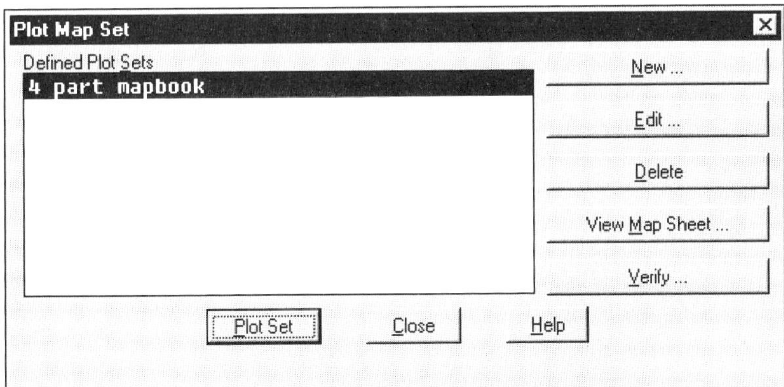

Figure 4–46

Click on the **Plot Set** button, and the printing of the four maps will begin. The entire area of the drawings will be divided into four sections as defined by the boundary lines in the PLOTFRAME drawing and will be displayed by a mosaic of four maps.

Either continue with the tutorials or exit AutoCAD Map 2000.

TUTORIAL: IMPORTING NON-DWG FILES

Often, you will want to make use of drawings that were created in a non-AutoCAD format. Most software packages used in conjunction with AutoCAD Map 2000 or GIS have the capability of exporting their vectors in the DXF format. AutoCAD Map 2000 has the capability of directly reading many of the other forms of GIS files. This exercise demonstrates how to access some of these types of data.

Step 1 AutoCAD Map 2000 can open a DXF file directly. Select **File** then **Open**. From under the **Files of Type** drop-down list box in the Select File dialog box choose the **DXF** option.

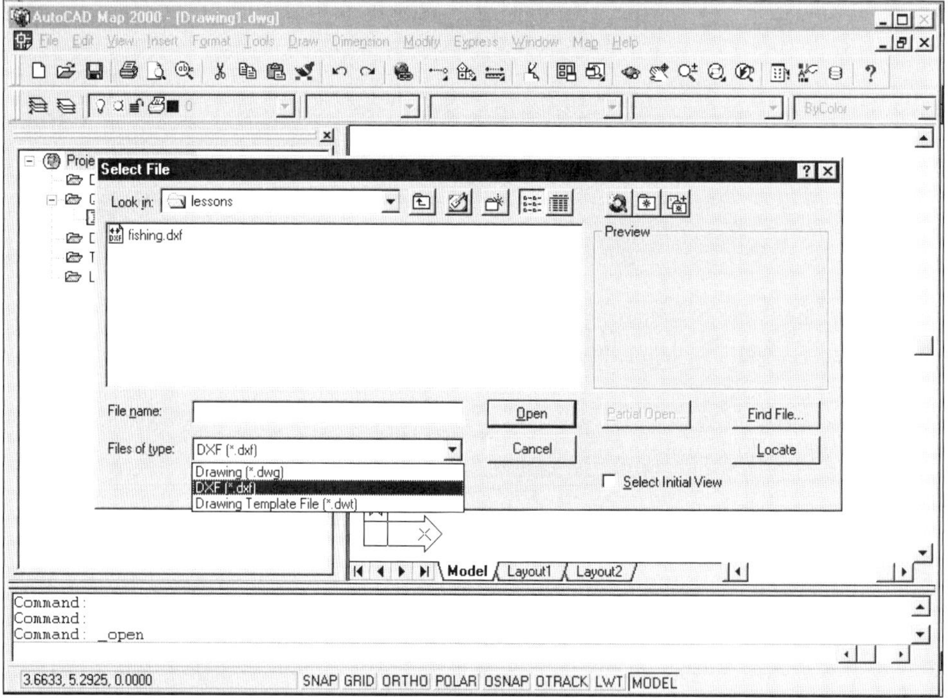

Figure 4–47

Next, select the directory and then the file you wish to open, in this case, **FISHING.DXF**. Click on **Open**. The file will be opened, and it can be used and saved as a DWG file when the work session is complete.

Step 2 Other file types can be imported into a work session. The following steps are a walk-through example because no actual data files have been provided for importing into AutoCAD Map 2000. Go to the **Map** menu then select **Map Tools** and **Import.**

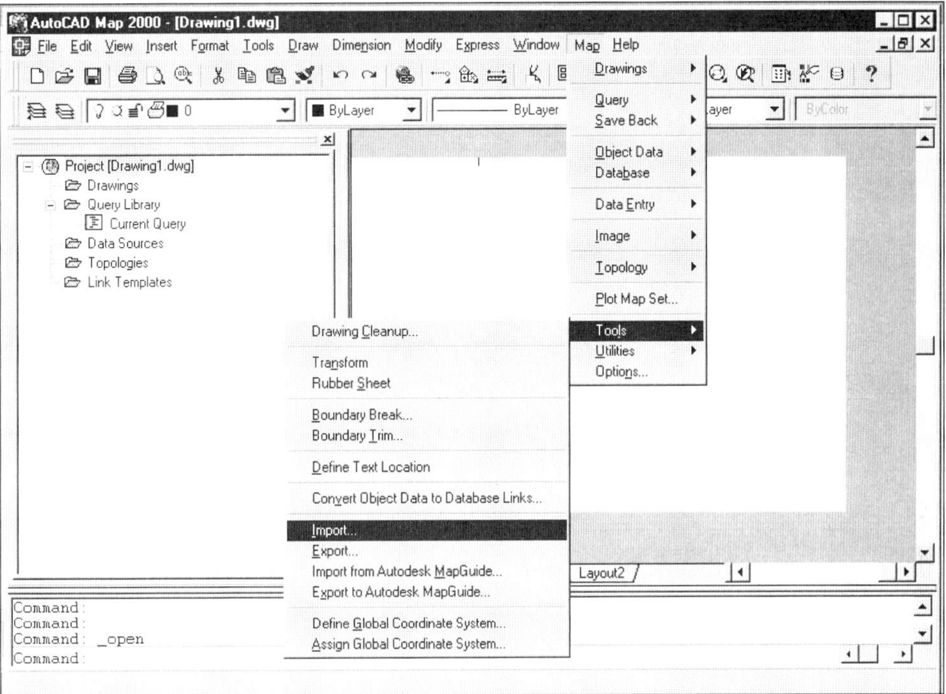

Figure 4–48

Step 3 From the Map Import dialog box select the file type you want to import from the list provided through the pull-down list box.

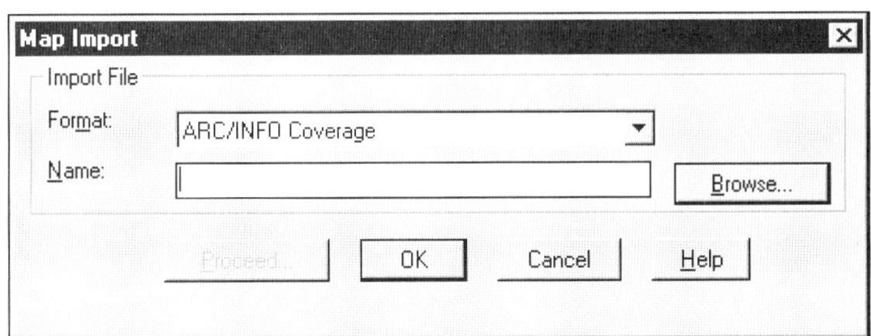

Figure 4–49

Step 4 You select the file you want to import and its location by using the **Browse** button.

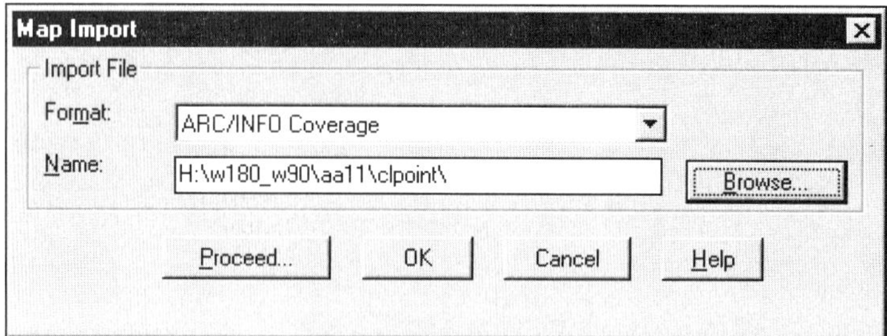

Figure 4–50

In this example shown, an Arc Info Coverage would be imported from the H: drive with a click on Proceed.

Step 5 An Import Data Options dialog box will appear, providing flexibility in the importing process.

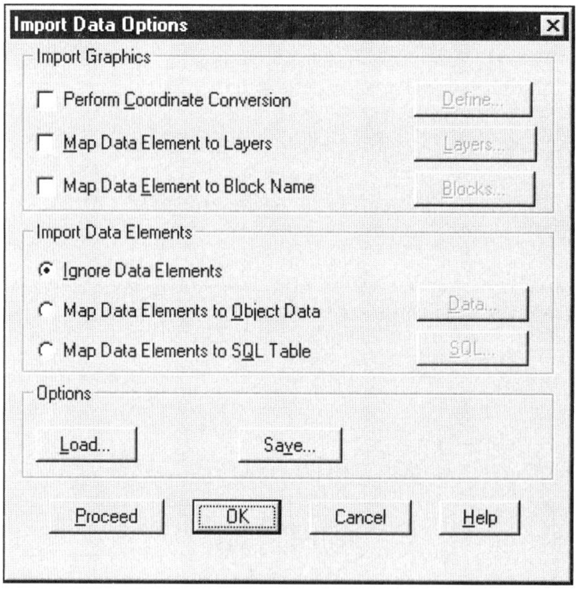

Figure 4–51

You can, for example, convert coordinate systems to match that which you are working in or have data being imported converted to object data . You would then click on **Proceed** and the item selected for import will be presented on the screen and could be saved as a DWG file for future use.

SUMMARY

Edge Matching, Rubber Sheeting and Importing Non-AutoCAD Map 2000 files will be common activities for most GIS users. With changing political boundaries or building expansions there can be a need to amalgamate drawing files. Edge Matching can take drawings of adjacent areas and merge them to become one. The procedure allows for small discrepancies between the drawings with an alignment procedure built into the process. Rubber Sheeting is a procedure which adjusts the location of known specified points to match the location of known points. This process is often used when old drawings are used in conjunction with drawing files from other sources deemed to be more accurate. The older files can be "Rubber Sheeted" to align with the newer map files. Importing non-DWG files is an important feature since most GIS software is capable of exporting data in one of the formats that is readable by AutoCAD Map 2000. Mapbook printing is an effective way to produce highly detailed, small page printouts of maps that would normally require a large size plotter to reveal the details in the drawings. The tutorial used the Plotframe.DWG drawing which partitioned the map into four sections but any number of sections can be established. To do this create a file similar to Plotframe.DWG with as many sections as desired and embed numerical object data in each partition.

REVIEW QUESTIONS

1. When would Edge Matching be employed?

2. What purpose does the Quick View function serve?

3. What would be a disadvantage to modifying original objects during a drawing cleanup?

4. What precautionary step should be taken with drawing files prior to performing a Save to Source Drawings function?

5. What is a fundamental concern when using Rubber Sheeted drawings?

6. What would be the advantage of a Mapbook compared to an E-size plot?

7. What file formats can be imported into Map 2000?

CHAPTER 5

Generating a Digital Map

OBJECTIVES

After completing this chapter you will know

- how the Universal Transverse Mercator system is used to georeference maps
- how to configure AutoCAD Map 2000 for use with a graphics tablet
- how to generate a digital map
- how to add object data to a drawing while it is being created

KEY TERMS

Projection	**COM port**
Graphics tablet	**Orthogonal projection**
Emulation	**Projective projection**
Limits	**Affine projection**
Object Snap	

A common requirement in the GIS workplace is the need to convert spatial data collected from the real world into a format that can be easily worked with and understood in the computer. This process requires the evaluation of the sources of data, the development of generalizations, and the organization and presentation of the information in a clear, concise manner.

Because the majority of spatial data sets will result in map displays being utilized to illustrate the content of the data, it is important that the base maps be as accurate as possible.

One of the most important cartographic concepts is that of *map projections*. A projection is a mathematical formula used to reduce the amount of inaccuracy that results when features on a spherical surface (the earth) are pressed onto a flat surface, a map. The projection formula adjusts the geographic coordinates of the data to minimize the distortion. Depending on the intended use of the map, some projections may be more suitable than others.

Selection of the best projection increases in importance with the extent of the area represented by the map. A map showing significant portions of the earth's surface requires the careful choice of an appropriate projection, whereas the projection for a street map may not be as critical.

The selection of an appropriate projection is reasonably straightforward .If you are using AutoCAD Map 2000 to digitize maps from paper-based maps, such as topographical maps, then you are likely to rely on the *Universal Transverse Mercator* projection (UTM) because sources like topographic maps have used this projection in the first place. As an alternative to the UTM system, the State Plane System of projections is a common choice.

Generally speaking, locations can be given on a conventional 1:24,000 topographic map by referring to a six-digit number and making use of the knowledge that the grid lines are 1,000 meters apart in either direction.

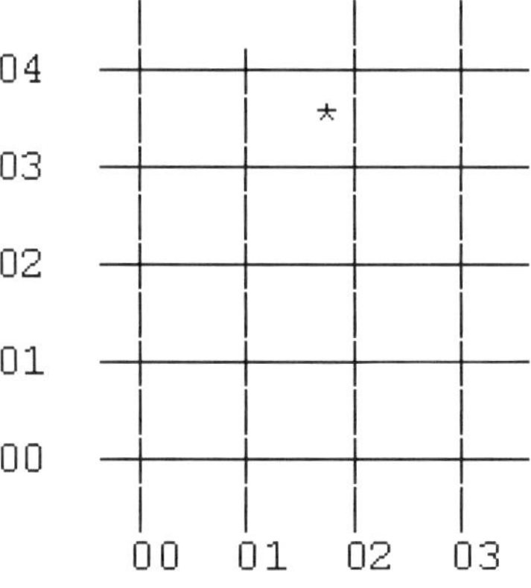

Figure 5–1

The six digits are determined by estimating ten divisions between each pair of numbered lines. These divisions exemplified don't show up on the map but must be estimated and read as tenths in order to determine the third digit for each direction. In the case above (Figure 5–1) the asterisk would be identified with the horizontal component 018. A similar approach for the vertical component in the grid above might yield a three digit number such as 035. The common grid number for the asterisk's location would be 018035.

When using the grid system with AutoCAD Map 2000, this procedure is taken to a more exact level. For example, each base map resides within a predetermined UTM zone. These zones, 6 degrees of longitude wide, are numbered starting at the International Date Line. Canada, for instance, is covered from Zone 7 on the Pacific Coast through Zone 22 on the Atlantic Coast. The zones are identified by number in the margin of topographic maps.

The UTM coordinates used in AutoCAD Map 2000 are expressed in meters and usually constitute 13 digits presented in the traditional (*X,Y*) ordered pair format. The *X* coordinate represents the "easting" relative to an arbitrarily picked "Central Meridian" for the zone you are working in. The value assigned to this arbitrary meridian is 500,000 meters. The second coordinate, or *Y* value represents the distance from the equator in meters.

To determine the location of a point on a topographic map, you can start by looking at the grid reference numbers located in the lower left corner of the margin area on the map. These numbers are usually pale blue in color and may be printed in superscript, so will appear smaller than the other grid-identifying numbers.

These numbers provide you with the actual distances that this lower left corner of the map is located relative to the central meridian and the equator. As mentioned previously, the grid lines on a 1:24,000 map are 1,000 meters apart.

In Figure 5–2, the vertical line numbers represent 301,000, 302,000 or 303,000 meters relative to the central meridian's value of 500,000 meters, and horizontal line values represent 4,800,000, 4,801,000, 4,802,000 meters from the equator.

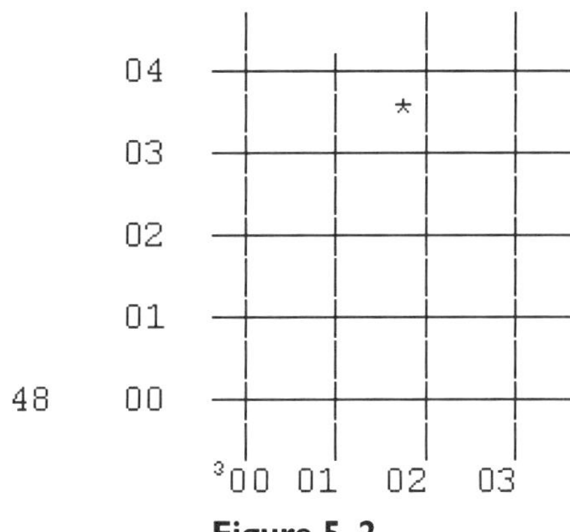

Figure 5–2

To determine the location of a point, in this case the asterisk, a scale or ruler should be used to measure the distance of the point from the grid lines accurately.

The coordinates for the asterisk now are as follows:

> across the bottom or easting, 301800
>
> alongthe side or northing, 4803500 for a combined ordered pair for the point being 301800,4803500.

This level of precision, to the nearest meter in theory, must be used when setting a map up in the UTM coordinate system in AutoCAD Map 2000. (You may want to refer back to this whenever you digitize maps in AutoCAD Map 2000.) It seems challenging to interpolate locations this way at first, but after a few times of using this approach, it becomes less onerous.

When preparing digital maps as described in the next lesson, the user will be required to utilize a minimum of two georeferenced points, identified by their 13-digit coordinates. The general rule of thumb for selecting locations to be used a reference points is to use one location in the lower left area of your region and a second point in the upper right region of the area to be mapped. When two ground control points are used to georeference the drawing, AutoCAD Map 2000 utilizes an *Orthogonal projection*. If three points are used, then a *Projective projection* is utilized, and if more than three points are used, an *Affine projectio*n is used. With the addition of more reference or ground control points comes a general improvement in geographic accuracy. The online help offers a detailed explanation of each projection.

TUTORIAL: CREATING DIGITAL MAPS USING A GRAPHICS TABLET USING UTM COORDINATES

Perhaps the most powerful and unique feature of a program like AutoCAD Map 2000 is the ability to draw maps. To draw a map that is geographically correct, one should use a coordinate system like the UTM system to establish reference points and then these are used to "digitize" the map information.

To digitize a map, the user requires a graphics tablet to be connected to the computer. These come in a variety of sizes starting at about 12 inches square (30 cm square) and ranging up to 48 by 36 inches or larger. In this process, one uses the tablet and its digitizing "puck" instead of the mouse to trace the paper map. The puck generally has four buttons on it (some models have 12 buttons or more) and includes a small magnifying glass with cross-hairs in it (Figure 5–3). The intersection point of the cross-hairs is the alignment point for digitizing data.

Typically, on a four-button puck, the yellow button or front center button behaves like the left mouse button and the red button (typically) or the left puck button behaves like the right mouse button.

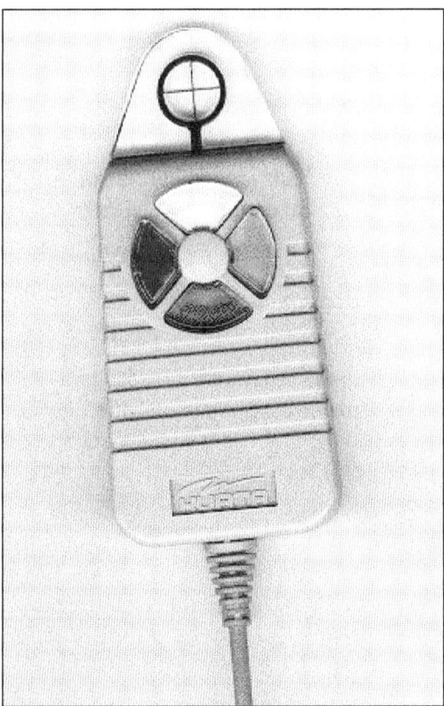

Figure 5–3

One must also realize that the puck operates in a weak magnetic field generated between the puck and the tablet. If you lift the puck up off the tablet, it will cease functioning. It has a sliding function rather than the rolling action that is common to the mouse, as you shall experience when you start using it.

Before one can accurately digitize a map the UTM (Universal Transverse Mercator) coordinates for at least two easily identifiable points on the original map must be obtained, one in the lower left region and one in the upper right region of the area to be digitized.

For this introductory exercise you have been provided with a paper map of Lennox and Addington County with townships outlined on it. (The copy of the map utilized in this tutorial is provided at the end of this chapter, see Figure 5-19.) Lennox and Addington County is a small county located in Ontario along the north shore of Lake Ontario. This exercise will get you started digitizing this map. Notice the annotations on this map that indicate the known points for the UTM coordinates. Also note the numbers identified as the "lower left limit" and "upper right limit." These numbers were taken from a topographic map following the procedure outlined above. Limits are used to set the size of the working area in world units are 13-digit UTM numbers taken from points that are opposite corners of an imaginary box that surrounds the area to be digitized.

Step 1 Ensure that the graphics tablet is **connected** to the computer and turned **on**, then start AutoCAD Map 2000.

Step 2 To digitize a map you must configure AutoCAD Map 2000 to use the graphics tablet instead of (or as well as) the mouse. There are several hot keys that can be useful when digitizing a map. The F6 key, for example, turns on the coordinate display so you can see the UTM or other coordinates. Others like the F3 key activates OSNAP or Object Snap.

Step 3 At the command prompt, type in **config**, then press ENTER. As an alternative, you can select **Tools**, then **Options**.

Step 4 An Options dialog box appears.

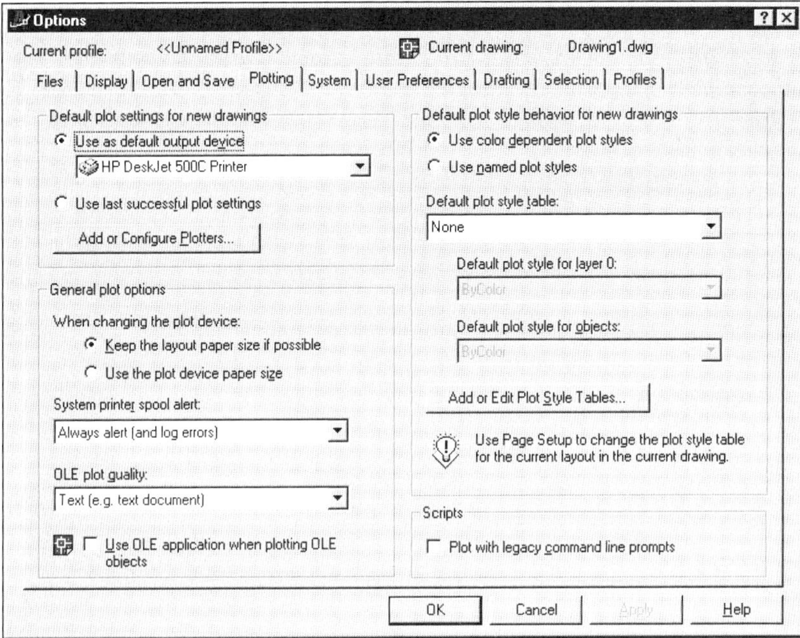

Figure 5–4

Click on the **System** tab.

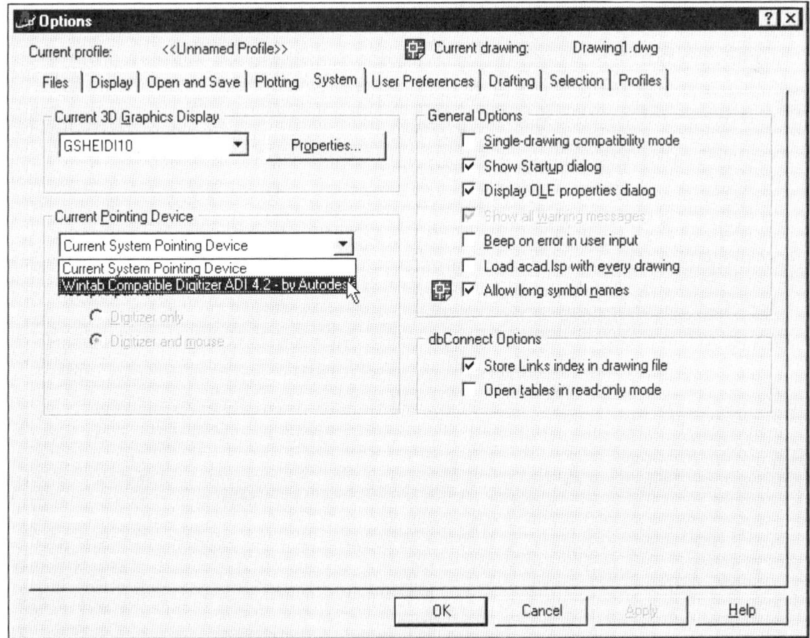

Figure 5–5

Step 5 The current pointing device is highlighted. In the case of Figure 5–5 it is the current system pointing device, or mouse. From the pull-down list find the digitizer you have installed a specific driver for or select the Wintab Compatible driver option. You have a choice of accepting input from the digitizer only or using both a mouse and a digitizer. Click on **OK**, and the digitizing tablet will be active.

Step 6 The mouse will still be active as will be the puck if you selected that option. Move the puck across the tablet surface to get the feel of it as it does not have a trackball in it, but works on the magnetic flux between it and the tablet.

Step 7 Now the tablet must be calibrated. Place your paper map of Lennox and Addington County under the protective plastic sheet on the tablet. If your tablet does not have the plastic overlay, you should consider fastening down the paper map with small pieces of tape in the corners.

Step 8 Go to the **Tools** menu and select **Tablet** then **Calibrate**.

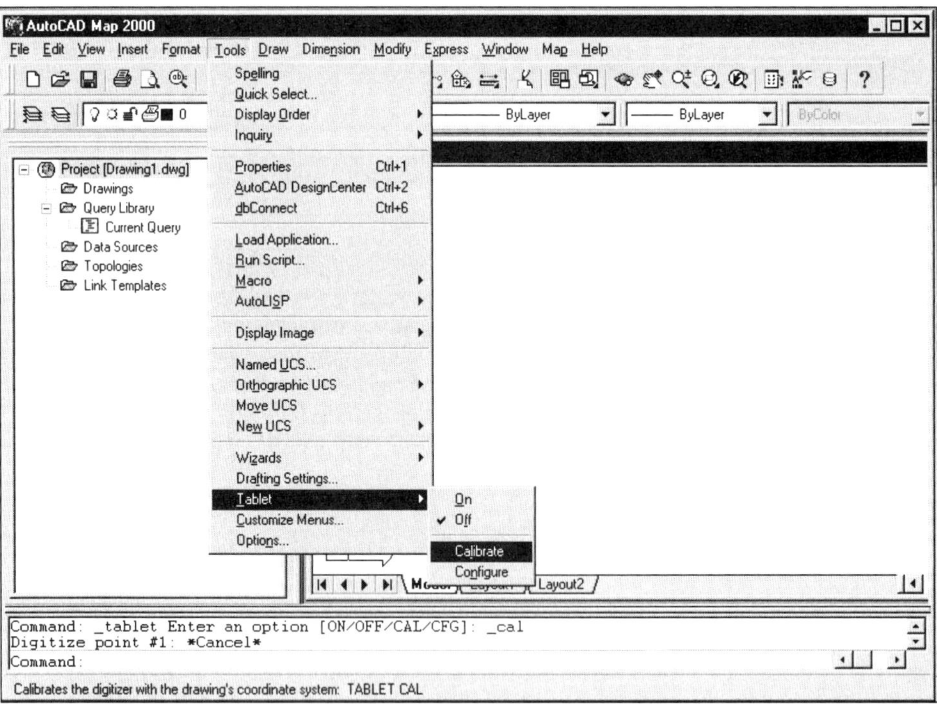

Figure 5–6

As described earlier, you must have at least two points identified on your map for which you know the location. These numbers measure distance east or west of a meridian and north of the equator in meters and were determined through the procedure outlined earlier dealing with UTMs. These are identified as the first and second digitizing points on the provided diagram (also see Figure 5–19) .

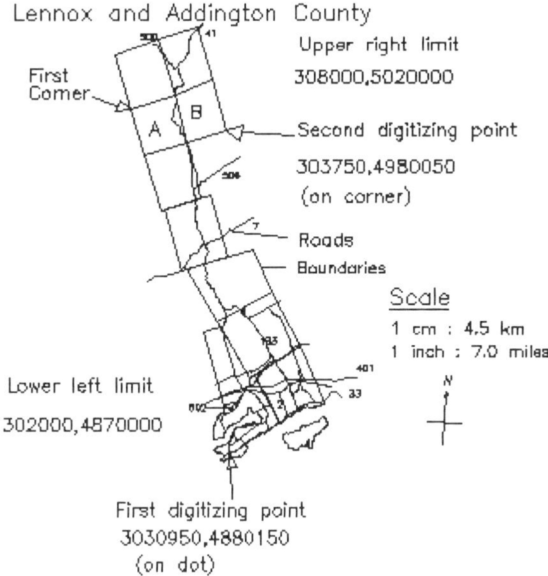

Figure 5–7

Step 9 When prompted to digitize the first point, locate the cross-hairs exactly over the lower left point on the paper map, identified as the "**first digitizing point**" and click the yellow puck button.

Step 10 When prompted to enter the coordinates of the first point, type in **303950,4880150**. Check the entry carefully. Press ENTER.

Step 11 When prompted to digitize the second point, locate the puck's cross-hairs over the point identified on the paper map as the "**second digitizing point**" and click the yellow button again.

Step 12 When prompted for the coordinates for the second point, type in **303750,4980050**. Press ENTER.

Step 13 You will be asked for a third point, press ENTER to skip the third repetition.

Tip: You should always keep a record of the coordinates you entered. If you don't get finished at one sitting, you can recalibrate using the same points and same coordinates, and your work will still be accurate.

You have now indicated to AutoCAD Map 2000 where on the planet you wish to draw. Now you must focus the computer screen over that area of the planet's surface. You will notice that you have difficulty finding the cross-hairs on your screen. You may periodically see them, but they seem to "blast" by your field of vision. The next step is to focus AutoCAD Map 2000 on the drawing area.

Step 14 Go to the **Format** menu and select **Drawing Limits.**

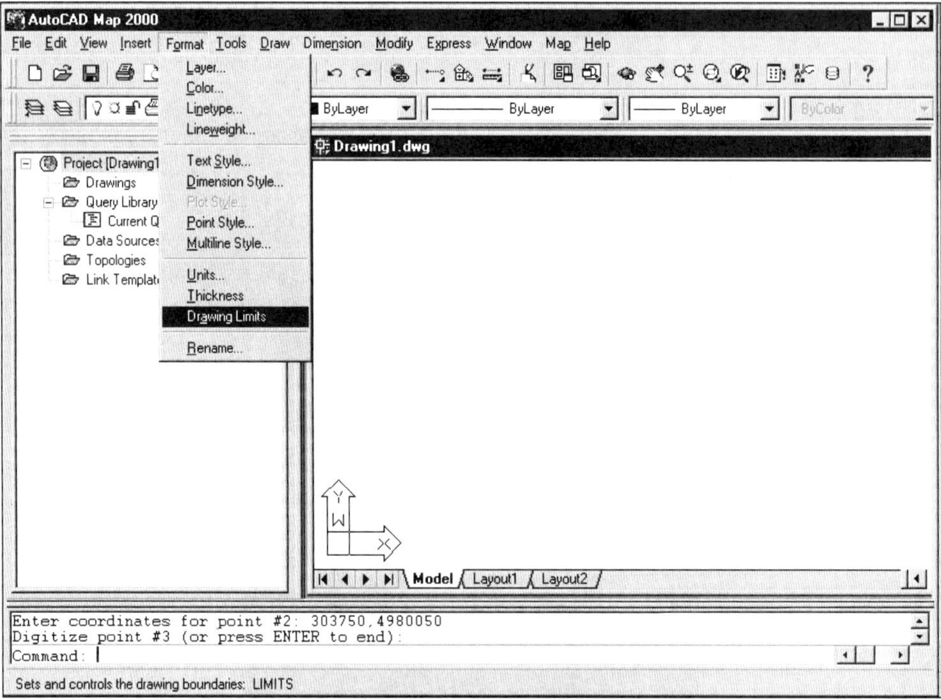

Figure 5–8

You will be prompted for the lower left corner's coordinates. These should be entered as an ordered pair (*X*,*Y*) or (easting, northing) and should be lower in value than the calibration coordinates. Type in **302000,4870000** for the lower left drawing limit, then press **enter.** These numbers identify a point more to the south and west than digitizing point 1.

Step 15 You will be prompted for the upper right -hand corner's coordinates. Type in **308000,5020000** then press **enter**. These numbers identify a location to the north and east of the area to be digitized. Steps 14 and 15 focus the screen on a drawing area only slightly larger than the area defined by the calibration numbers.

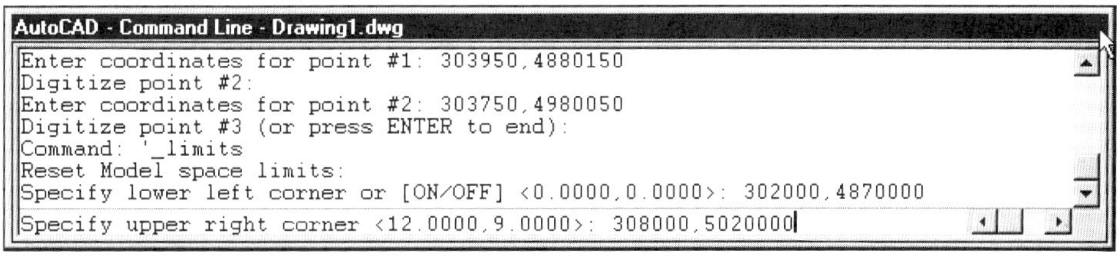

```
AutoCAD - Command Line - Drawing1.dwg
Enter coordinates for point #1: 303950,4880150
Digitize point #2:
Enter coordinates for point #2: 303750,4980050
Digitize point #3 (or press ENTER to end):
Command: '_limits
Reset Model space limits:
Specify lower left corner or [ON/OFF] <0.0000,0.0000>: 302000,4870000
Specify upper right corner <12.0000,9.0000>: 308000,5020000
```

Figure 5–9

Step 16 To align the tablet and screen type in **z** and press ENTER. From the list of "zoom" options, type **e** (for extents) and press ENTER.

Step 17 You are now ready to start digitizing a map. However, often the cross-hairs seem to be off to one side of the screen. Now is a good time to center the screen. If the cross-hairs are off to one side for example, type in **pan** and press ENTER.

Step 18 A "hand" will appear on the screen. This panning tool can be used to drag the drawing area to one side or another (or up and down) the screen. Place the hand close to the left side of the screen and click the yellow button. Move the cross-hairs to the right to center your work area. Click on the red puck button to finish the pan function. A pop-up menu will appear and provide the opportunity to exit this function. Click on the **Exit** button. The screen will have been shifted. If you need to repeat the process, do so. This will work for any direction. As an alternative you can go to the **View** menu and select **Pan**, then highlight **Right** or you can try adjusting the scrollbars along the bottom and side of the screen.

When you have finished adjusting the screen to coincide with the location of the paper map on the graphics tablet, you should be able to place the puck on any portion of the paper map and still see the cross-hairs of the puck on the screen.

Step 19 Maps are generally drawn using a continuous line drawing option called polyline. You can access this tool by typing **pline** and pressing ENTER or you can go to **Draw** and select **Polyline.**

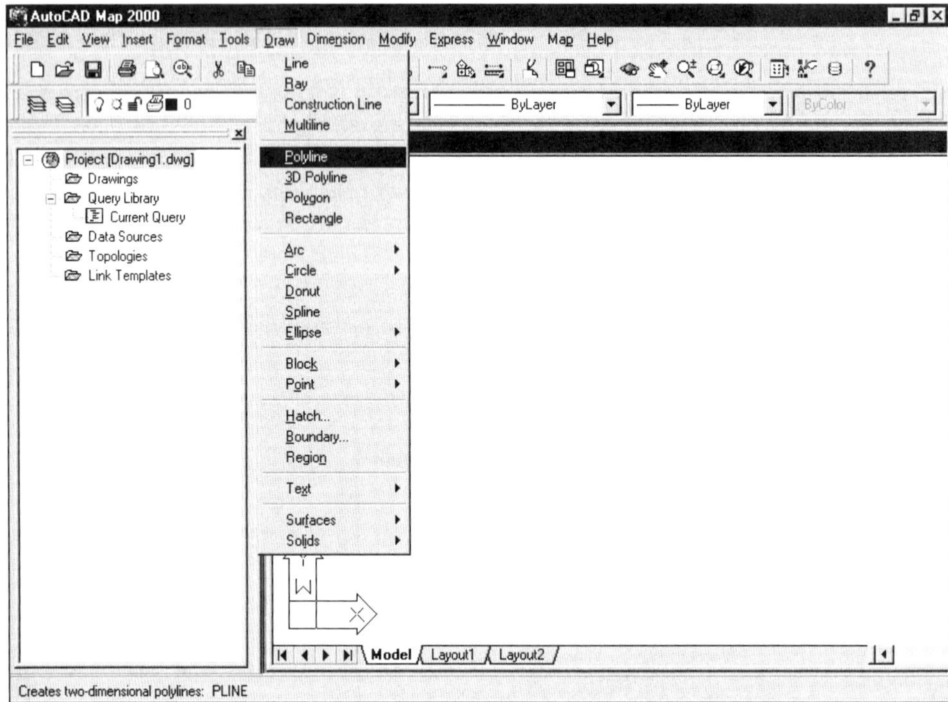

Figure 5–10

Step 20 After starting the pline command you are prompted for a first point. Place the cross-hairs of the puck over the point on the map indicated by arrow (it says start here for Step 20 and is located on the lower left corner of a township identified as "A") and click once with the yellow button.

Step 21 Slide the puck up to the first corner (as indicated on the paper map) and click again with the yellow button. When you click, you anchor the line to the screen.

Step 22 Now slide the puck across the map to the next corner of the township and click again with the yellow button.

Step 23 The outline of the township is starting to be recognizable. Move the puck down to the next corner and click again with the yellow button.

Step 24 ***Important:*** You must draw *closed* polygons. Read this before proceeding.

Move the puck cross-hairs near to the starting point used in Step 20 but **DO NOT CLICK. INSTEAD** type in "**c**" (for close) and press **enter**. AutoCAD Map 2000 will automatically close the polygon.

Step 25 Move the puck to the lower right corner of Township B. (This is the same point as the second digitizing point and where you will start to draw this township.)

Step 26 Move the cross-hairs up to the top menu strip. Go to **Tools** and select **Drafting Settings**.

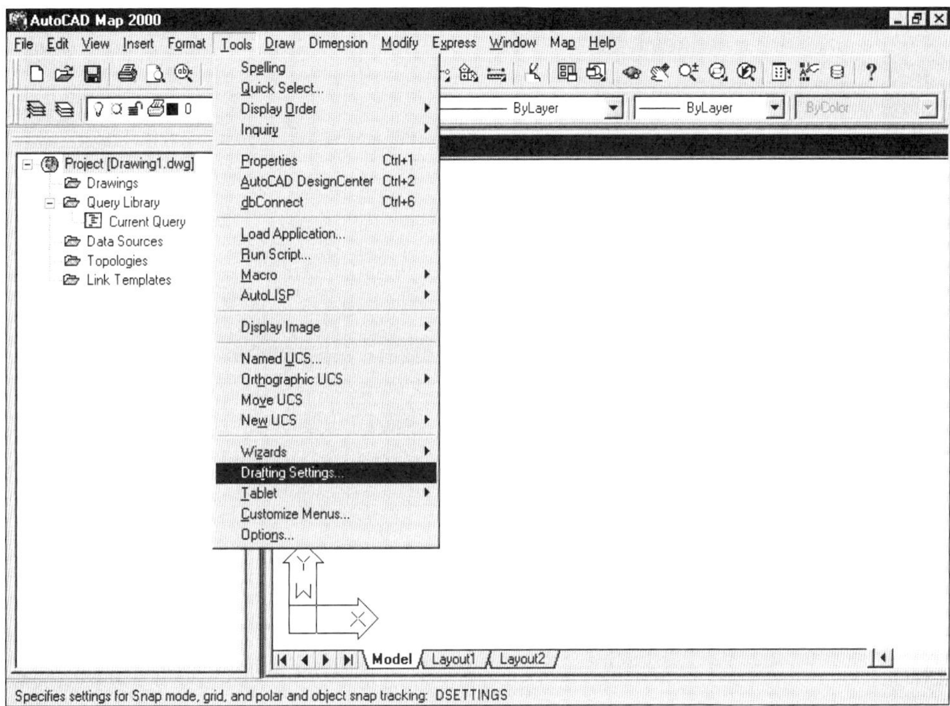

Figure 5–11

Step 27 The Drafting Settings dialog box will appear. Select the **Object Snap** tab. This tool will allow AutoCAD Map 2000 to automatically adjoin the lines of the shared borders of the townships so that double lines don't appear, making for a far superior finished product.

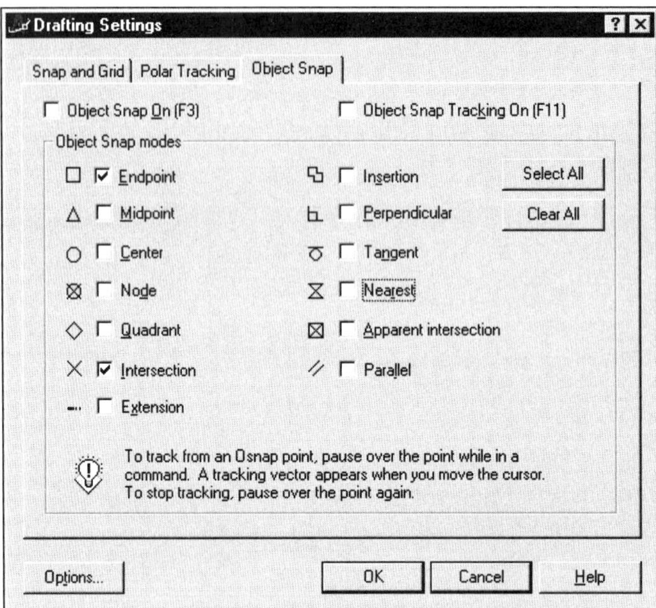

Figure 5–12

Step 28 In the dialog box select the **Nearest** option in the Object Snap Modes section. A checkmark will appear in the box.

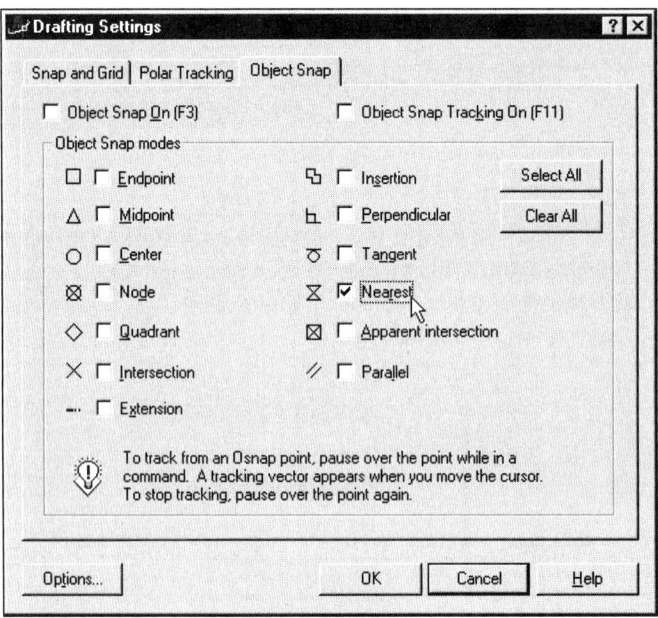

Figure 5–13

Click on **OK**.

This feature will allow AutoCAD Map 2000 to "lock onto" the nearest point you already have digitized if you place the box that is located at the center of the crosshairs around that point.

Step 29 Type in **pline** at the command line and press ENTER.

Step 30 Locate the crosshairs over the point earlier referred to as the "**second digitizing point**" and click the yellow button.

Step 31 Slide the puck cross-hairs up until they reach the corner of the township and click the yellow button again.

Step 32 Move the puck cross-hairs to the left until the corner of the previously digitized township is inside the box located at the center of the cross-hairs, and click the yellow button.

Step 33 Move the cross-hairs down to the bottom left corner of the township, so that the lower corner of the previously drawn township is within the box at the center of the cross-hairs, and click the yellow button again.

Step 34 Move the cross-hairs toward the point at which you started digitizing this township until it is inside the box located at the center of the cross-hairs. In the command line, type **c** for close and press ENTER.

Step 35 Repeat the process for the other townships in order to obtain a complete map consisting of closed polygons. The uneven boundary along the bottom represents a shoreline of a lake and will require digitizing by drawing a continuous polyline from a multitude of very short line segments.

ADD A LAYER FOR "ROADS"

An accepted practice in mapping is to generate a separate layer for each map feature. This would imply that features like roads, swamps, power lines, lakes, and so on would each be drawn on a dedicated layer.

Step 36 Go to the **Format** menu and select **Layer** and click again. As an alternative you can activate the Layer Property Manager screen by clicking on the icon in the Object Properties Toolbar.

Figure 5–14

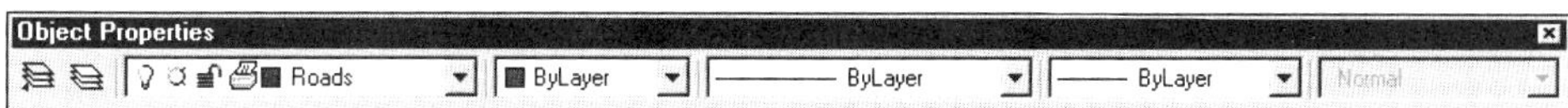

Figure 5–15

Step 37 In the Layer Properties Manager dialog box, click on the **New** button. A new default layer will be created. Click again on the **Layer1** entry or in the text box located to the right of Name.

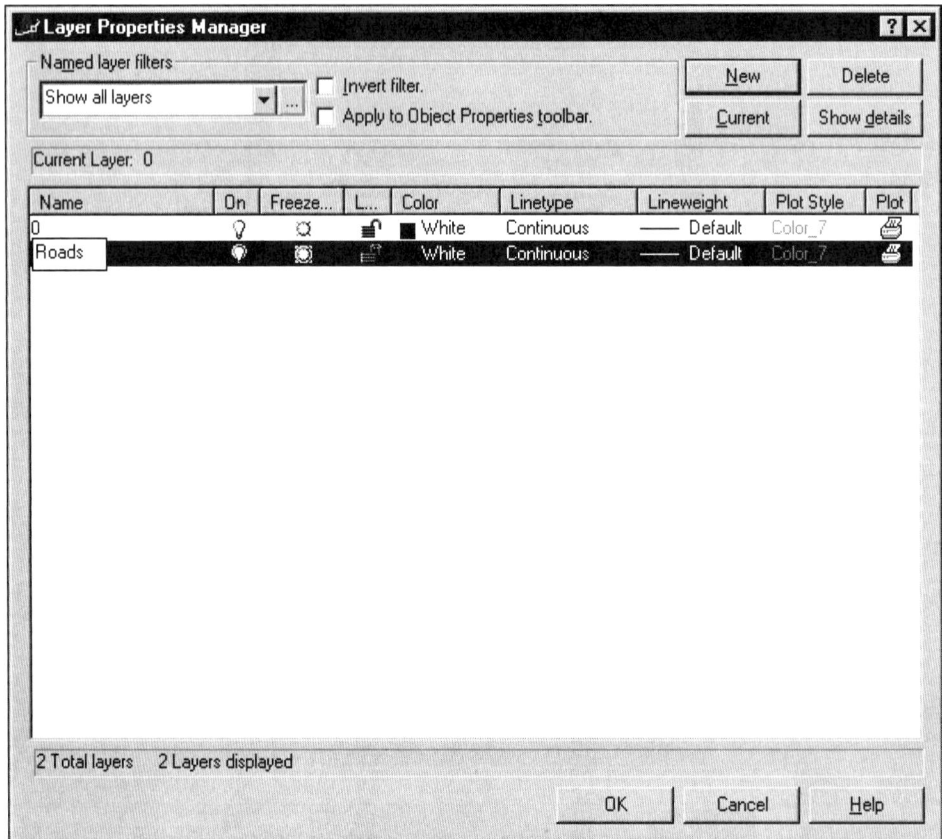

Figure 5–16

Type in **Roads** as the new name.

Step 38 Place the puck pointer on the small Color box and click once. Place the puck pointer on the color of your choice (in this example, red) and click, then click on **OK**. The color you selected will appear in the dialog box.

Step 39 To draw in the new layer we must make it the current one. The Roads layer you created should still be highlighted. If is not, click on it. Now click on the **Current** button and then on **OK**.

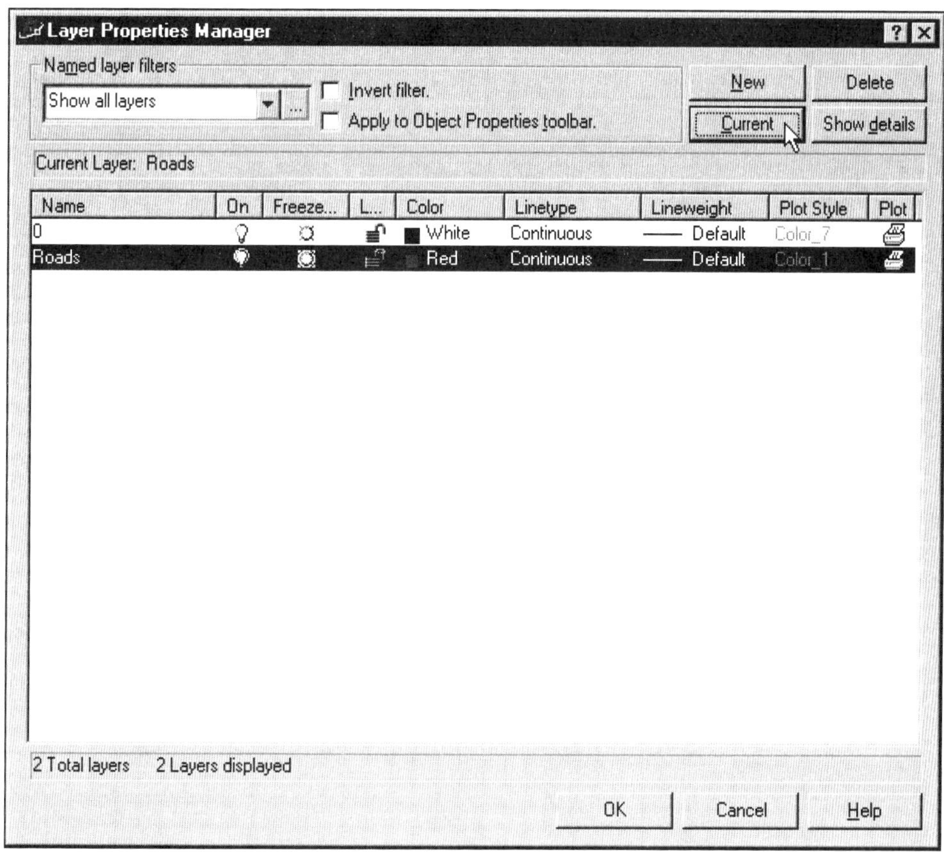

Figure 5–17

Step 40 You will see your new layer and drawing color identified in the Object Properties tool bar.

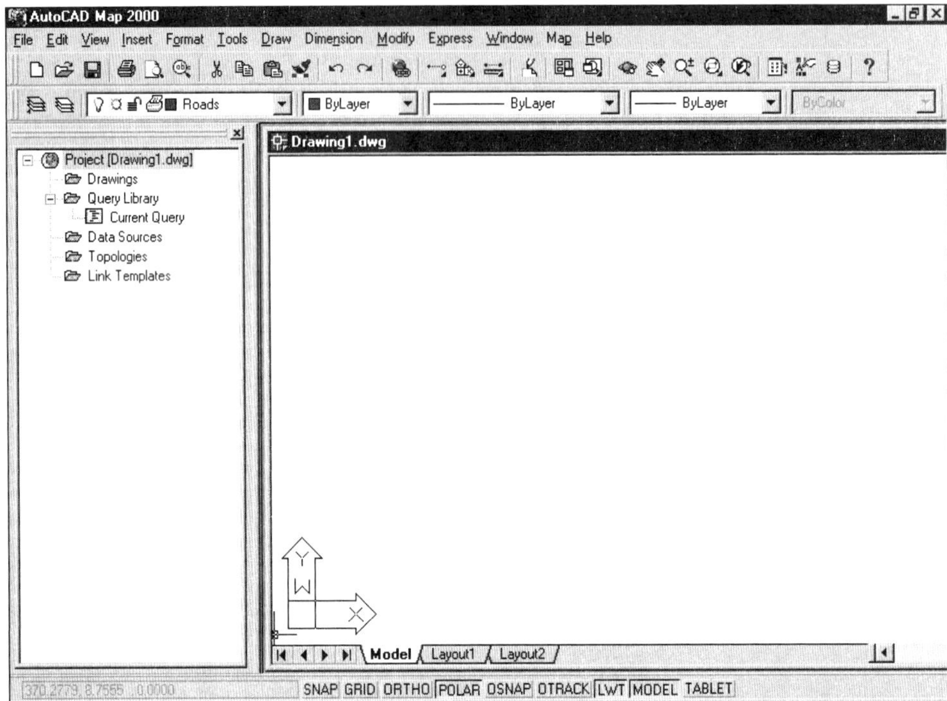

Figure 5–18

Step 41 Go to **Tools** and select **Drafting Settings**. Click on the **Object Snap** tab. This time, click on the box with the checkmark in it to turn off the snap procedure. Click on **OK**.

Step 42 Type in **pline** and press ENTER. This time you are drawing a line feature, not an area, so you will not close the polyline when you reach the end of the road. Instead press ENTER to stop drawing.

Step 43 Locate the cross-hairs over the road shown on the paper map where it enters Township B and click the yellow button.

To digitize a curved line like a road with a straight line drawing tool you must make a series of very short straight lines. Click the yellow button at every point that the road changes direction. When you finish digitizing the road as it passes through the two townships tap the ENTER key to end the drawing process.

Complete the drawing by digitizing all the road features.

Step 44 When finished drawing you should reconfigure AutoCAD Map 2000 back to the mouse.

Go the **Tools** menu and select **Options**. Click on the **System** tab to view the **Pointer** configuration. After reviewing the current configuration, scroll up the list of supported pointing devices until you see the **current system pointing device**. Highlight this choice, then click on **OK**.

The digitizer is now disabled and the mouse is the only pointing device.

Step 45 Often the entire drawing project doesn't get completed at one sitting. To finish a working session type **quit** or select File then **exit**. When asked about saving changes, click on **OK**. You will be prompted to provide a location for the file and a name for it. Set a location for saving the file to and type in a file name, such as **LNA2** and click on the **Save** button. The file will be saved and AutoCAD Map 2000 will close.

CONTINUING THE DRAWING IN A FUTURE WORK SESSION

To continue with the drawing assignment at a future session, open the file you had partially completed. You must then use the Tablet and Calibrate feature to register the UTM coordinates of the paper map to the on-screen drawing. The limits will be intact from the previous work session and will not require setting again. After inputting the digitizing points, complete a Zoom, All to focus and align the drawing.

Either exit AutoCAD Map 2000 or continue with the tutorials. A large copy of the Lennox and Addington County map is provided below (Figure 5–19).

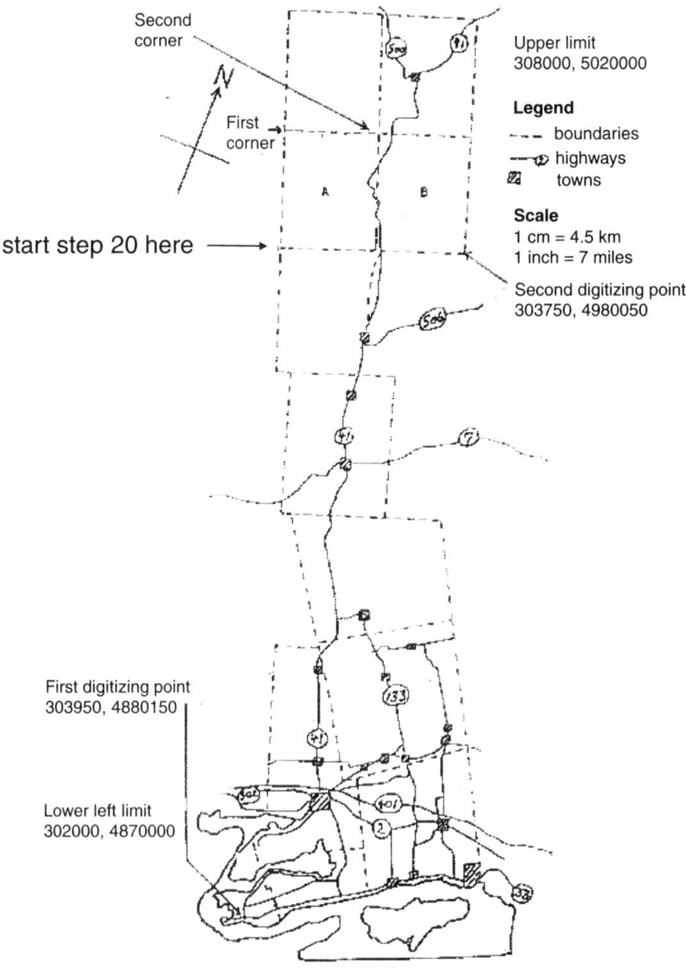

Figure 5–19

ADDING OBJECT DATA WHILE DIGITIZING

Object data can be added to the features being digitized while they are being created. To add the data the Object Data Table must be created prior to digitizing. You can add object data to both nodes and links. Use the Map menu and select Data Entry, then Digitize Setup. As the Object type, select either Linear or Node. Highlight the Attach Data option, then select Data to Attach, pick the table you wish to place data in, then click on OK. Return to the Map menu, highlight Data Entry and Digitize to begin digitizing. As you digitize each feature you will prompted to enter object data values in the table.

SUMMARY

Digitizing a paper map can often be the only way of acquiring a high quality, accurate base map for a region. The original paper map can be a photograph, a topographic map or an original drawing. The tutorial introduces UTM coordinates because they are commonplace but other coordinate systems could be used. The key to having a geographically correct digital map after the digitizing has been completed, lies in the selection of locations or features on the paper original that can be identified on topographic maps through the use of the coordinate system. The calibration process uses these locations to establish a projection for the map, either Orthogonal, Projective, or Affine. The other numerical values used, the Limits, establish the size of the screen display relative to the drawing or map area. Once digitizing has commenced, it is very important to create closed polygons if regions such as counties or states are being created. To minimize errors, the Object Snap function should be used whenever possible. It is also important to digitize different features on their own unique layers. Whenever one is digitizing, a record of the digitizing or ground control points and the limits should be kept to facilitate the completion of the exercise at a later date.

REVIEW QUESTIONS

1. How many digits are used to make up the ordered pair of numbers used to pinpoint a location in the UTM system?

2. Which direction does the first value represent, latitude or longitude?

3. What is the difference between configuring the tablet and calibrating the tablet?

4. How many data or ground control points are required to calibrate a tablet?

5. How do the coordinate values for the limits generally compare to the coordinates of the Ground Control Points?

6. At what level of theoretical accuracy do the 13-digit coordinates allow the map to be georeferenced to?

7. What does UTM stand for?

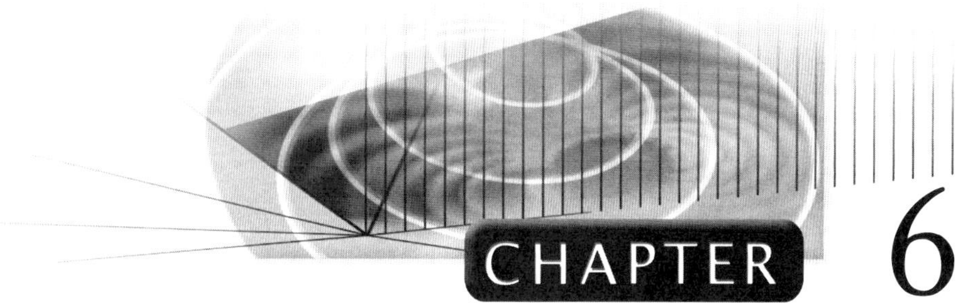

External Databases

OBJECTIVES

After completing this chapter you will know how to

- activate the ODBC capability of your software
- create a link between AutoCAD Map 2000 and a database of descriptive data
- access the data from AutoCAD Map 2000

KEY TERMS

ODBC	**Environment**
Link Path Name	**Catalog**
SQL	**Key column**

BACKGROUND INFORMATION AND NOMENCLATURE

AutoCAD Map 2000 has the capability to recognize data that has been prepared in an external database, such as dBase, Excel, or Access, and to have links created between that data and your drawing.

This chapter will guide you through the use of dBase3 and Access files in conjunction with a map of property ownership. This tutorial is designed to show you how to link your maps to external data and to provide some insight, into the many ways you can use this capability.

To understand the steps involved in establishing the linkage between AutoCAD Map 2000 and the database, there is some nomenclature to deal with.

The link between the two software packages is made with a *link template*, which is generated using an automated procedure in which you identify a *key column* from the database table, and a *catalog* for the data.

The *table* is a set of values to be linked to the drawing. The key column will be a column from the table that is used to make the link with the drawing. This column generally contains some type of information that can also be found in the drawing, such as identifying numbers or names. The catalog is the directory and path for the database table.

TUTORIAL: EXTERNAL DATABASE CONNECTIONS TO DBASE3

Prior to starting this exercise, ensure that the PROPERTI.DBF and the DBASE3.DWG files have been copied into a specified directory inside AutoCAD Map 2000. For the purpose of this exercise the location of these files is in the Acadmap4\Sample\DBF\ directory.

PART I ACTIVATING ODBC

Step 1 Open the **Control Panel** and locate the **ODBC** icon. (If you do not have this icon, you must install this feature to your operating system. If necessary, refer to your operating system manual for directions.)

Figure 6–1

Step 2 Activate the ODBC icon by double-clicking, and a list of potential data sources will be presented in the ODBC Data Source Administrator dialog box. The actual list may vary slightly between operating systems and with the software installed on a computer.

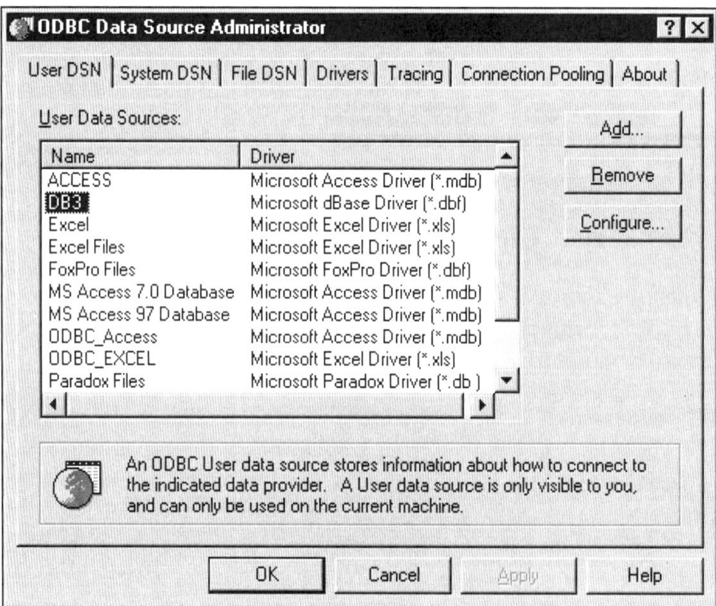

Figure 6–2

Highlight the DB3 line and then click on the **Configure** button. The ODBC dBASE Setup dialog box will appear. In this case, the system will use dBase files for data of a format up to version 5.

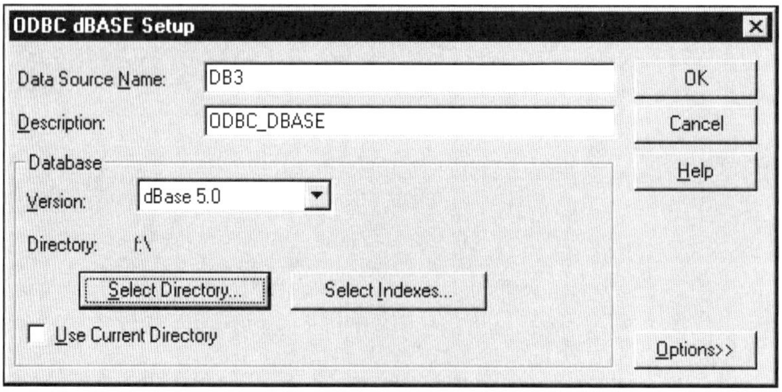

ODBC dBASE Setup
Data Source Name: DB3
Description: ODBC_DBASE
Database
Version: dBase 5.0
Directory: f:\
Select Directory...
Select Indexes...
Use Current Directory
OK
Cancel
Help
Options>>

Figure 6–3

Click on the **Select Directory** button and **Browse** directories until you locate the **PROPERTI.DBF** file required for this tutorial.

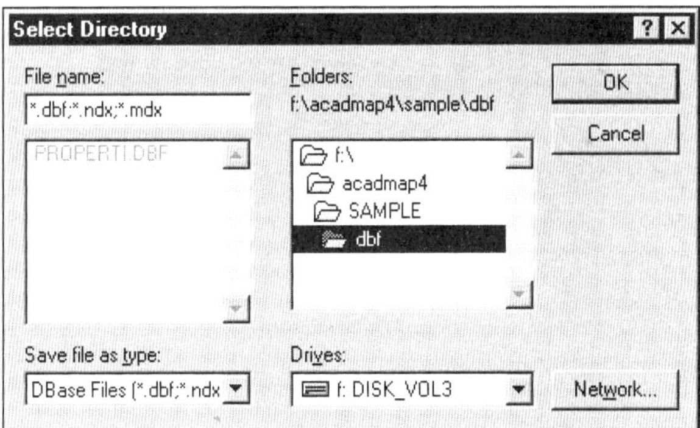

Figure 6–4

Click on **OK** and then at the ODBC dBASE Setup dialog box select **OK** to return to the Data Source Administration screen. Click on **OK** again. Close the Control Panel.

PART 2 ACCESSING THE DATA FROM AUTOCAD MAP 2000

Step 1 Open AutoCAD Map 2000 and open the drawing called **DBASE3** in the Sample directory.

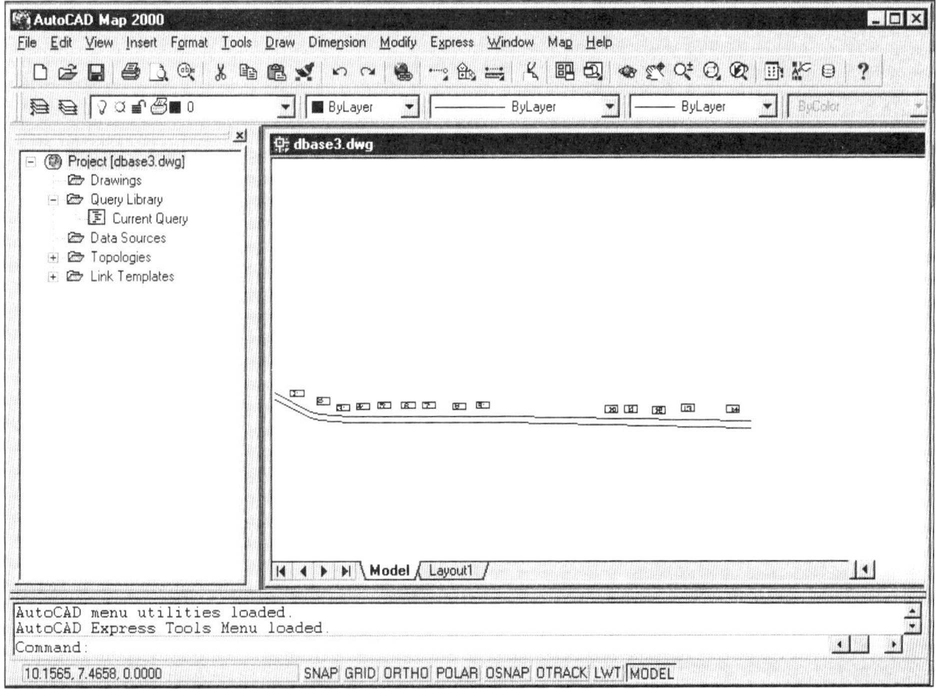

Figure 6–5

Step 2 From the **Map** menu, select **Database,** then **Data Sources** and **Configure.**

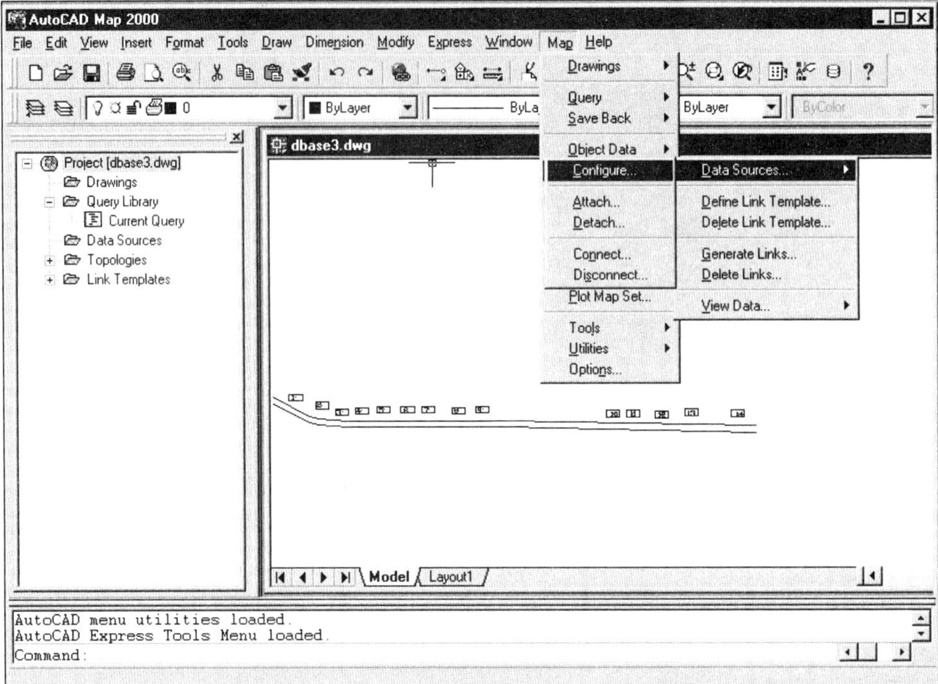

Figure 6–6

The Configure Data Source dialog box will be presented. Type in **db3** as the data source in the Data Source Name edit box.

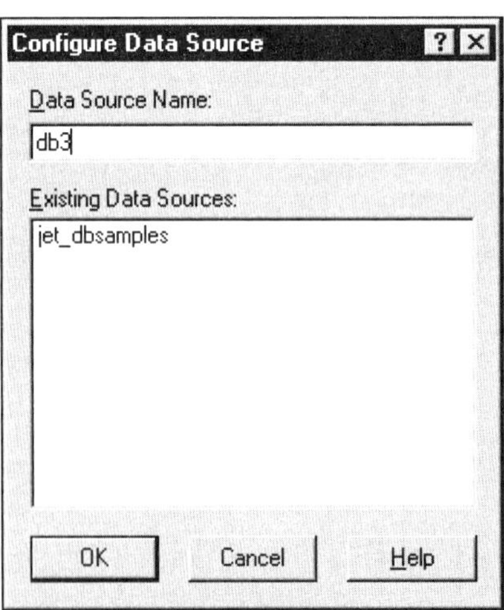

Figure 6–7

Click on **OK** and the Data Link Properties dialog box will be presented. Choose the **Connection** tab.

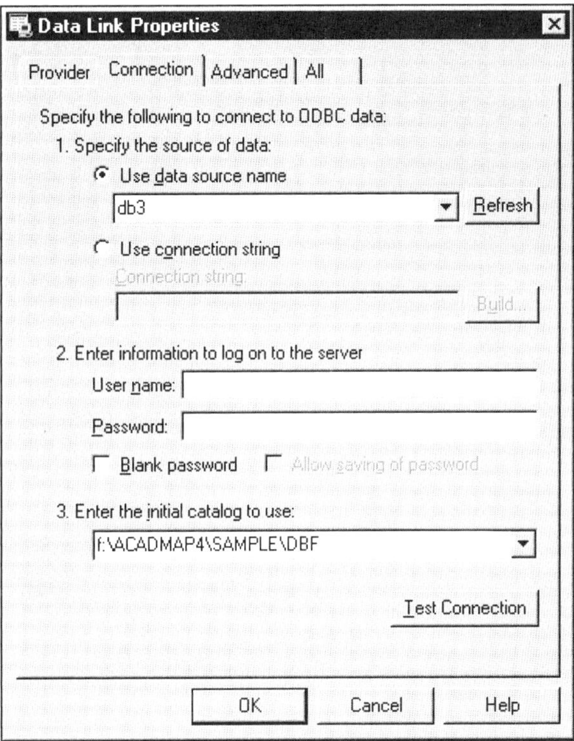

Figure 6–8

Step 3 Type in **db3** in the Use Data Source Name edit box. In the central portion of the screen, provide any password and user ID that is required for your particular system. In the lower section, select the catalog or directory for the data. In this case, it is F:\ACADMAP4\SAMPLE\DBF. Click on the **Test Connection** button to check if AutoCAD Map 2000 will be able to connect with your database. A message will indicate success.

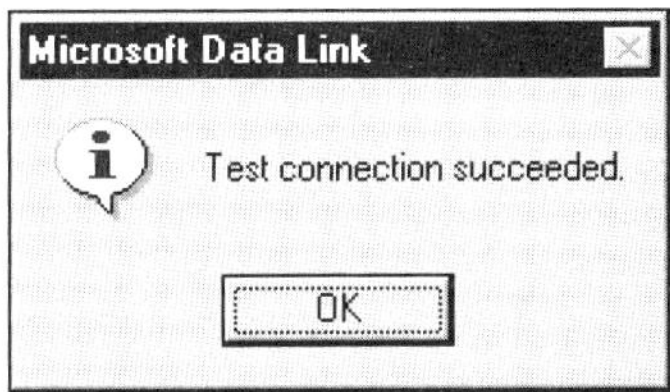

Figure 6–9

Click on **OK** to close the connection message and again to close the Data Link Properties dialog box.

These steps have basically indicated to AutoCAD Map 2000 which database software you will be using. The next series of steps make the actual connection to the database.

Step 4 Go to the **Map** menu and select **Database** then **Data Source** and **Attach.**

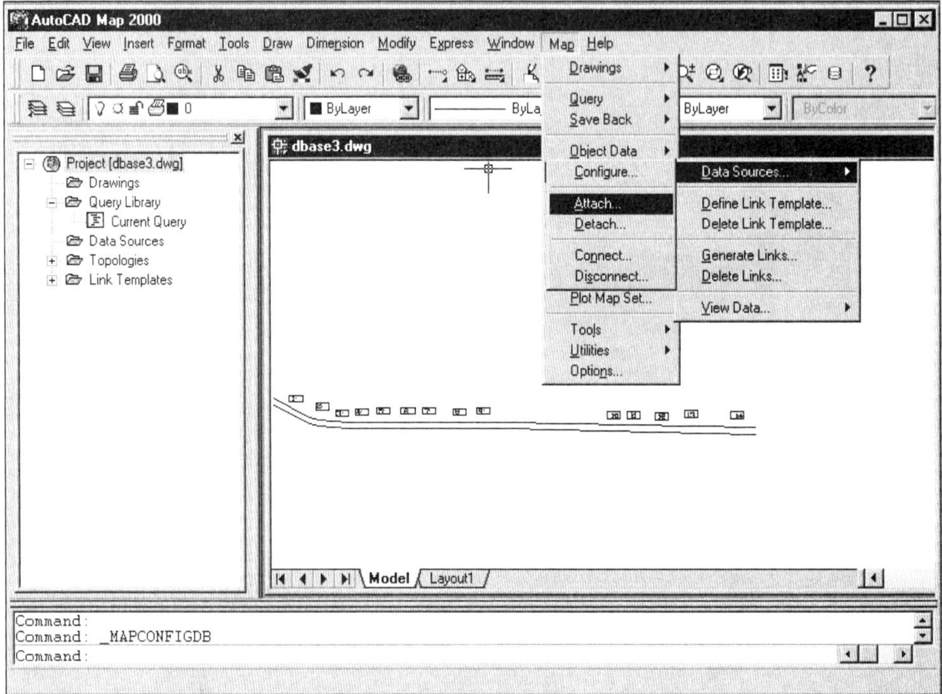

Figure 6–10

The Attach Data Source dialog box will appear. Select the **DBASE3.UDL** item, then click on **Open**.

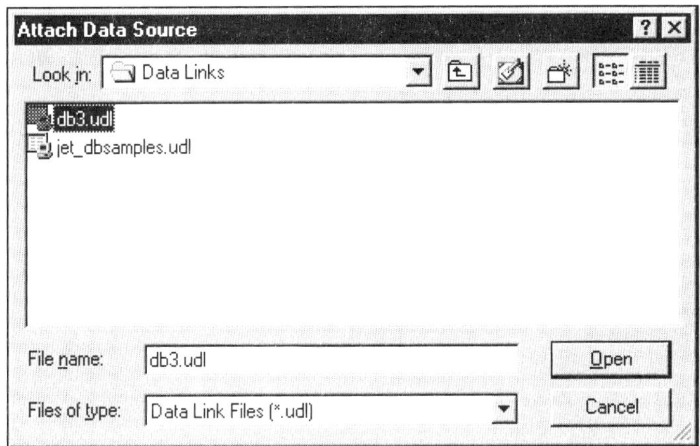

Figure 6–11

Step 5 Go to the **Tools** menu and select **dbConnect**.

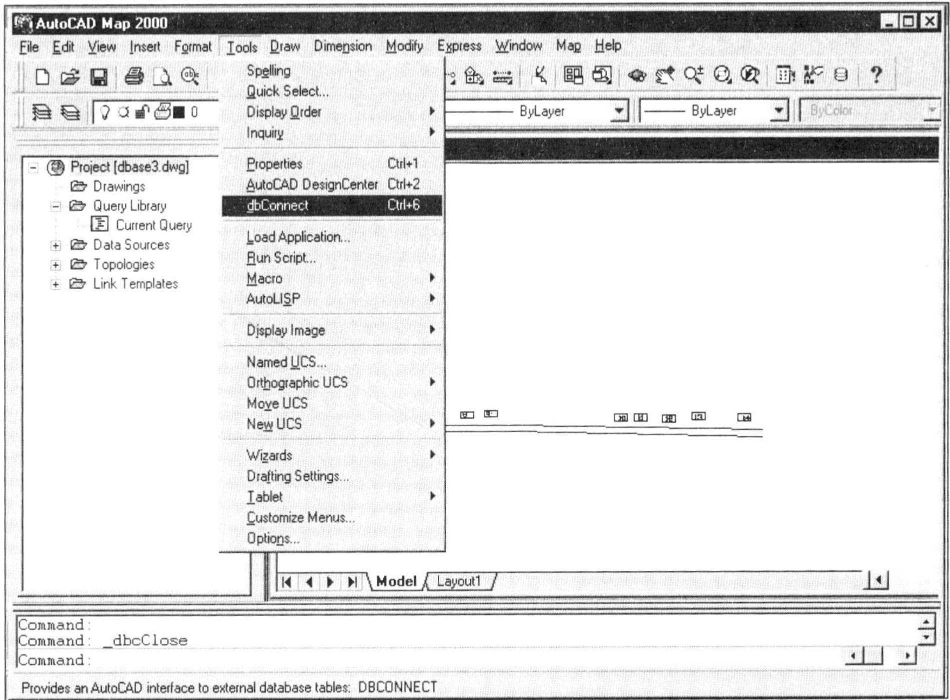

Figure 6–12

A new menu item will appear along the top of the AutoCAD Map 2000 screen and a new work session area will also appear along the left side of the screen.

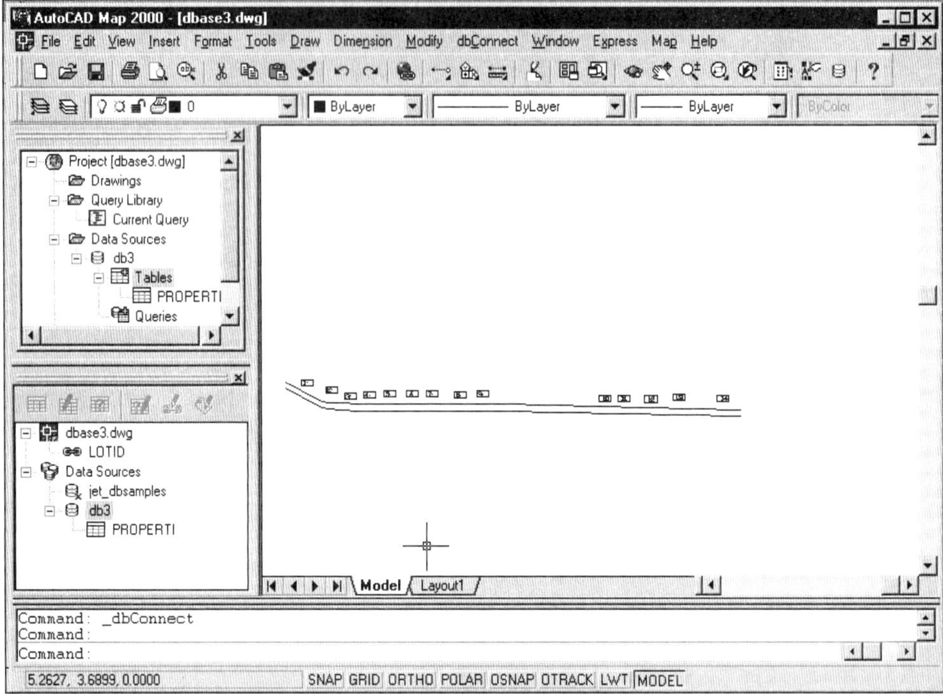

Figure 6–13

Step 6 A linkage must now be created between features common to the drawing and the database table. Go to the **Map** menu and select **Database** and **Define Link Template**.

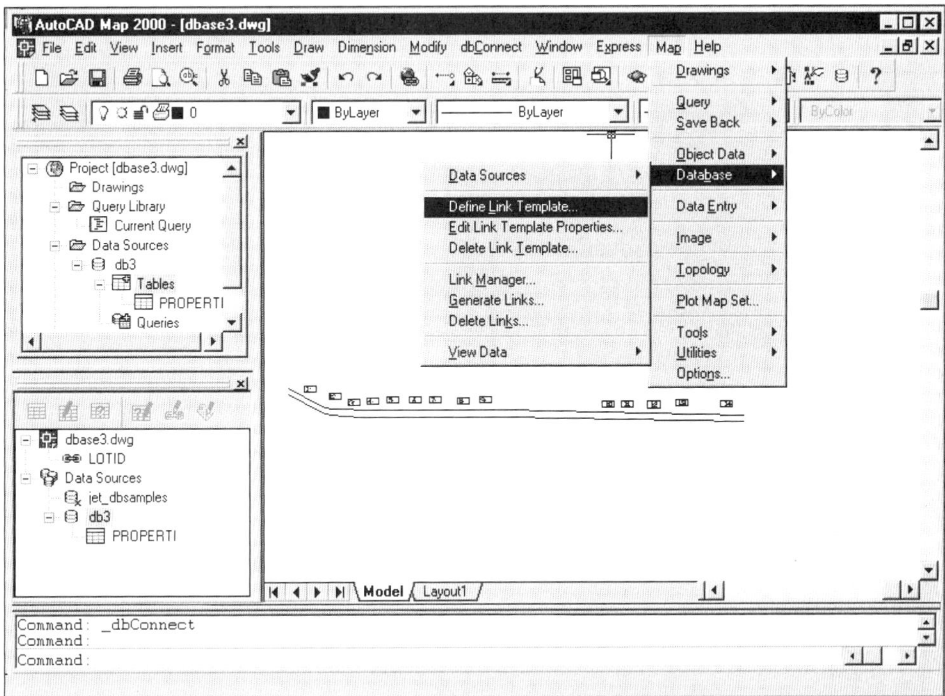

Figure 6–14

Step 7 In the Define Link Template dialog box, enter a name for the link template. In this case, the name was **Lotlink**. Select the **Lot** item as the Key, then click on **OK.**

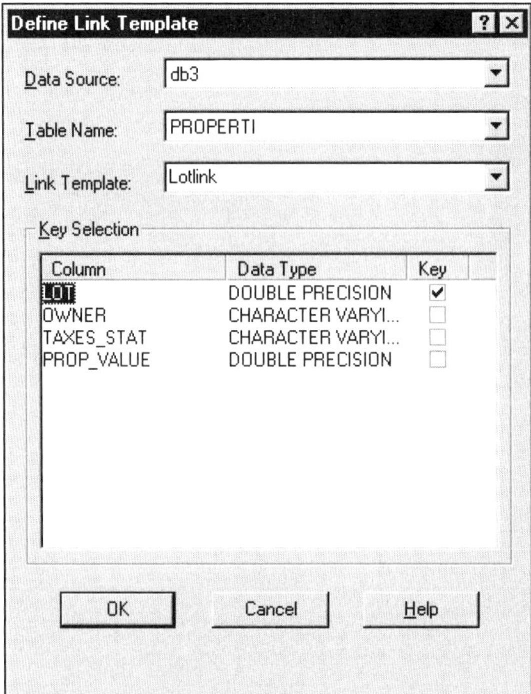

Figure 6–15

Now the actual links will be created under the name of Lotlink. Go to **Map**, then **Database** and **Generate Links**. In the Generate Data Links dialog box, choose **Text** as the linkage type by clicking on the radio button. This will have the links to the database created through the numeric text used for lot identification. In the Data Link section ensure that the Link Template is **Lotlink** and that the **Create Database Links** radio button is selected.

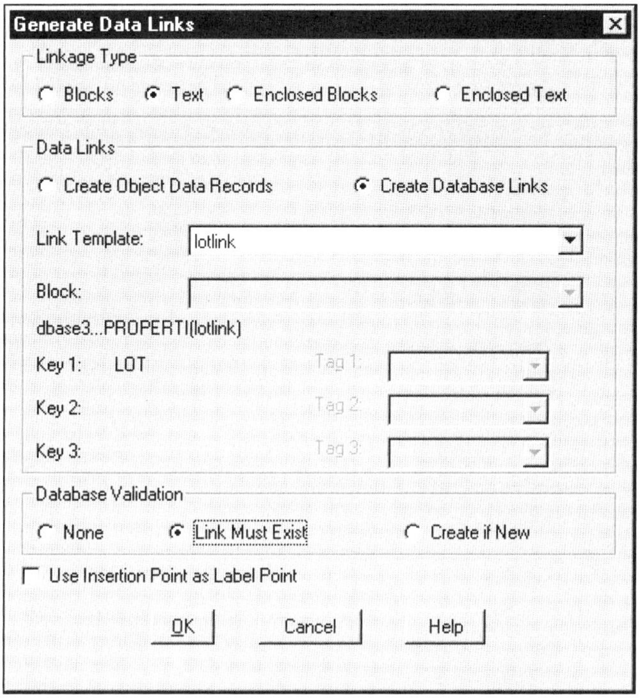

Generate Data Links
Linkage Type
Blocks Text Enclosed Blocks Enclosed Text
Data Links
Create Object Data Records Create Database Links
Link Template: lotlink
Block:
dbase3...PROPERTI(lotlink)
Key 1: LOT Tag 1:
Key 2: Tag 2:
Key 3: Tag 3:
Database Validation
None Link Must Exist Create if New
Use Insertion Point as Label Point
OK Cancel Help

Figure 6–16

In the Database Validation section, select the **Link Must Exist** radio button, then click on **OK**. In the Command prompt text box you will be prompted to select objects. Type **All** and press ENTER. Objects will be identified and links created with the results being tabulated in the command prompt text box.

AutoCAD - Command Line - dbase3.dwg

```
Text objects to generate from: Select/<All>:
14 object(s) to process.
Checking unresolved links 14...done.
Processing 14...done. 14 links created.
Command:
```

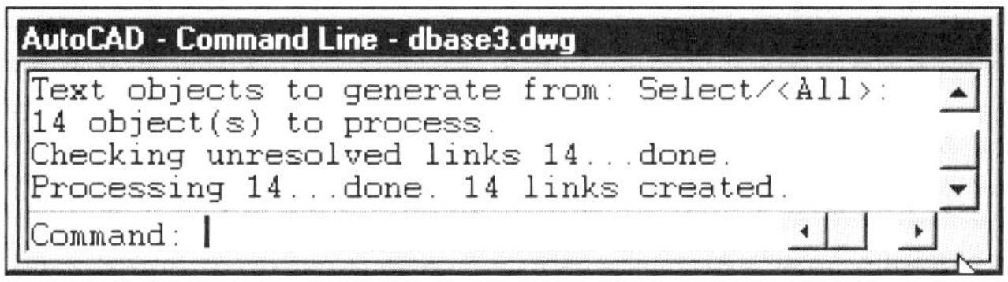

Figure 6–17

Step 8 To access the data in the PROPERTI.DBF file, go to the **dbConnect** menu and select **View Data**, then **View External Table**. The Select Data Object dialog box allows you to choose the PROPERTI table.

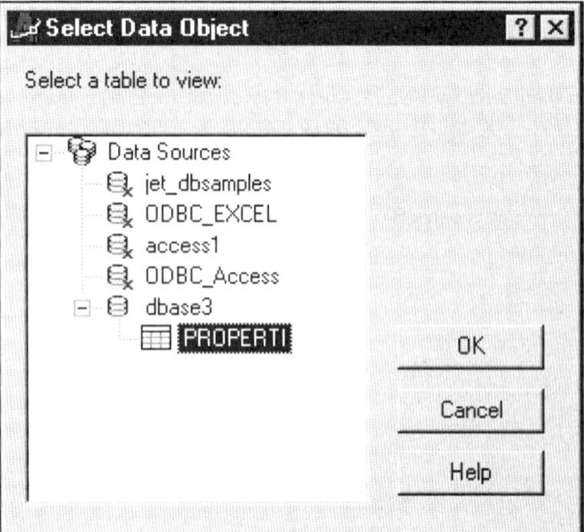

Figure 6–18

Click on **OK** and the table will be presented on screen.

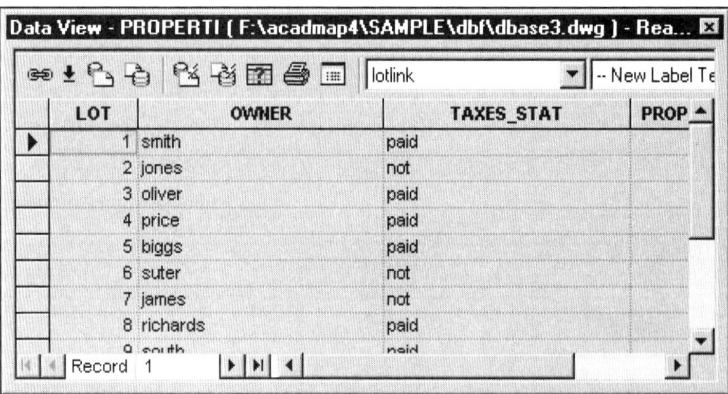

Figure 6–19

Step 9 A query can be performed on the data by going to the **dbConnect** menu and selecting **Queries** then **New Query on an External Table** or by using the dbConnect work session toolbar.

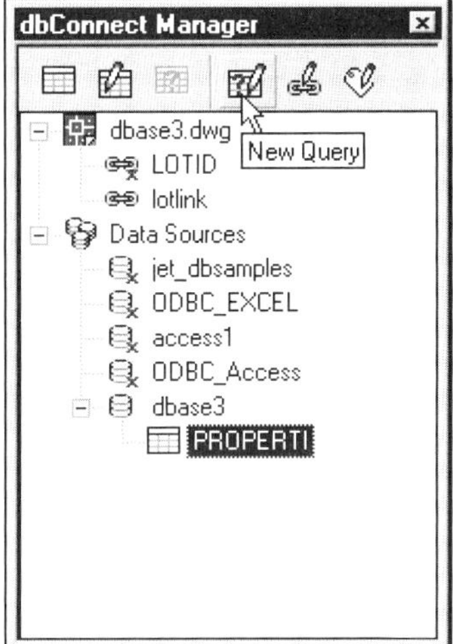

Figure 6–20

Highlight the **PROPERTI** item, then click on the **New Query** icon. The New Query Wizard guides you through the process. Assign a name to the query. In this case, the name used was a default name of **PROPERTIQuery1**. Click on **Continue**.

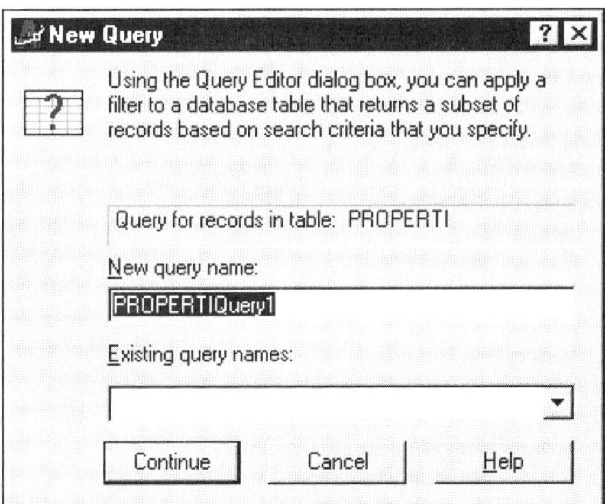

Figure 6–21

Step 10 The Query Editor screen offers a variety of approaches to conducting a query. Select the **SQL Query** tab.

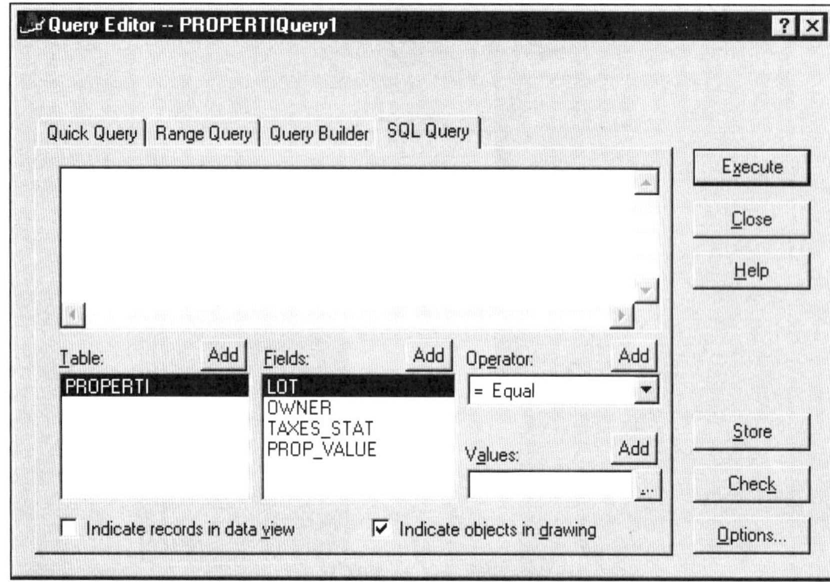

Figure 6–22

Working from left to right, click on the **Table** item **PROPERTI** then select **Add.** The query will start to take shape in the Query Editor. In the **Fields** section select **PROP_VALUE** then **Add.** For the **Operator** select = and then **Add** and for the **Value** type in **100000** and click on **Add.**

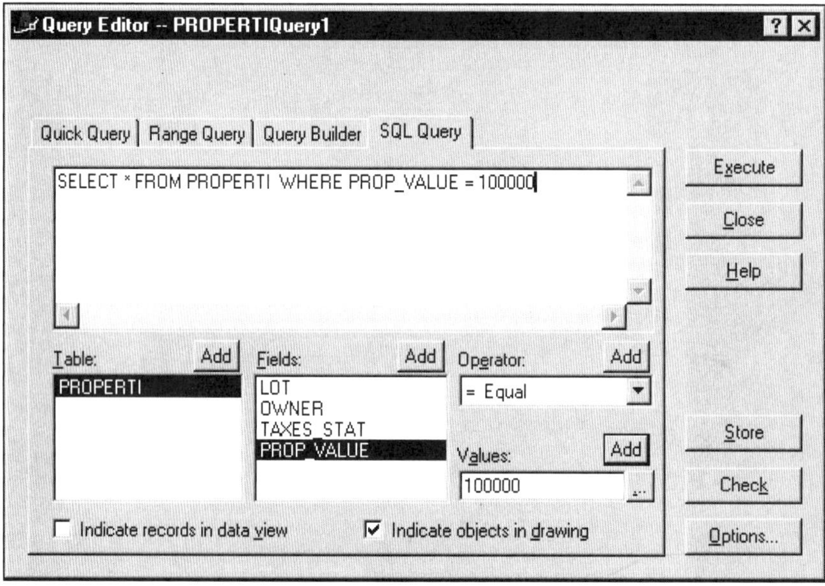

Figure 6–23

This will search for all the properties that have values of $100,000.00. Near the bottom of the Query Editor dialog box, there are two options for displaying the results of the search. Select the **Indicate Objects in Drawing** radio button. Click on **Execute**.

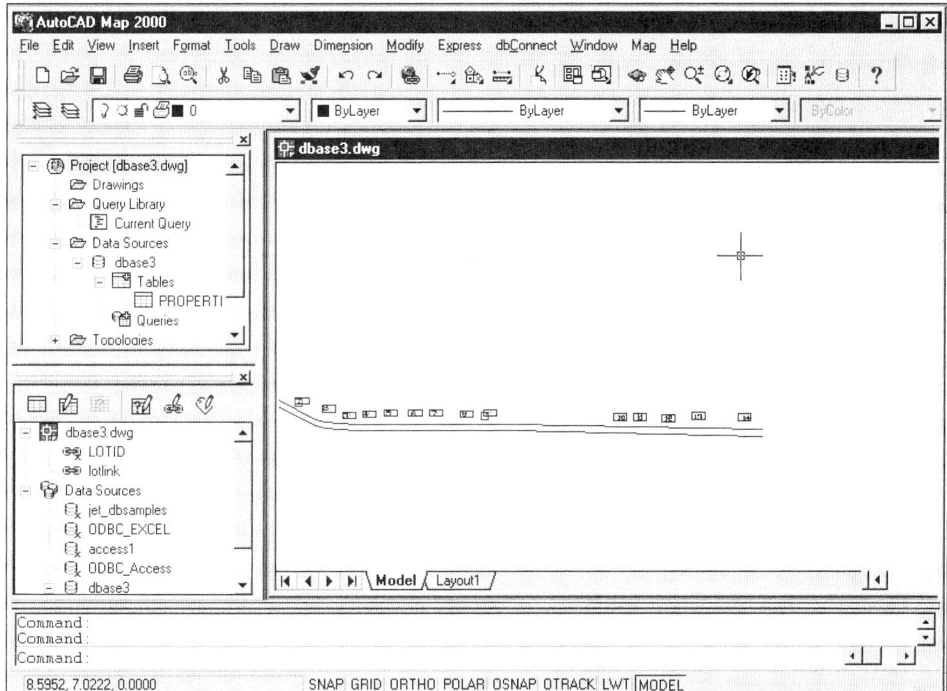

Figure 6–24

The two properties will be marked in the drawing. By selecting the other display option the data can be revealed in tabular form.

Step 11 It is also useful to know that you can edit and print the database table from AutoCAD Map 2000. To access the table for editing purposes, select **dbConnect** and choose **View Data** and **Edit External Table**.

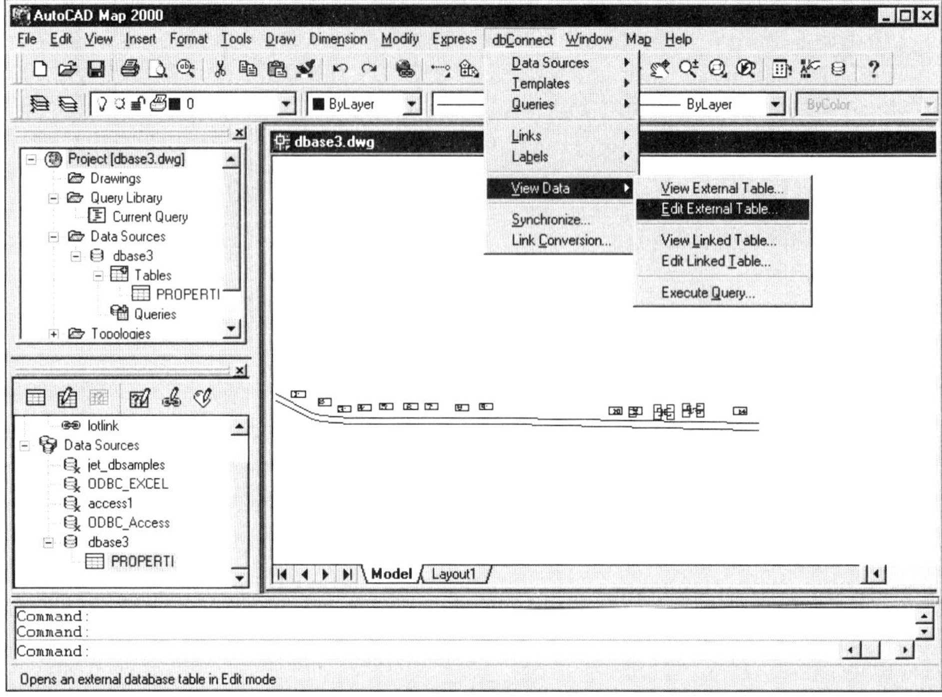

Figure 6–25

In the **Select Data Object** screen choose the **PROPERTI** item and **OK**. The table will appear in an editable form.

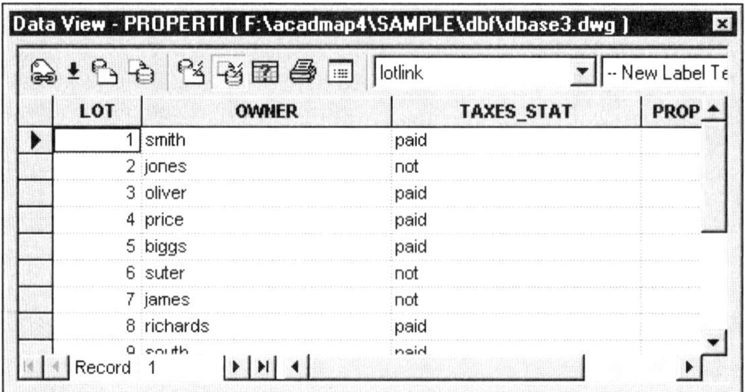
Data View - PROPERTI (F:\acadmap4\SAMPLE\dbf\dbase3.dwg)

lotlink | -- New Label Te

LOT	OWNER	TAXES_STAT	PROP
1	smith	paid	
2	jones	not	
3	oliver	paid	
4	price	paid	
5	biggs	paid	
6	suter	not	
7	james	not	
8	richards	paid	

Record 1

Figure 6–26

Step 12 The activation of the Data View menu item in the AutoCAD MAP 2000 menu bar is concurrent with this procedure.

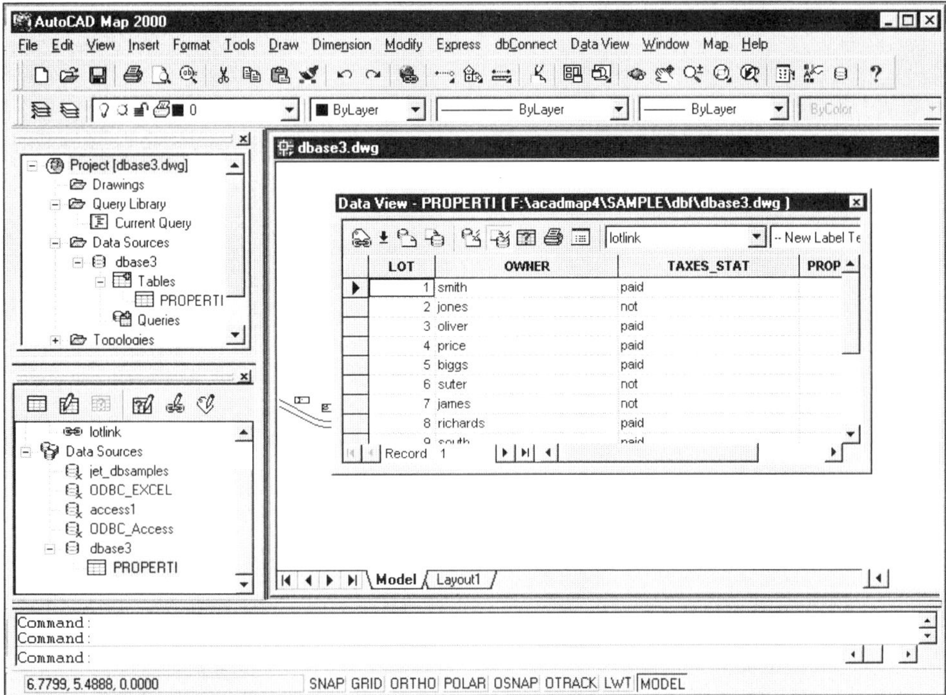

Figure 6–27

With the Data View menu you can query individual properties from the table. To do this, select the **View Linked Records** option from the Data View menu, then use the pick box to select one or more properties for which you wish the database information. In this example, Properties 8 and 9 were selected; when the selection process was ended using the right mouse button, the data for those records was provided in the table.

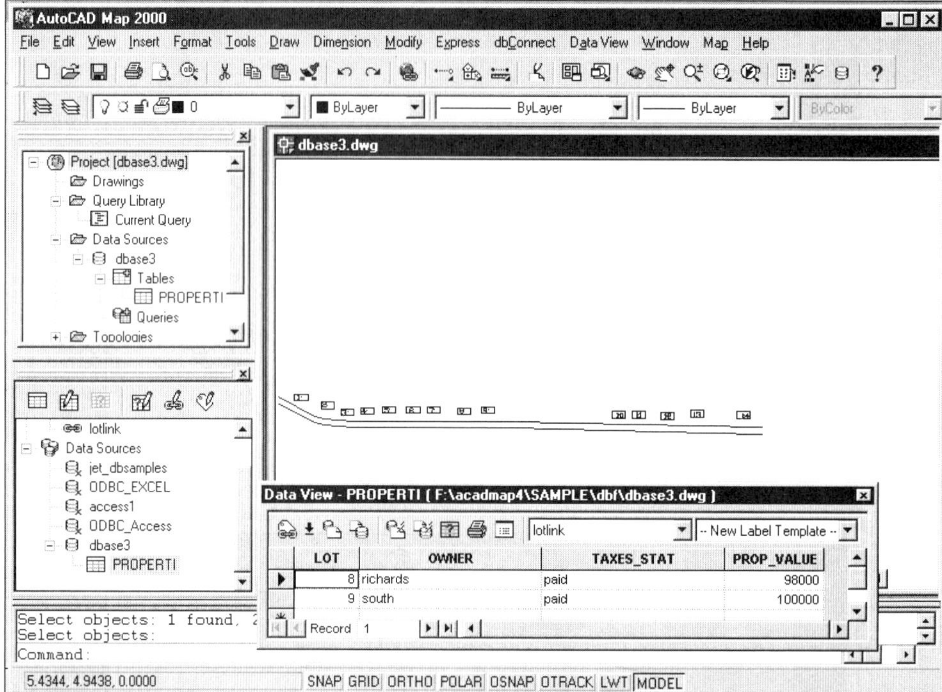

Figure 6–28

Step 13 Printing the table is routine. You can print either the entire database table or parts of it that may have resulted from a query or viewing linked data. When you close the table, the Data View menu is removed.

Step 14 To Detach the database from the drawing, go to the **Map** menu and select **Database**, then **Data Sources** and **Detach**. Select the database you have been working with as the one to detach. Then go to **Tools** and select the **dbConnect** item to remove the checkmark and deactivate it.

Either continue with the tutorials or exit AutoCAD Map 2000. The drawing does not need to be saved.

TUTORIAL: EXTERNAL DATABASE CONNECTIONS TO ACCESS

Prior to starting this exercise, ensure that the ACCESSMAP.DWG and the ACCESS1.MDB files have been copied to your computer. For the purpose of this exercise, the location of the ACCESS1.MDB file isPROGRA~1\MICROS~2\OFFICE\SAMPLES\ACCESS1.MDB.

PART 1 ACTIVATING ODBC

Step 1 Open the **Control Panel** and locate the **ODBC** icon.

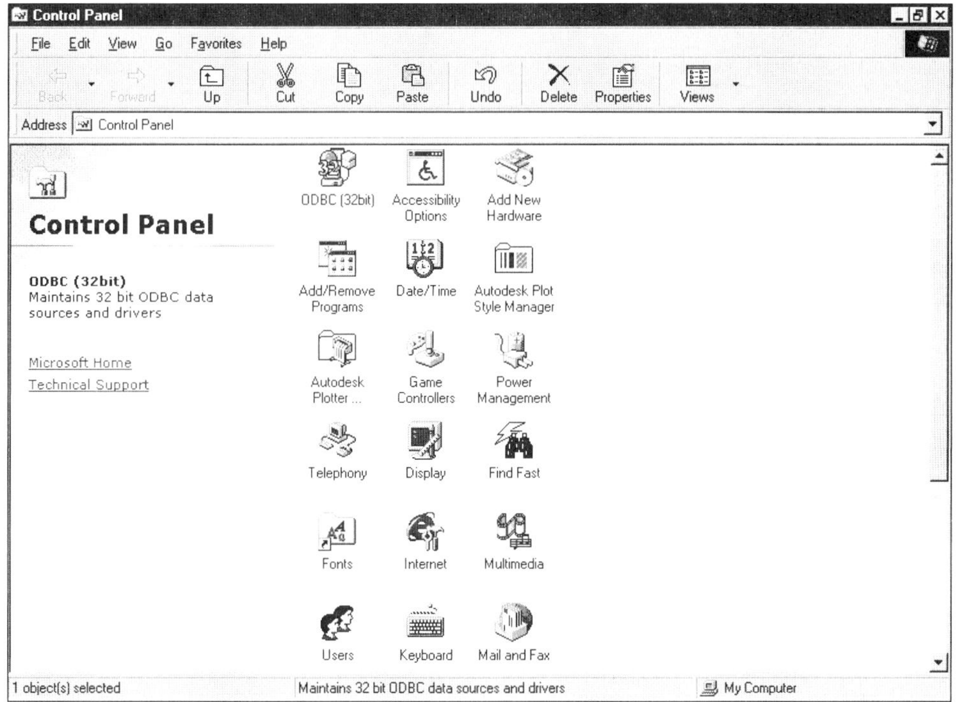

Figure 6–29

Step 2 Activate the **ODBC** icon by double-clicking, and a list of potential data sources will be presented. The actual list may vary slightly between operating systems and with the software installed on a computer.

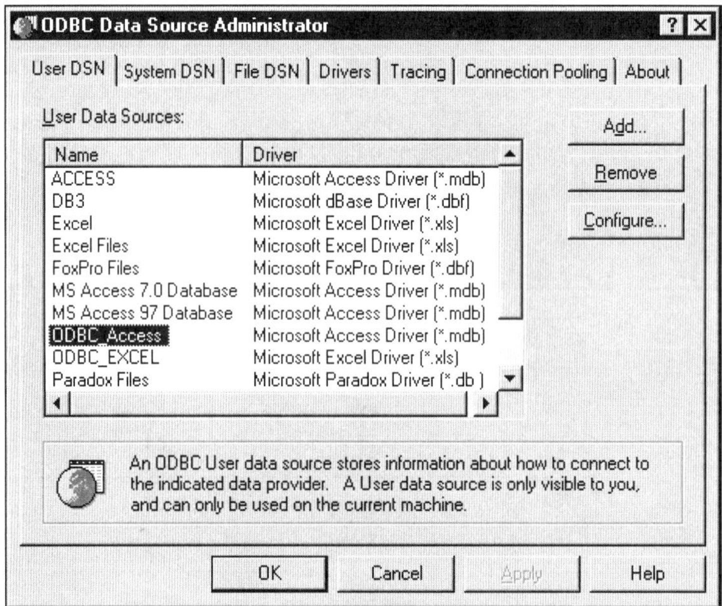

Figure 6–30

Click on **ODBC_ACCESS**, then click on **Configure** and the ODBC Microsoft Access 97 Setup dialog box will appear. Click on the **Select** button and find the **ACCESS1.MDB** file.

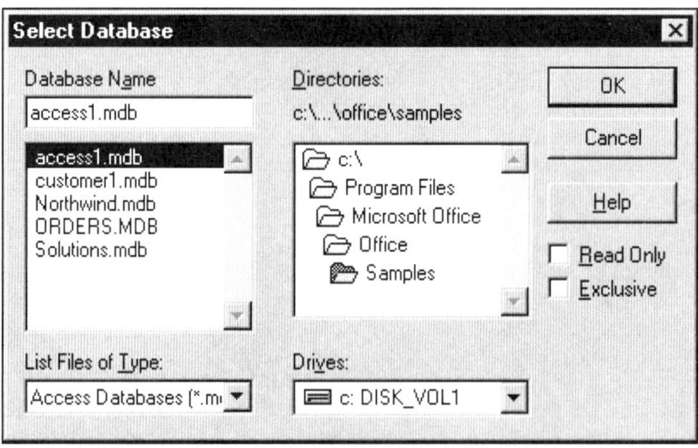

Figure 6–31

Click on **OK** to set the database and to return to the ODBC Microsoft Access 97 Setup dialog box. The database selection will be identified in the central portion of the Setup screen.

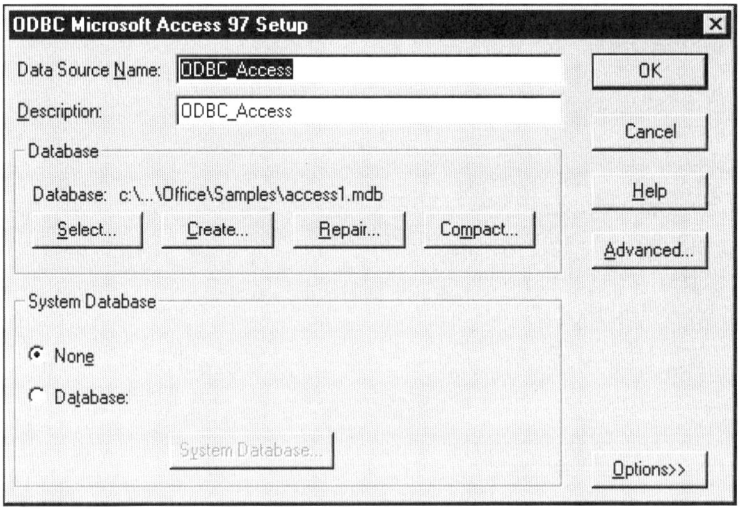

Figure 6–32

Click on **OK** to return to the **Administrator** screen, then click **OK** again. Close the **Control Panel.**

PART 2 ACCESS DATA AND AUTOCAD MAP 2000

Step 1 Open AutoCAD Map 2000 and open the drawing called **ACCESSMAP**.

Step 2 Go the **Map** menu and select **Database**, then **Data Sources** and choose **Configure**.

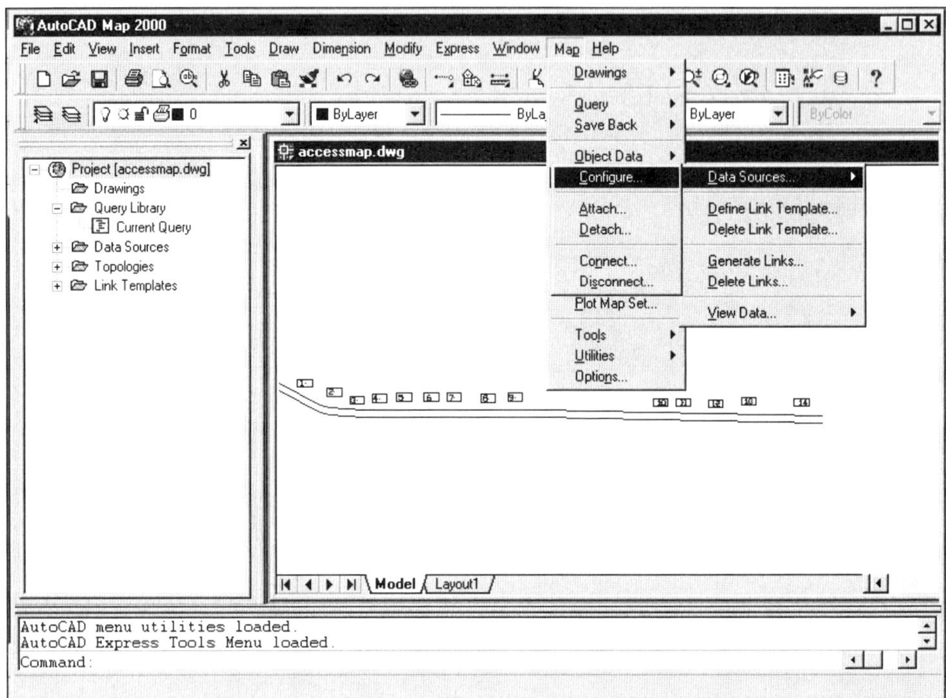

Figure 6–33

Step 3 The Configure Data Source dialog box facilitates the selection of the data type you are connecting to. Type in **ODBC_Access** as the Data Source Name, then click on **OK**.

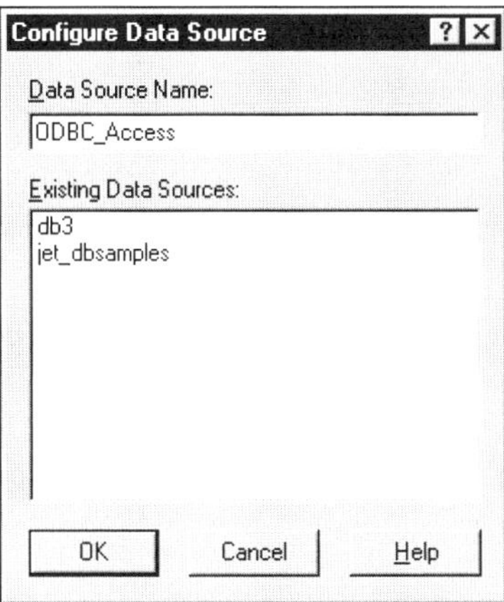

Figure 6–34

Step 4 The Data Link Properties dialog box is presented. Select the **Connection** tab. For the Data Source Name type in **ODBC_Access**. In the central part of the screen provide a username and password if your system requires it. In the lower part of the screen select the catalog for the database file by referring to the pull-down list.

Data Link Properties
Provider Connection Advanced All
Specify the following to connect to ODBC data:
1. Specify the source of data:
Use data source name
ODBC_Access
Refresh
Use connection string
Connection string:
Build...
2. Enter information to log on to the server
User name:
Password:
Blank password
Allow saving of password
3. Enter the initial catalog to use:
c:\Program Files\Microsoft Office\Office\Samples\access
c:\Program Files\Microsoft Office\Office\Samples\custome
c:\Program Files\Microsoft Office\Office\Samples\Northwin
c:\Program Files\Microsoft Office\Office\Samples\ORDERS
c:\Program Files\Microsoft Office\Office\Samples\Solutions
OK
Cancel
Help

Figure 6–35

To check whether or not AutoCAD Map 2000 will be able to make the connection with your database table click on the **Test Connection** button. A message will provide you with the test results. Click on **OK** twice to close the screens.

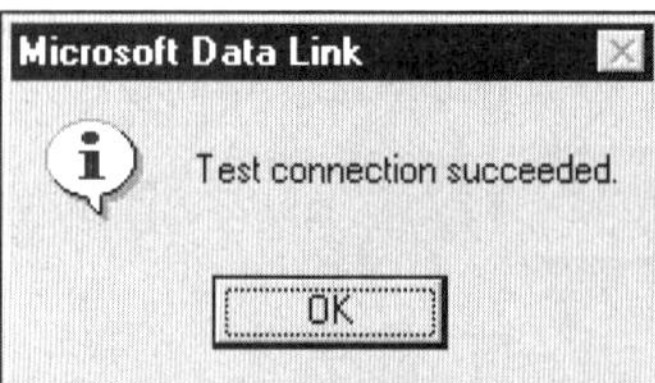

Figure 6–36

The previous steps have prepared for the connection to be made between AutoCAD Map 2000 and the database. Now the actual linkage has to be made.

Step 5 Go to the **Map** menu and select **Database** then **Data Sources** and **Attach**. The Attach Data Source dialog box will appear.

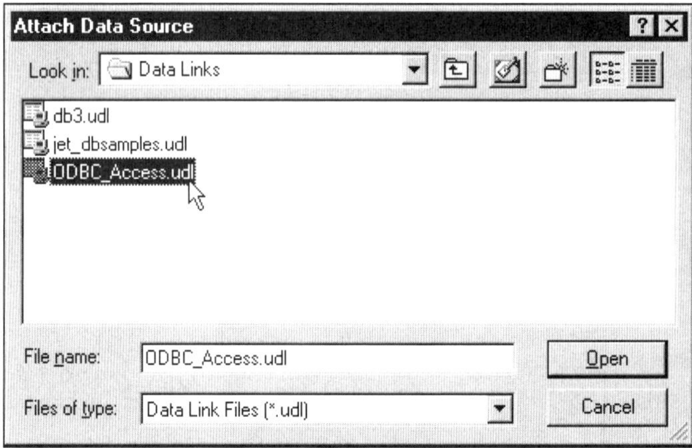

Figure 6–37

Select the **ODBC_Access.udl** item and click on the **Open** button. This change will be reflected in the work session window of the screen.

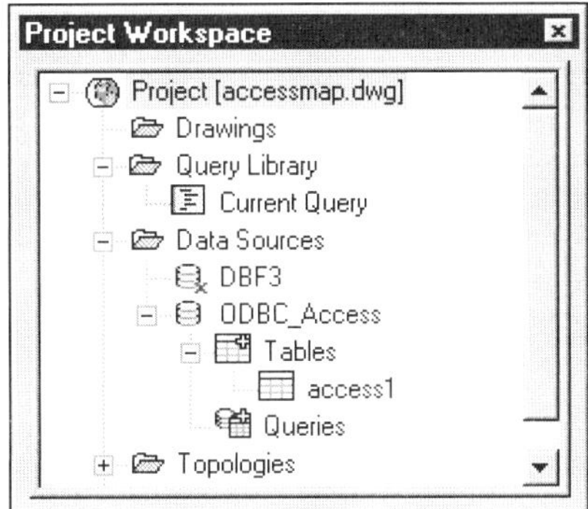

Figure 6–38

Step 6 Go to the **Tools** menu and select **dbConnect**. A new work session window will appear as will a new menu item, dbConnect.

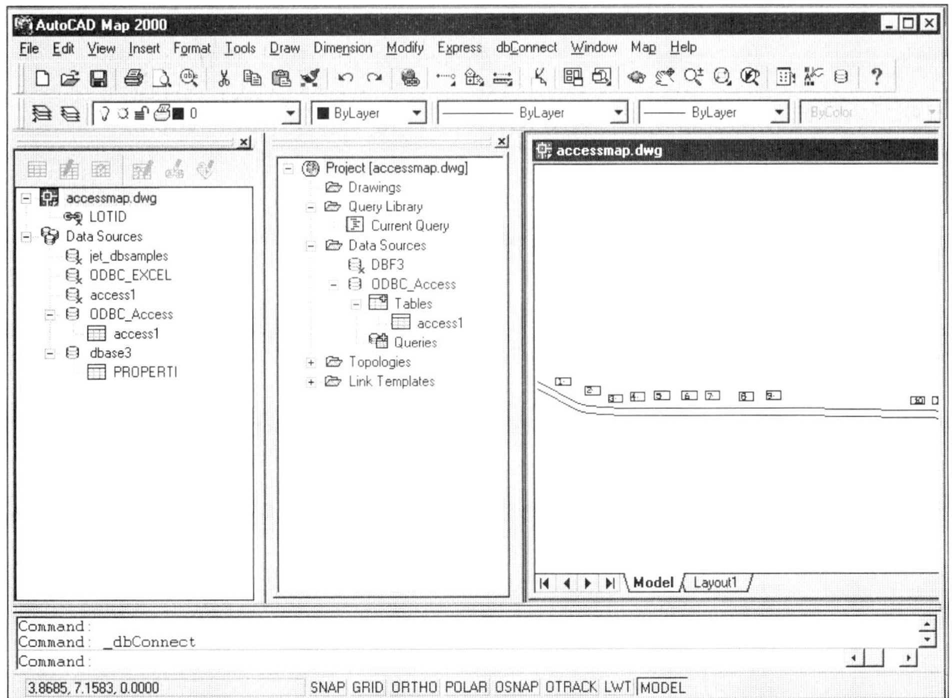

Figure 6–39

Step 7 A linkage between the database and the drawing must now be established. Go to **Map**, then select **Database** and **Define Link Template.**

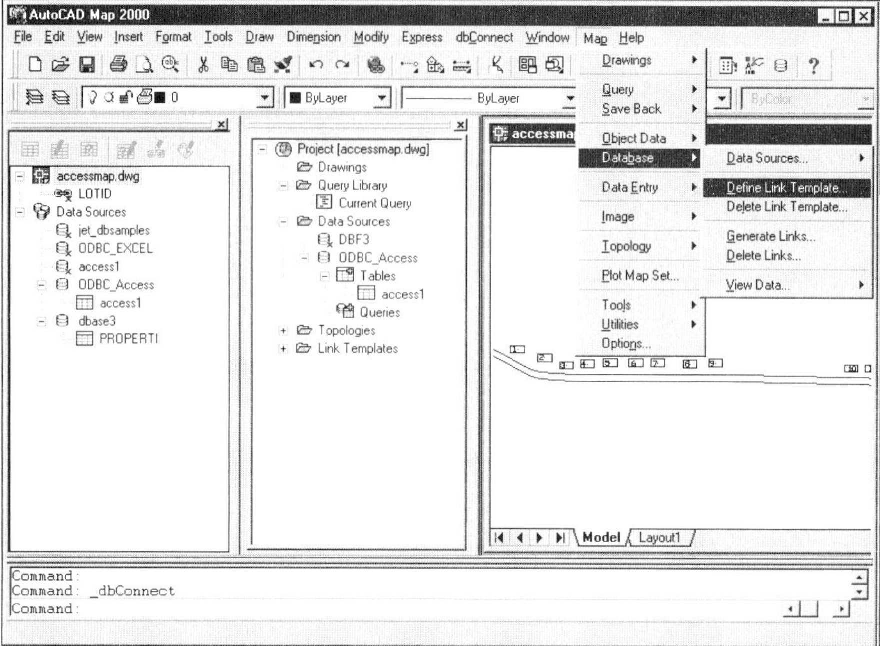

Figure 6–40

At the Define Link Template dialog box type in a link template name. In this example the name used is **lotidlink**. Select the **Lotid** item as the key item. Click on **OK**. Now the actual links will be created using the Lotidlink template.

Define Link Template

Data Source: ODBC_Access

Table Name: access1

Link Template: lotidlink

Key Selection

Column	Data Type	Key
ID	INTEGER	
lotid	INTEGER	✓
"electric service"	INTEGER	
"natural gas service, 1...	INTEGER	
"last assessment"	INTEGER	

OK Cancel Help

Figure 6–41

Step 8 Select **Map** then **Database** and Generate Links. The Generate Data Links dialog box is presented. For the Linkage type select **Text**.

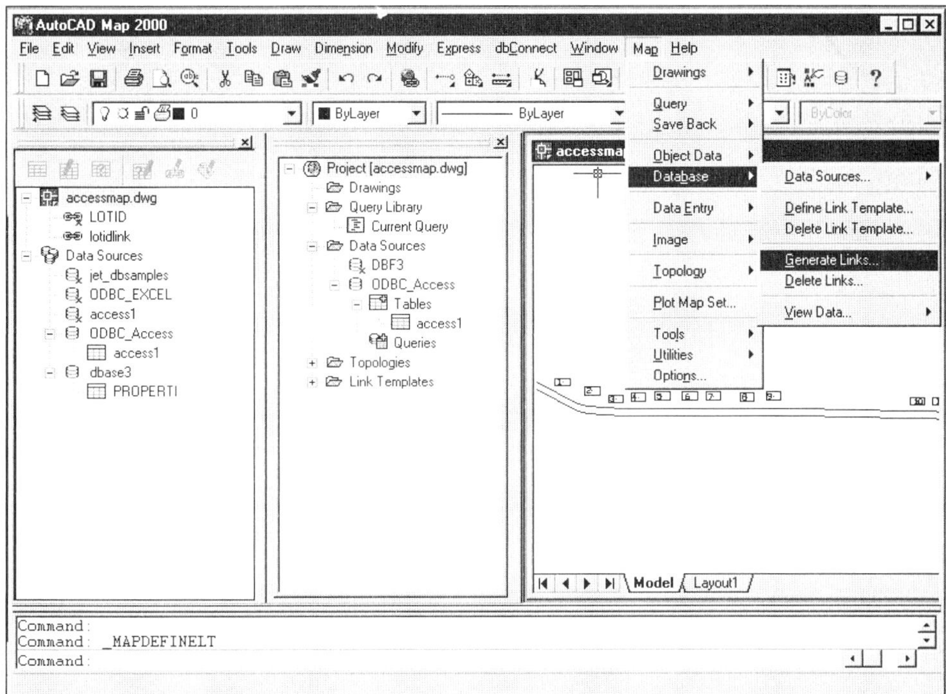

Figure 6–42

In the Database section of the screen select **Create Database Links** and set the Link Template to **Lotldlink**. Set the Database Validation to **Link Must Exist** by clicking on the radio button.

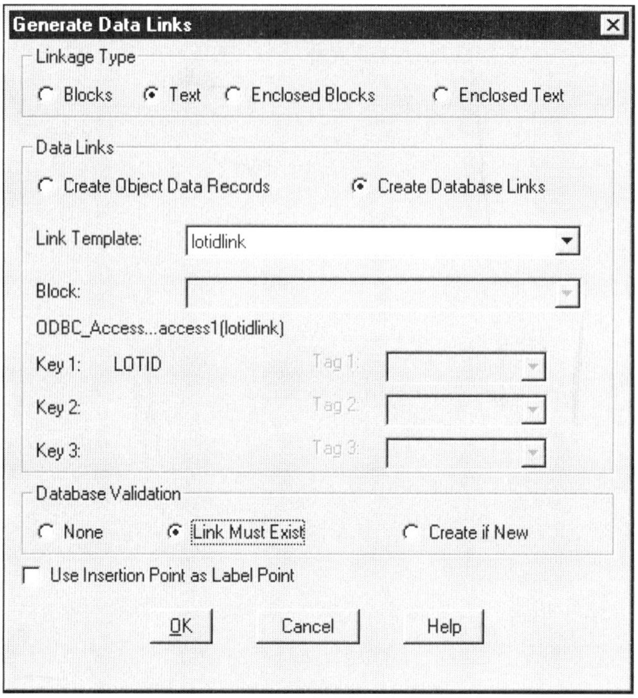

Figure 6–43

Click on **OK** to proceed. The command prompt text box will provide a request to select objects. Press ENTER to select **All** objects. Objects will be selected and links created.

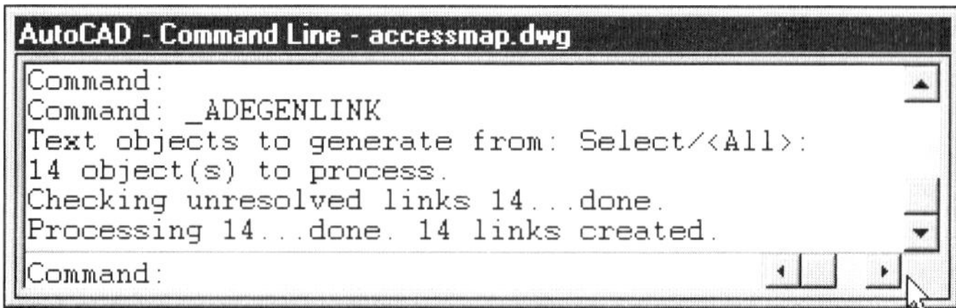

Figure 6–44

Step 9 To see the database table that has just been attached to the drawing go to the **dbConnect** menu and select **View Data** and **View External Table**. The Select Data Object dialog box will accommodate the selection of the database table. Highlight **Access1** and click on. **OK**

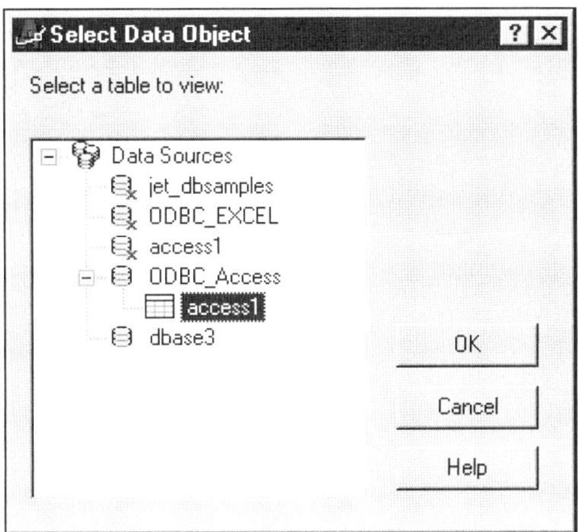

Figure 6–45

The table is presented.

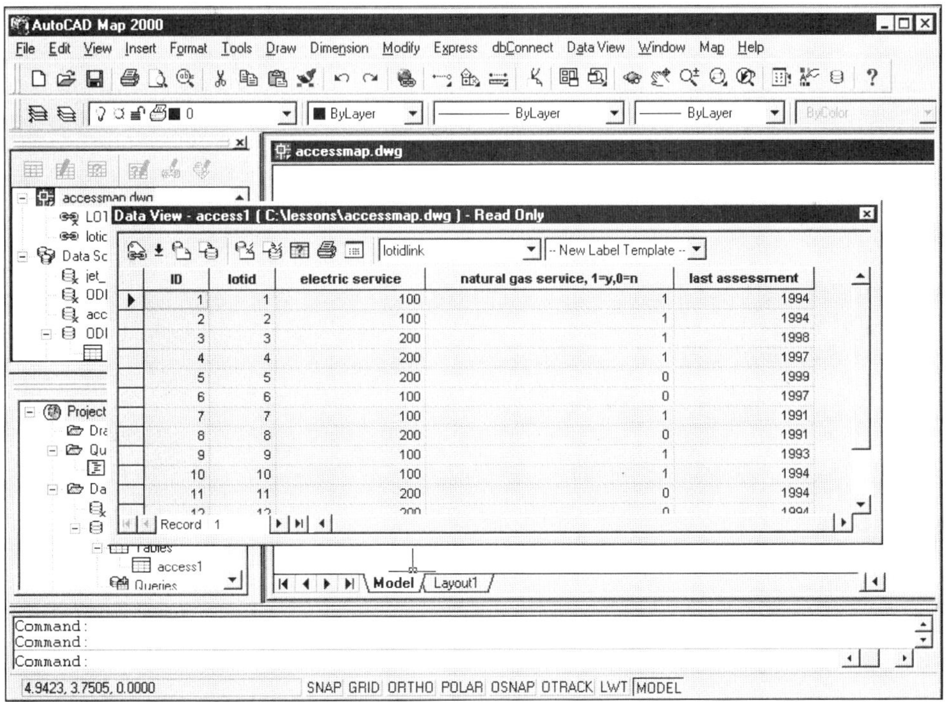

ID	lotid	electric service	natural gas service, 1=y,0=n	last assessment
1	1	100	1	1994
2	2	100	1	1994
3	3	200	1	1998
4	4	200	1	1997
5	5	200	0	1999
6	6	100	0	1997
7	7	100	1	1991
8	8	200	0	1991
9	9	100	1	1993
10	10	100	1	1994
11	11	200	0	1994

Figure 6–46

Step 10 To run a query on the data, close the table. Go to **dbConnect** and select **Queries**, then **New Query on an External Database**.

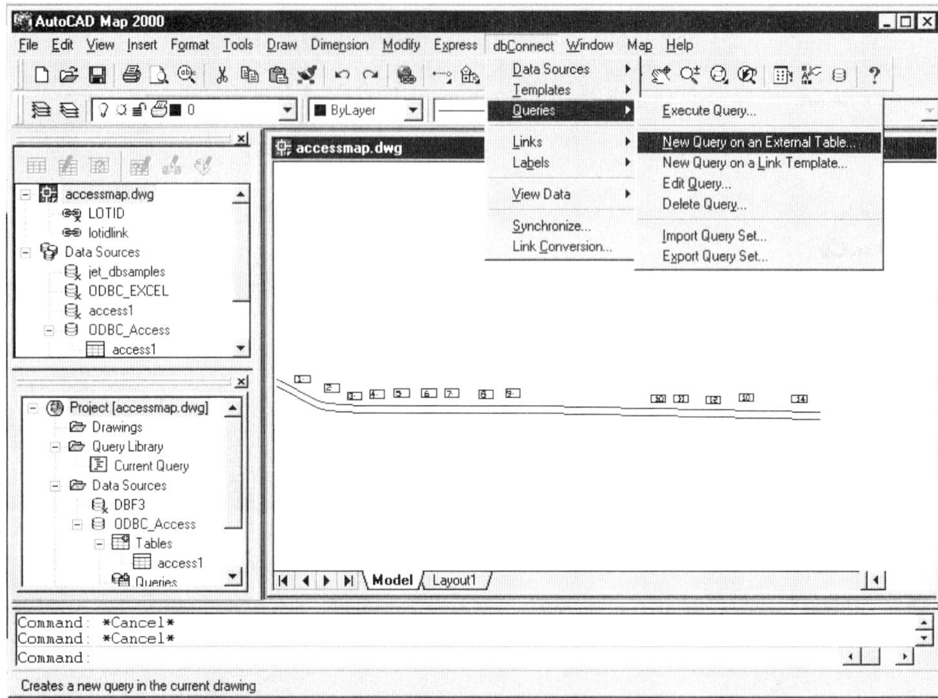

Figure 6–47

Step 11 In the Select Data Object dialog box, highlight **Access1** then select **Continue**. In the New Query dialog box, establish a new name for the query. In this case the name **access1Query1** was utilized.

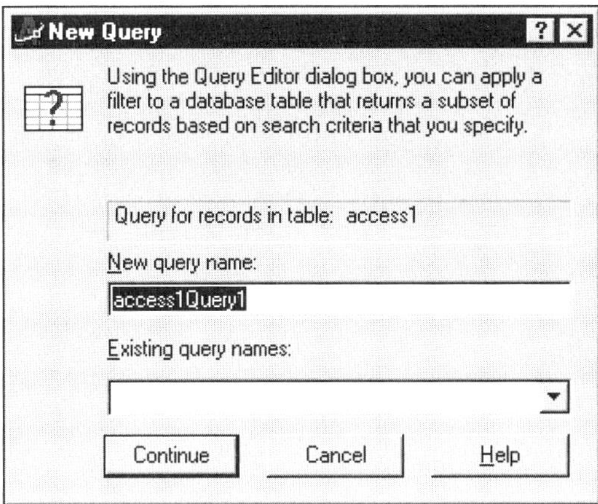

Figure 6–48

Click on **Continue.** The Query Editor screen is presented.

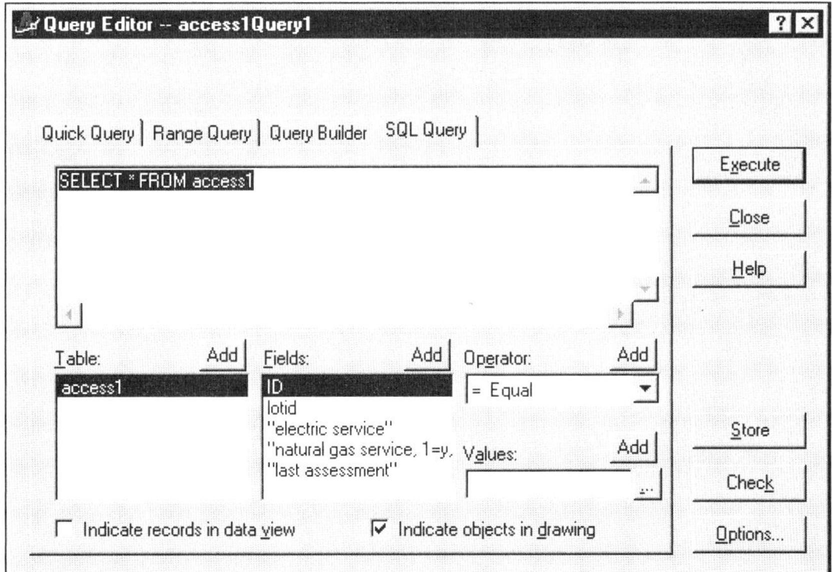

Figure 6–49

To identify the properties that have a 200-amp electric service, select the "**electric service**" field and click on the **Add** button. For the Operator choose **= Equal**, then click on **Add** and finally type in a value of **200** in the Values edit box and click on **Add** again. The query statement should read

SELECT * FROM access WHERE "**electric service**" **= 200**.To see the properties on the map, ensure that the **Indicate Objects in Dawing and Indicate Objects in Record View** checkboxes are selected. Click on **Execute**. The results will be displayed as a table and marked on the map as well.

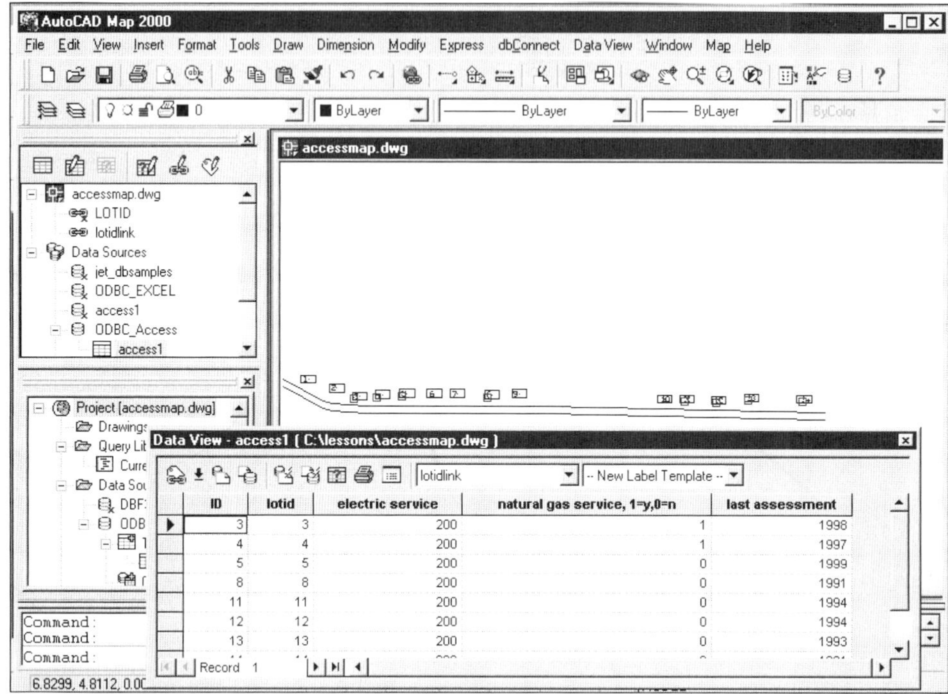

Figure 6–50

More enhanced queries might have parameters that, for example, specify a range of property values or that identify properties with unpaid taxes. The sophistication of the query is dependent only on the data available in the tables. The table being queried in this example is shown below.

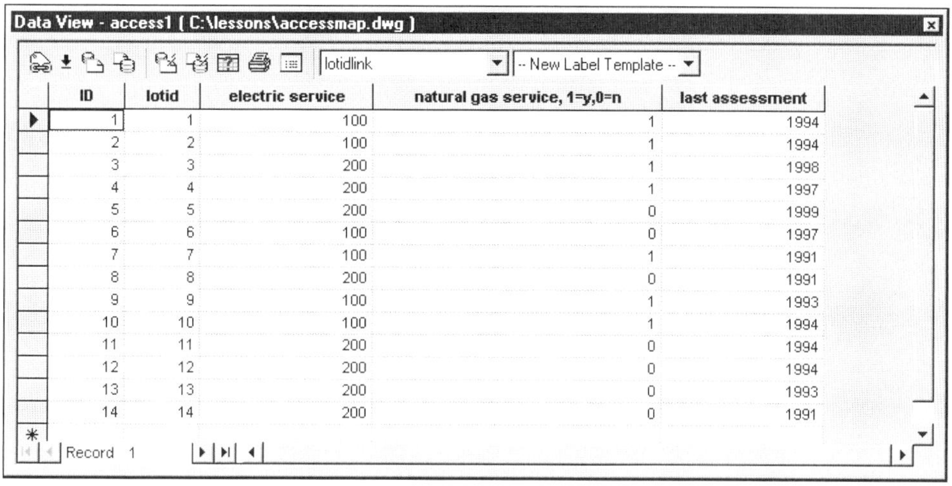

Data View - access1 (C:\lessons\accessmap.dwg)

lotidlink | -- New Label Template --

ID	lotid	electric service	natural gas service, 1=y,0=n	last assessment
1	1	100	1	1994
2	2	100	1	1994
3	3	200	1	1998
4	4	200	1	1997
5	5	200	0	1999
6	6	100	0	1997
7	7	100	1	1991
8	8	200	0	1991
9	9	100	1	1993
10	10	100	1	1994
11	11	200	0	1994
12	12	200	0	1994
13	13	200	0	1993
14	14	200	0	1991

Record 1

Figure 6–51

Either continue with the tutorials or exit AutoCAD Map 2000. This drawing does not need to be saved.

SUMMARY

AutoCAD Map 2000 offers the unique ability to connect to a variety of industry leading databases. This chapter introduced the process for linking drawings to two commonly used databases, Access and dBase. The linkage between the data in the database and features in the drawing is completed through a link template. This generally requires that identifying entries from a key column in the database also are featured in the drawing being linked. In the tutorials, these features are text entries represented as numbers. The same numbers appear in the key column of the database. Prior to attempting to attach a database to a map file, ensure that your computer has the ODBC driver option installed. The procedure for activating the ODBC driver for a database is outlined for Access and dBase. For other database packages the process is similar. When a database connection is made, a new AutoCAD Map 2000 menu item appears, dbConnect. The tools provided through this menu facilitate access to the data. The ability to generate maps from data that is stored in conventional databases is a feature that makes AutoCAD Map 2000 extremely useful in countless settings that range from municipal offices to health care provision to business analysis applications.

REVIEW QUESTIONS

1. What is a Link Template used for?

2. What function does the key column serve?

3. What menu item is added to AutoCAD Map 2000 after the dbConnect command is activated?

4. Can the database table from Access or dBase be viewed from within AutoCAD Map 2000?

Applications in GIS

Civil Engineering Application

OBJECTIVES

After completing this chapter you will know how to

- conduct queries based on specific polygons
- prepare data sets for queries relating to identified risks

In this application, the problem faced is the determination of the number of residential properties that would be negatively affected by a 100-year-storm flood level.

Generally, in such a case, the drawings used in the determination would be provided by a variety of sources. The Conservation Authority in Canada or the Federal Emergency Management Agency in the United States might provide flood plain data, whereas the Public Utilities or other municipal offices could provide elevation information and building location by class data. Flood zone determination is an excellent example of an application that makes AutoCAD Map 2000 such a valuable tool.

TUTORIAL: FLOOD ZONE PROPERTIES

In this application, drawings are provided that show streets, houses, contours, and flood zone areas. The first part of this exercise requires you to attach the pertinent drawings to the work session.

Step 1 Open the **NAPBASE.DWG** file. This drawing provides a backdrop of streets, street names, and the river. Select the **Map** menu, then **Drawings** and **Define/Modify Drawing Set** menu to **Attach** the **FLOODZONE.DWG** file. Complete a **Quickview** of the attached drawing.

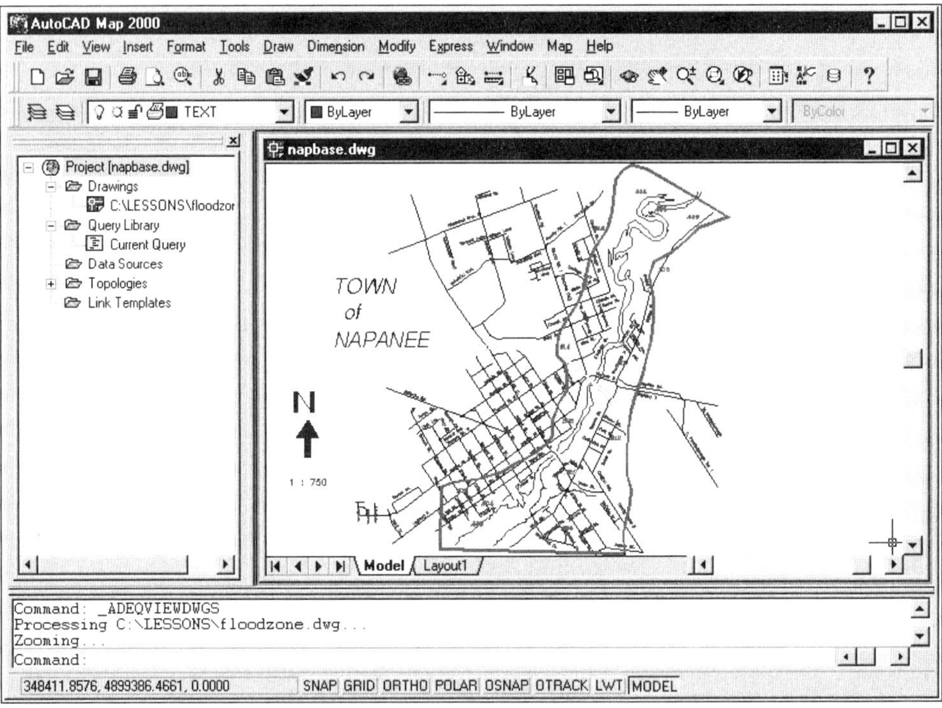

Figure 7–1

Step 2 Attach the **HOUSESTOP.DWG**. This drawing has a polygon topology added to the polygons that represent houses. A data table named **Danger** was created for this drawing and a data value of **1** was attached to each polygon. In this case, the arbitrary value 1 indicates a residential building.

Step 3 Go to the **Map** menu and select **Query** then **Object Thematic Query**. At the Limit to Location item, choose the **Define** option. In the Location Condition dialog box, select **Polygon** for the boundary type and **Inside** for the selection type. Then click on **Define**.

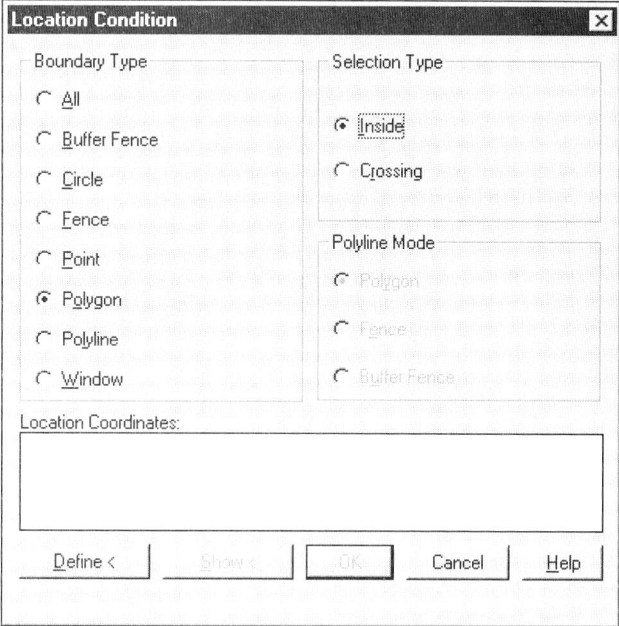

Figure 7–2

Draw a polygon by tracing over the Floodzone outline. For the **Data** select the **Danger** table.

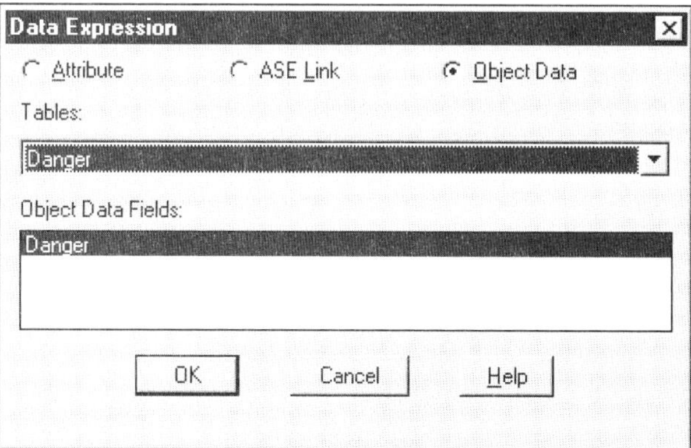

Figure 7–3

Set the Display Pattern to **Fill. Edit** the query to look for the houses inside the specified polygon (drawn to identify the query location) with an Object Data value of **1**, the value previously attached to all houses.

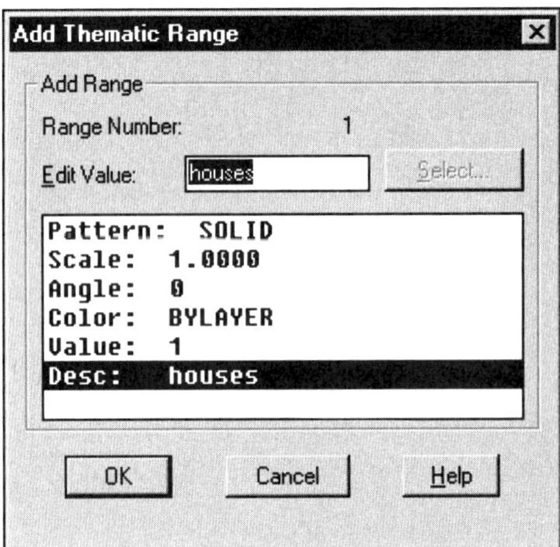

Figure 7–4

Proceed with the query. The houses located inside the location polygon will be drawn on the screen, and the number of affected residences will be displayed in the command prompt area of the AutoCAD Map 2000 screen. You can zoom in to view the individual structures.

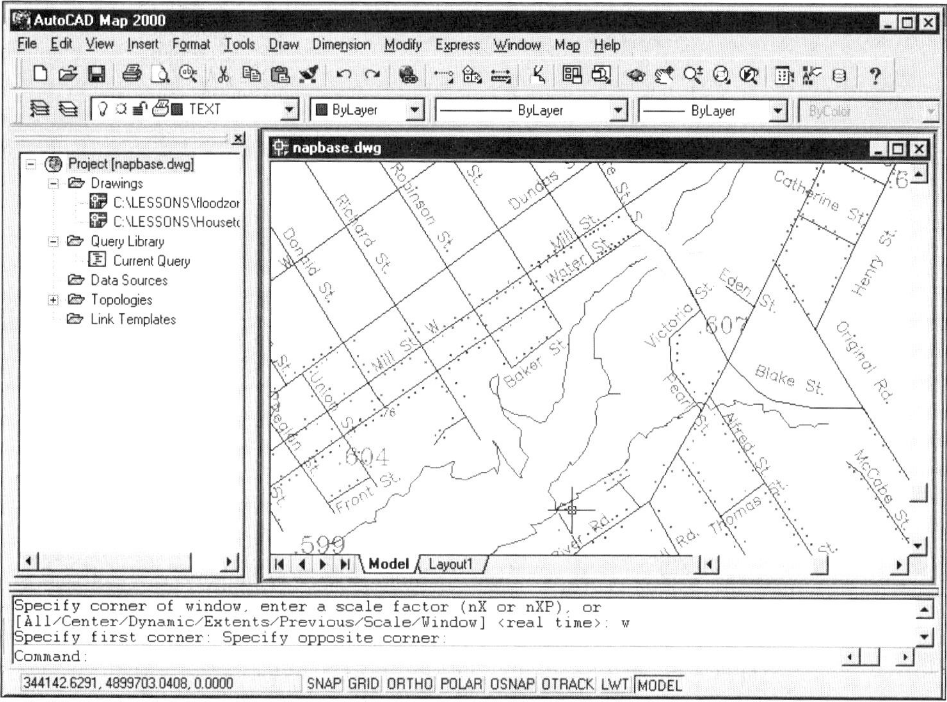

Figure 7–5

If at any time you wish to see the outline of the flood zone you can conduct a **Quickview** for that drawing. You can save the resulting drawing if you wish, then exit AutoCAD Map 2000 or continue on with the tutorials.

FLOOD ZONE PROPERTY EXERCISE

Use the Napbase, Floodzone, Multi-resident, and the Contour drawings to determine the number of apartment buildings (multi-residential units) that are inside the flood zone. Add identifying object data to the multi-residential buildings and conduct a query to identify those that lie in a danger zone.

SUMMARY

Engineering applications abound with AutoCAD Map 2000. In this tutorial, a high rainfall situation requires that the houses in a flood plane be identified. Maps prepared in accordance with procedures outlined in Chapter 1 are used to facilitate the use of an Object Data Query. The follow-up exercise reinforces the procedures for querying drawings for a specific purpose. Similar applications could include topics ranging from identifying properties in avalanche zones to areas in hurricane paths.

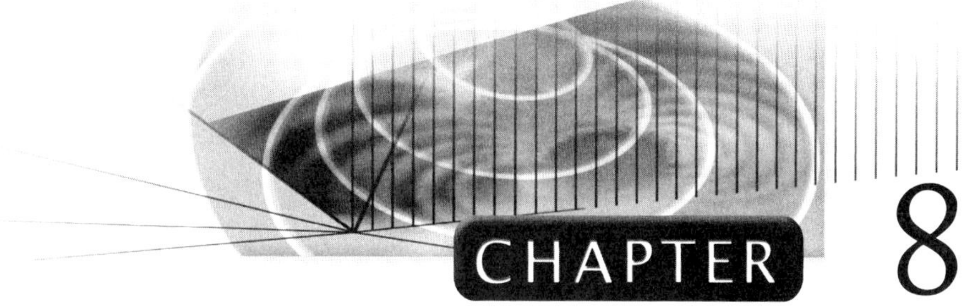

CHAPTER 8

Environmental Application

OBJECTIVES

In this chapter you will learn how to:

- access multiple drawings
- perform spatial analysis studies

Environmental uses for GIS technology are numerous because the technology can analyze both the spatial and temporal components of environmental issues.

ACID RAIN

Acid rain occurs whenever precipitation falls with a pH lower than 5.6. Acid rain causes a variety of environmental and economic damages exemplified by building deterioration, car finish deterioration, fish depletion from thousands of lakes and other serious consequences.

It is generally accepted that acid rain is caused by water vapor combining with nitrogen oxides and/or sulfur dioxide in the atmosphere before falling as rain or snow.

This exercise uses information derived from the National Atmospheric Deposition Program and the U.S. Environmental Protection Agency's Web sites. The data maps show the states with high levels of NO_x, SO_2, and some pH levels of rainfall in the eastern half of the continental United States.

TUTORIAL: ACID RAIN

Step 1 Open the **USA** drawing and **Attach** the **USASO2** drawing. This drawing has data attached to several states indicating that they generate SO_2 emissions of at least 100 kilotons (100,000 tons) per year.

Step 2 You will now conduct a Fill-based Object Thematic Query to reveal which states are involved. Go to the **Map** menu and select **Query**, then **Object Thematic Query**. Click on **Limit to Location** and. Select a boundary type defined as the area **Inside** a **Circle**.

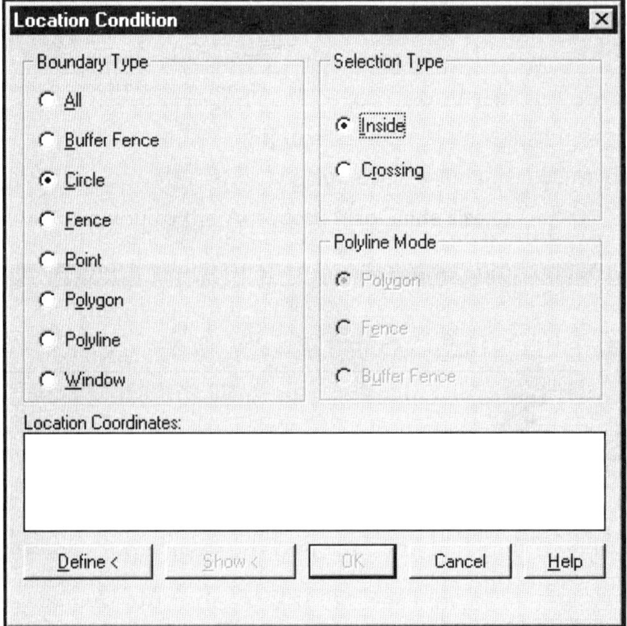

Figure 8–1

Define the circle as one that surrounds the USA.

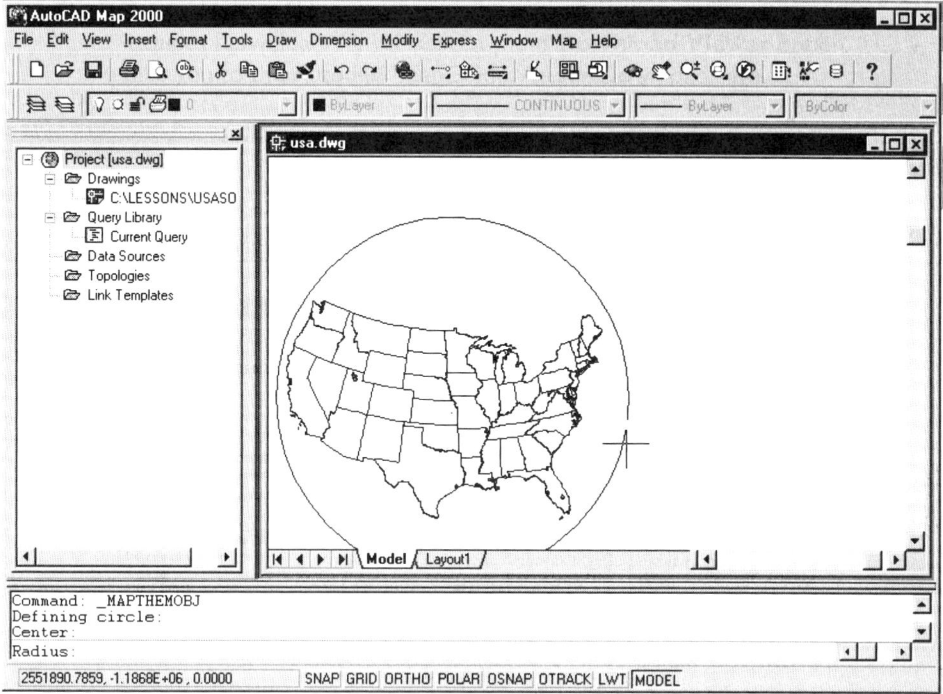

Figure 8–2

Step 3 Set the Thematic Expression to search for **Data** in a table named **SO2**.

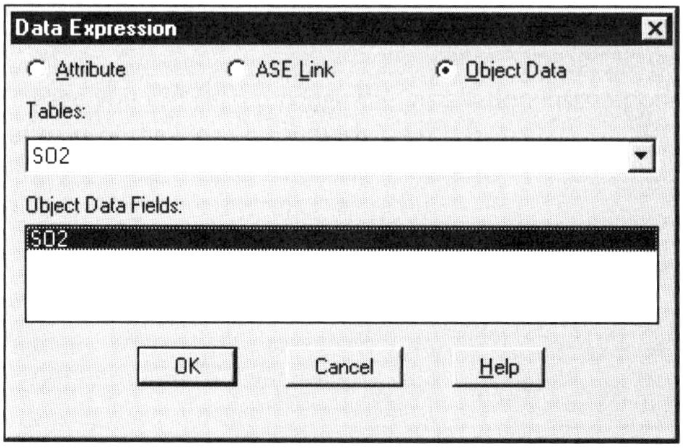

Data Expression

Attribute ASE Link Object Data

Tables:

SO2

Object Data Fields:

SO2

OK Cancel Help

Figure 8–3

Use the **Fill** option in the Display Parameters to **Define** a Thematic map. In the Edit Thematic Range dialog box, **Add** the parameters to search, and display the data by assigning a **Line** pattern with a spacing of **350000** at an angle of **0**. Set the color as **blue (5)** and establish the value as **100** with a description of **100,000 tons per year**.

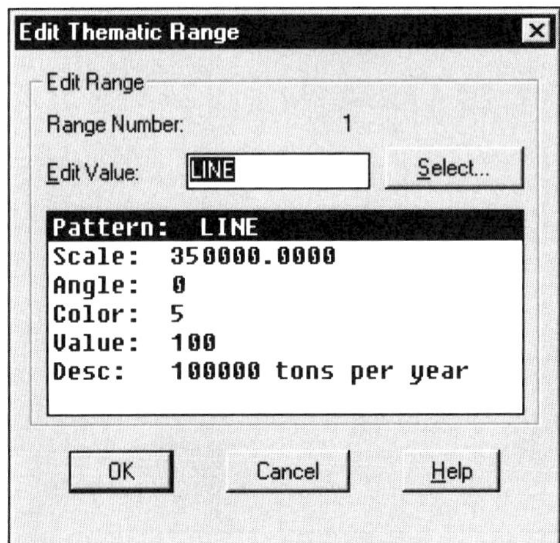

Figure 8–4

Step 4 After establishing the presentation qualities, select the **Legend** option from the Thematic Display options dialog box. Confirm the **Legend Creation**, then **Pick** a location on the map for the legend with boxed symbols and set text heights for *X* and *Y* values at **200000.** Click on **OK** and then on **OK** again at the Thematic Display Options dialog box. At the Object Thematic Mapping screen, click on **Proceed**. The involved states will be displayed.

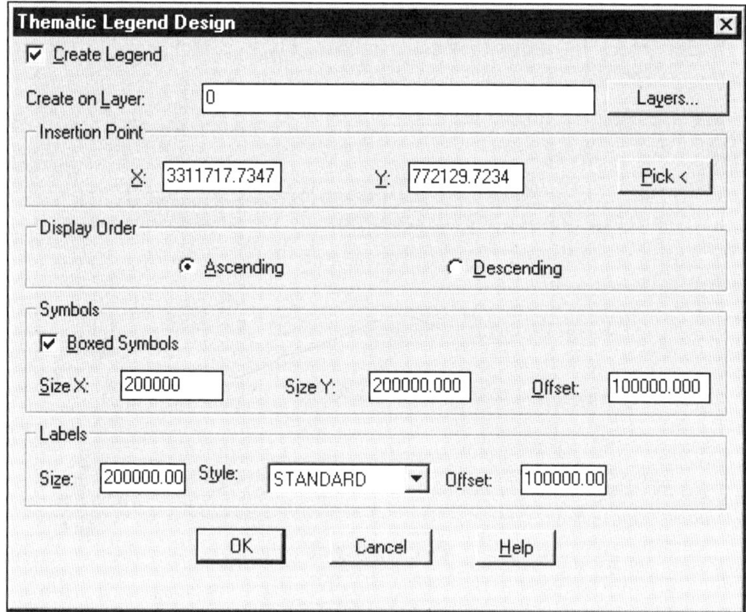

Figure 8–5

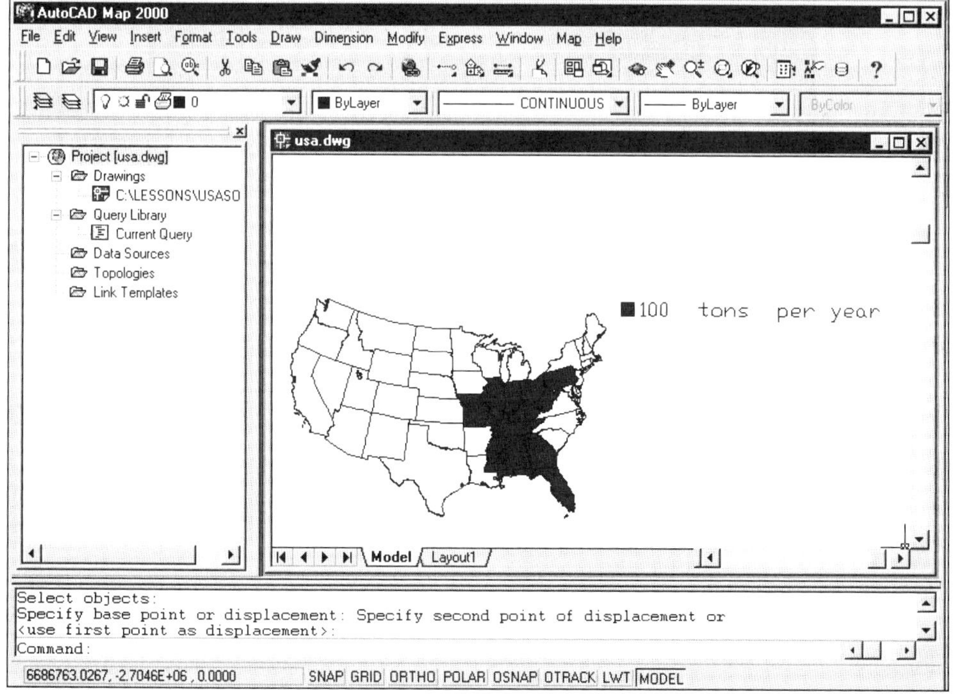

Figure 8–6

Step 5 Now attach the **USANO2** drawing. Conduct a second Object Thematic Query to reveal states that are associated with 30 kilotons (30,000 tons) of nitrogen oxides emissions per year.

Step 6 Use the same circular search location to search for the data in the table identified as **NO2**.

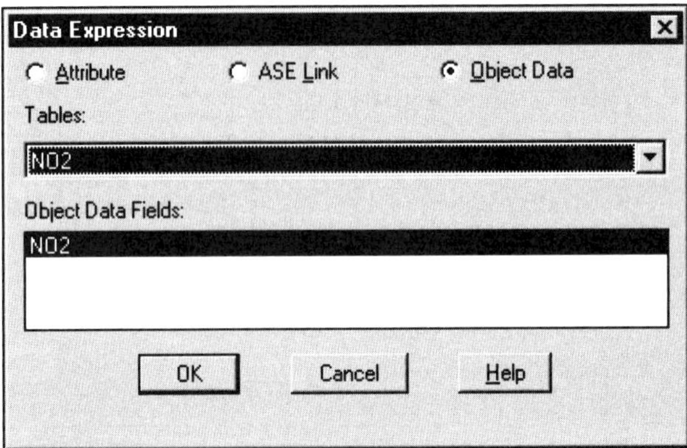

Figure 8–7

Define the **Fill** display to include a **Line** pattern with spacing **300000** at an angle of **90** degrees. Use **red (1)** for the color, and set the value at **30** with a description of **30000 tons per year**.

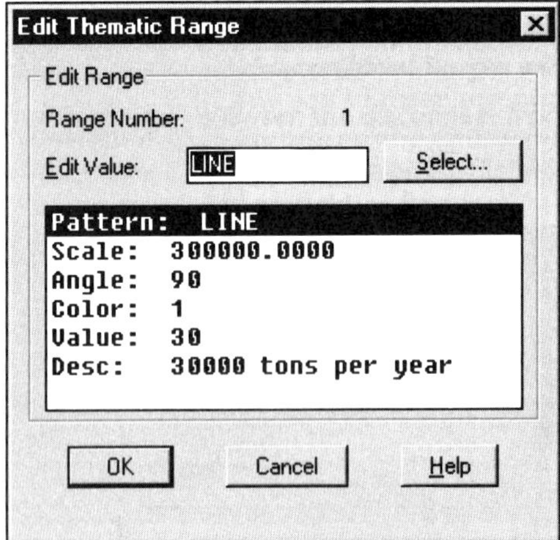

Figure 8–8

Add a suitable legend. The resulting map will show a considerable overlap between the states that generate high levels of SO_2 and those that generate high levels of NO_x.

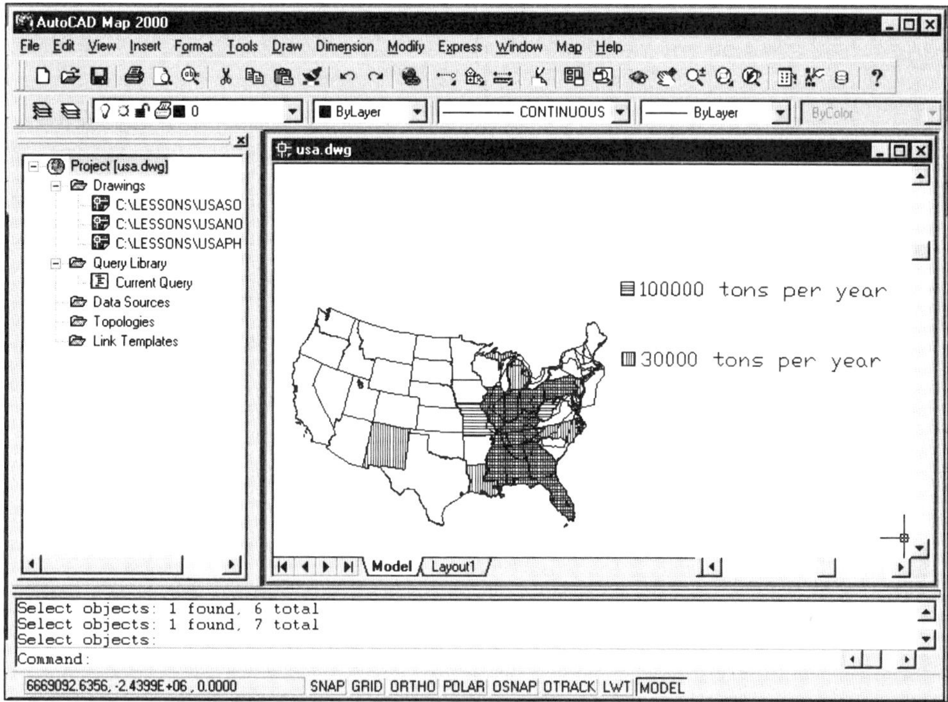

Figure 8–9

Step 7 Attach the **USAPHRAIN** map. Conduct a third Object Thematic Query based on the same search location, using the **Fill** option to reveal the areas that have precipitation pH levels of **4.3, 4.4, 4.5**, and **4.6.** Set the query to search for **Data** in the **pH** table.

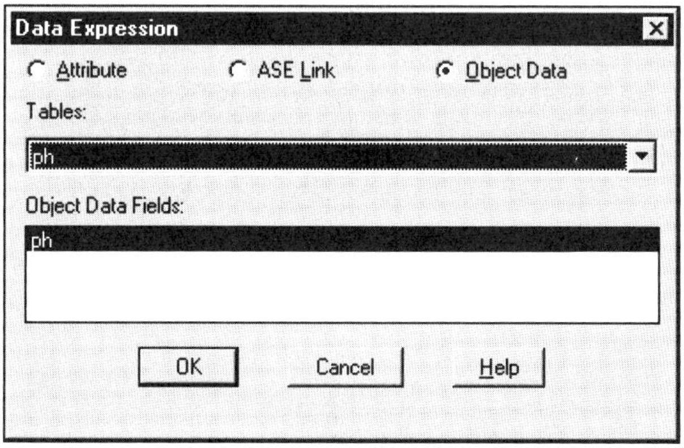

Data Expression

Attribute　　ASE Link　　Object Data

Tables:

ph

Object Data Fields:

ph

OK　　Cancel　　Help

Figure 8–10

Establish the display parameters as follows:

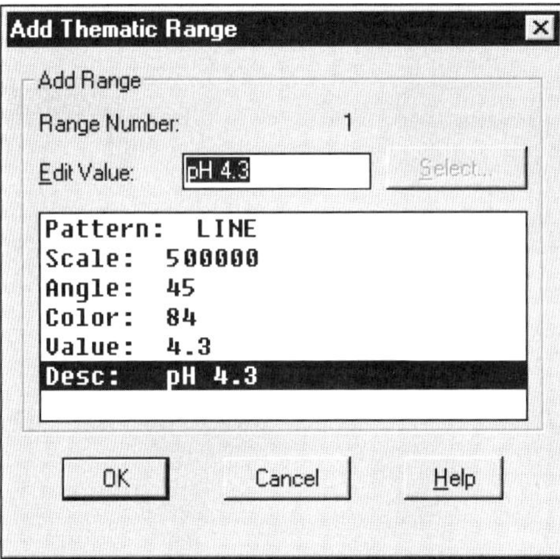

Figure 8–11

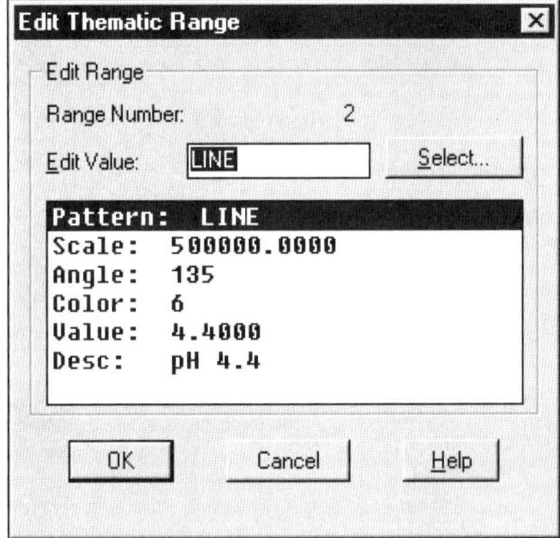

Figure 8–12

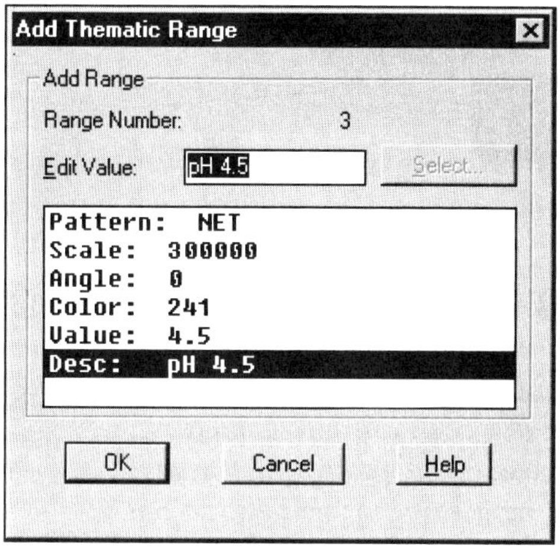

Add Thematic Range

Add Range

Range Number: 3

Edit Value: pH 4.5 Select...

```
Pattern:  NET
Scale:   300000
Angle:   0
Color:   241
Value:   4.5
Desc:    pH 4.5
```

OK Cancel Help

Figure 8–13

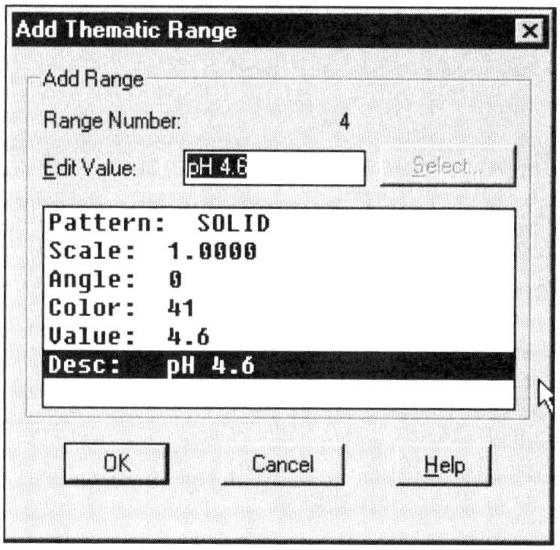

Add Thematic Range

Add Range

Range Number: 4

Edit Value: pH 4.6 Select...

```
Pattern:  SOLID
Scale:   1.0000
Angle:   0
Color:   41
Value:   4.6
Desc:    pH 4.6
```

OK Cancel Help

Figure 8–14

The Thematic Display Options dialog box will appear with four items in it, which should be **arranged in order** according to the **Value**.

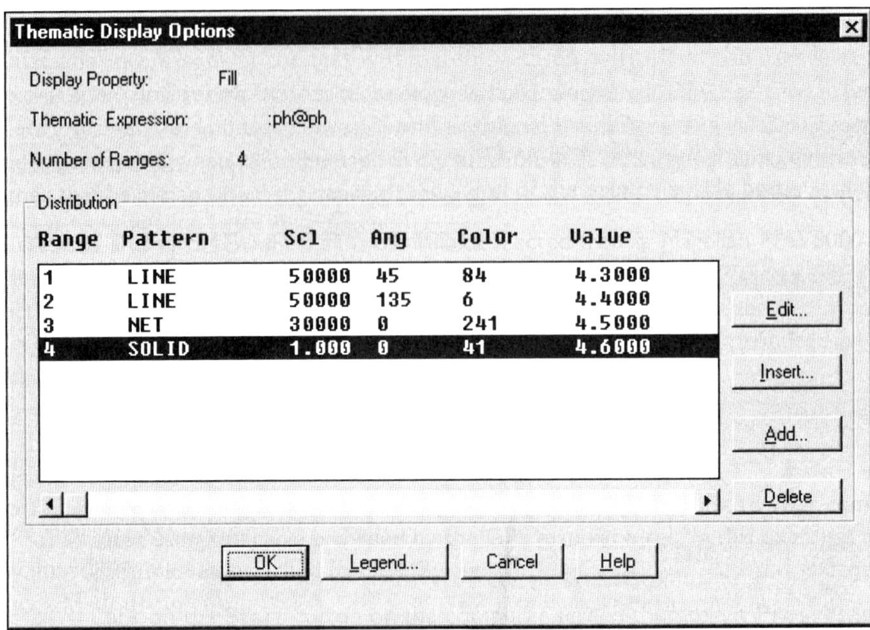

Figure 8–15

Step 8 Add a legend, then **Proceed** to produce a finished map that will reveal the locality of high pH rain levels as related to the atmospheric gas output.

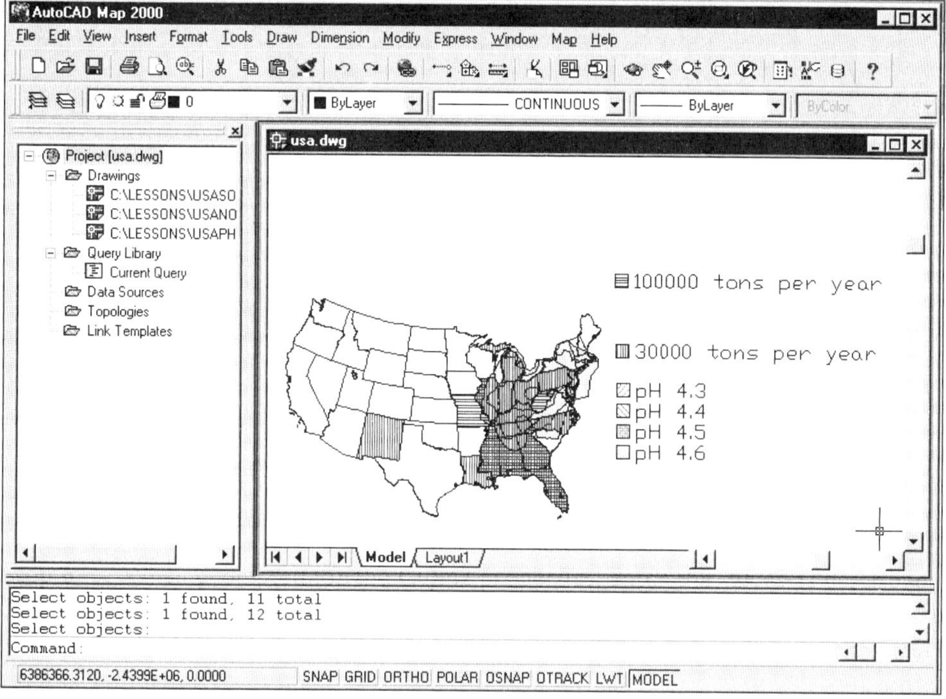

Figure 8–16

The drawing can either be saved or discarded, and you can either exit AutoCAD Map 2000 or continue on with the tutorials.

SUMMARY

This tutorial studies the relationship between two variables; emissions into the atmosphere and acid precipitation. The provided drawings contain Object Data relating to the two types of variables. The query illustrates the procedure for producing a thematic map with several classes of data being identified as well as having a legend applied to the finished map. Similar applications could vary from analyzing and displaying crime rates in a district to illustrating ocean surface temperatures.

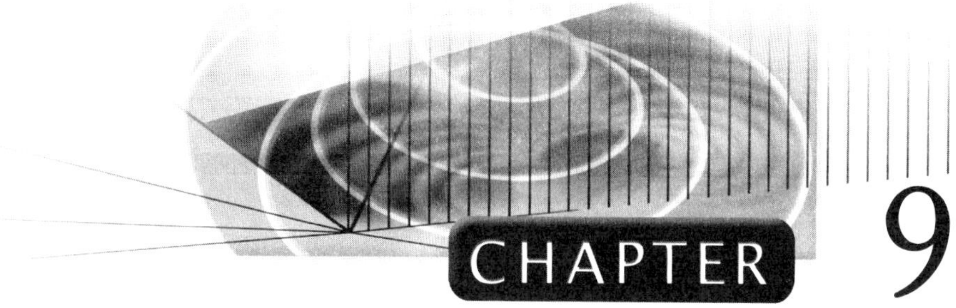

A Tourism Application

OBJECTIVES

When you have finished this chapter you will know how to:

- conduct multiple queries
- view multiple drawings simultaneously
- prepare drawing data sets
- prepare data-enriched drawing sets
- delete layers from drawings

Tourism is a rapidly growing field in the global economy. It is also heavily reliant on map-based information!

This chapter is presented in three parts. The first illustrates an application of many of the techniques covered in earlier chapters, and the second reveals how one can set up the drawing files for such an application. The final component is an exercise designed to reinforce both these techniques.

TUTORIAL: QUERYING MULTIPLE DRAWINGS

For the purpose of this exercise, let us suppose that a small group of people were going camping for a short period of time to an area that offered a variety of activities, such as fishing, canoeing, and hiking, that would meet the group's diverse interests. The group has settled on a general area for the excursion and now wants to select specific sites for their activities.

Step 1 Open AutoCAD Map 2000 and select the **TOURBASE.DWG** file, shown in Figure 9–1. This drawing will serve as the base map and covers the region that the group has decided to visit. The first task is to determine which campground to stay in based on proximity to the activity sites.

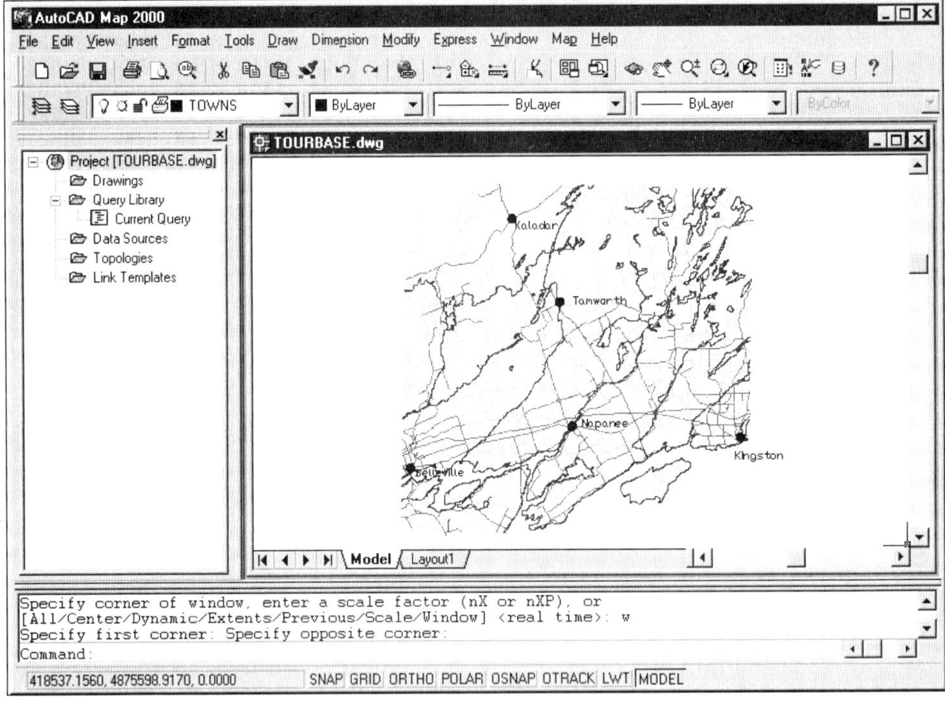

Figure 9–1

Step 2 To assist in the decision making, it would be helpful to be able to view all of the available information. Select **Map**, then **Drawings** and **Define/Modify Drawing Set**. Use the **Attach** function to bring up the Select Drawings to Attach dialog box, and then add the **Camp, Canoeroute, Fish**, and **Walk** drawings. Click on **OK** and at the Define/Modify Drawing Set dialog box click on **OK** again.

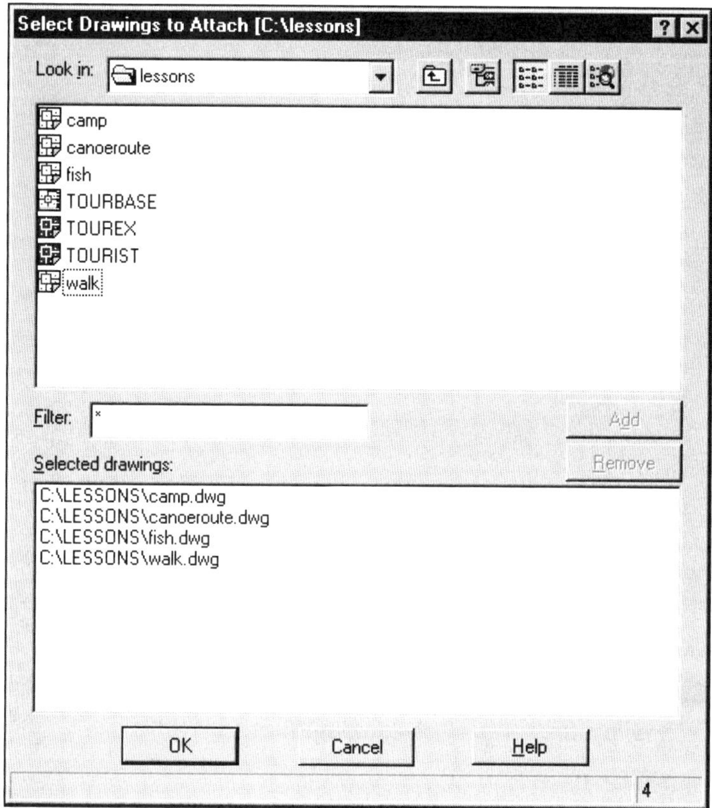

Figure 9–2

Step 3 Go the **Map** menu and select **Drawings** then **Quick View Drawings**.

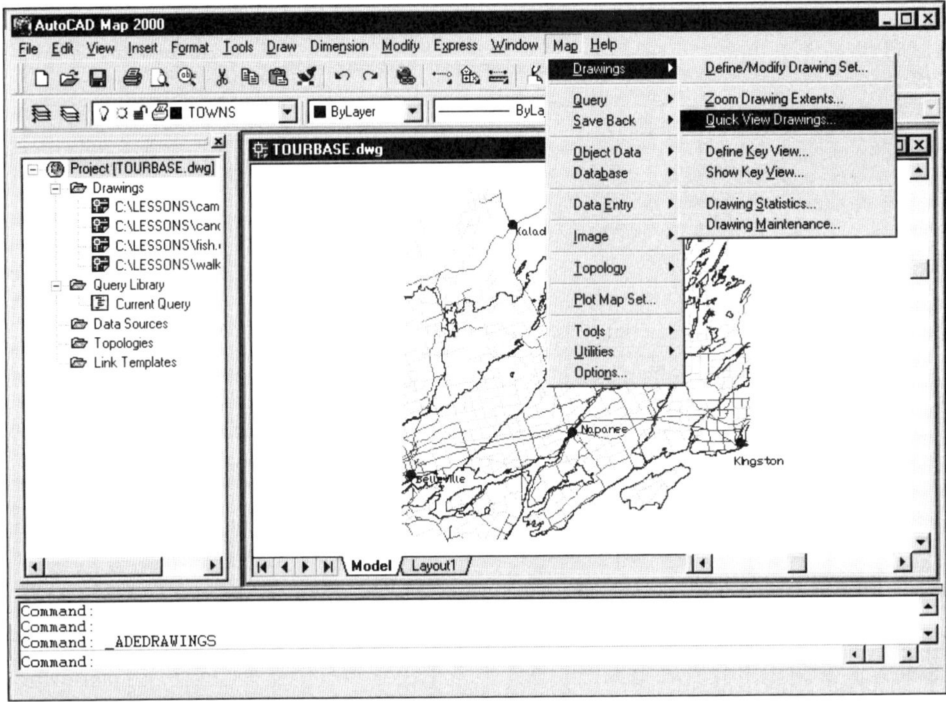

Figure 9–3

You can select the drawings to be displayed either one at a time or all at once as shown in the Quick View Drawings dialog box.

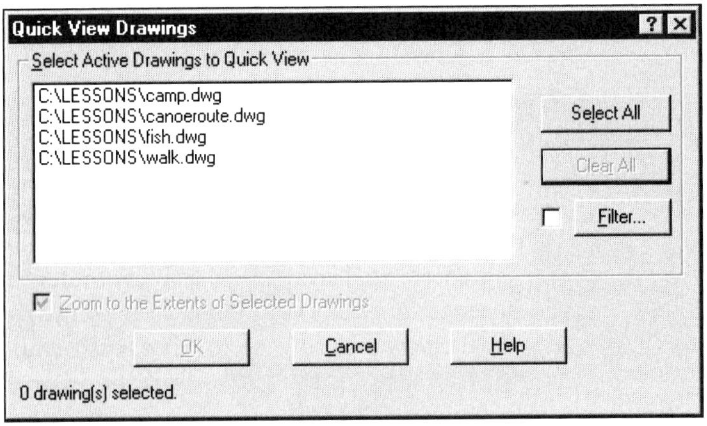

Quick View Drawings
Select Active Drawings to Quick View
C:\LESSONS\camp.dwg
C:\LESSONS\canoeroute.dwg
C:\LESSONS\fish.dwg
C:\LESSONS\walk.dwg
Select All
Clear All
Filter...
Zoom to the Extents of Selected Drawings
OK
Cancel
Help
0 drawing(s) selected.

Figure 9–4

Selecting these drawings displays the campgrounds, hiking trails, fishing locations, and canoe routes in the region. The touring group decides that they want to find the locations for walleye fishing, an easy canoe route, and a wilderness hike and select a campground closest to sites for those activities.

Step 4 Before continuing, press **r** and ENTER to refresh the screen and remove the Quick View.

Step 5 The attached drawings have data embedded in them as follows:

FISH :	Walleye or Bass
CAMP:	Government or Private
WALK :	Park or Wilderness
CANOE:	Length in km (1,000 meters)

Go to the **Map** menu and select **Query** then **Object Thematic Query**. At the Object Thematic Mapping dialog box, select **Limit to Location**, then click on **Define**. Select **Circle, Inside**, and click on **Define**; draw a circle that encompasses the map. For the Thematic Expression make sure that **Data** has been selected then click on **Define**. In the Data Expression dialog box, select the **Fish** item from the list of tables with **Species** as the data field.

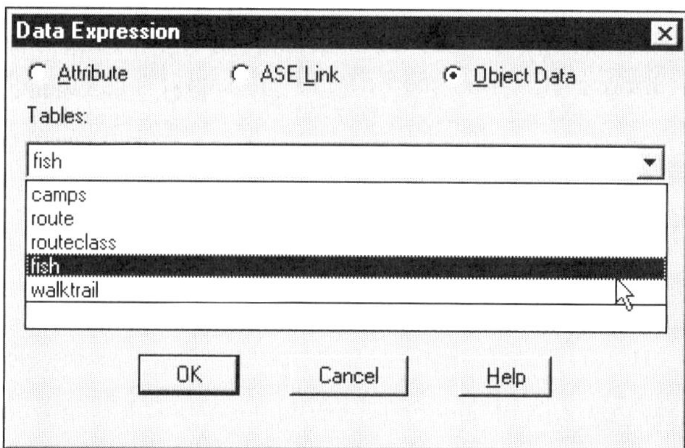

Figure 9–5

Step 6 In the Display Parameters section of the dialog box, set the Display Property to **Fill** and click on **Define**. Use the **Add** button in the Thematic Display Options dialog box to establish a Thematic Range to define the search for **Walleye**.

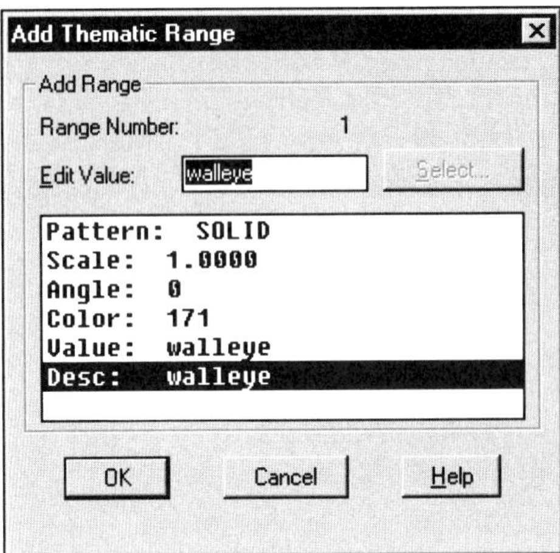

Figure 9–6

Click on **OK** to return to the Thematic Display Options_dialog box, then click on **OK** again and click on **Proceed**. The sites for walleye fishing will be identified as filled blocks.

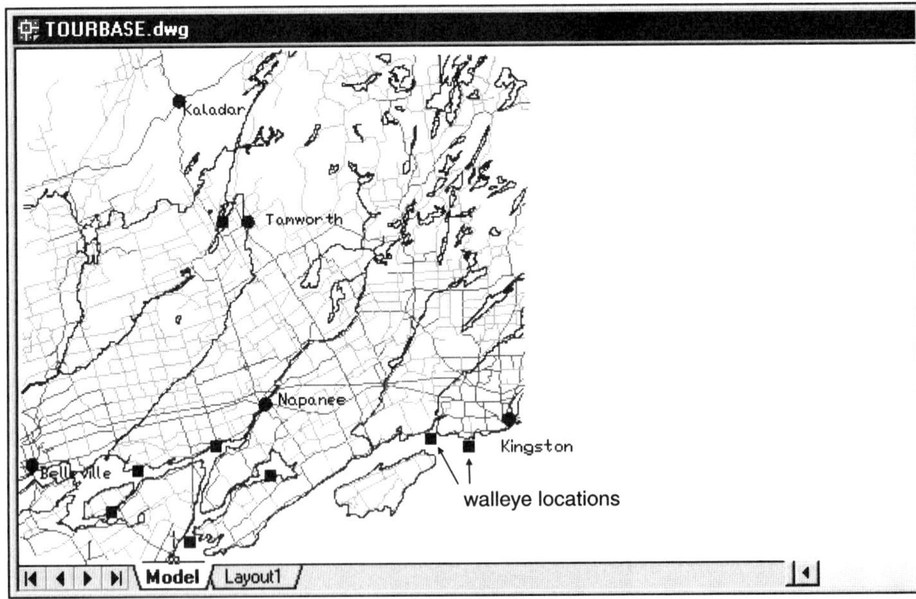

Figure 9–7

Step 7 A similar procedure will be followed to find a canoe route of a moderate length. In this case, there are three routes identified on the Canoeroute map, shown originally in the Quick View, one 7 km, one 12 km, and the other 17 km in length. The group wants to find the 12-km route. Go to the **Map** menu and select **Query** then **Object Thematic Query**. Use the same location circle as was used in the fishing query. For the Thematic Expression ensure that **Data** is selected, and then define the object data to search the **Route** table and the **Length** data field and click on **OK**.

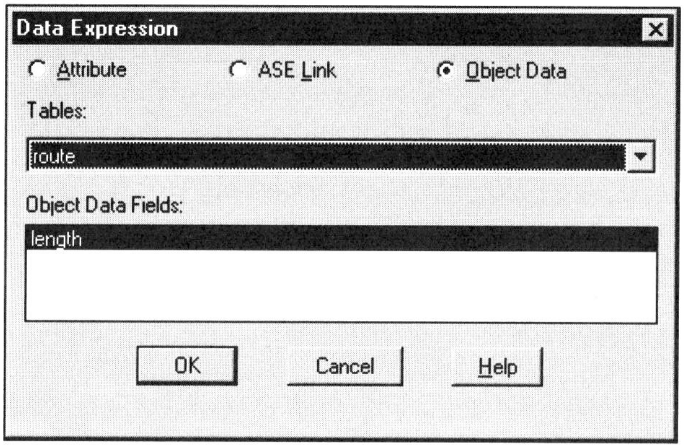

Data Expression

Attribute ASE Link Object Data

Tables:

route

Object Data Fields:

length

OK Cancel Help

Figure 9–8

Step 8 Set the Display Parameters to **Fill**, and click on **Define**. Set the Thematic Range values to a **Solid** with a Scale of 1, an Angle of **0**, a distinctive **Color** (**84** in this case), a value of **12**, and a description of **12 km.**

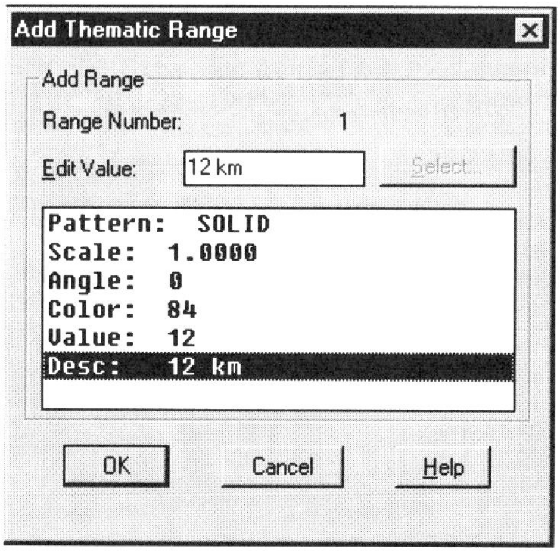

Add Thematic Range

Add Range

Range Number: 1

Edit Value: 12 km Select...

Pattern: SOLID
Scale: 1.0000
Angle: 0
Color: 84
Value: 12
Desc: 12 km

OK Cancel Help

Figure 9–9

Click on **OK** and then on **OK** again at the Thematic Display Options dialog box, and then click on **Proceed** at the Object Thematic Mapping dialog box. The three canoe routes will be highlighted on the map, but the 12-km route will be shaded in.

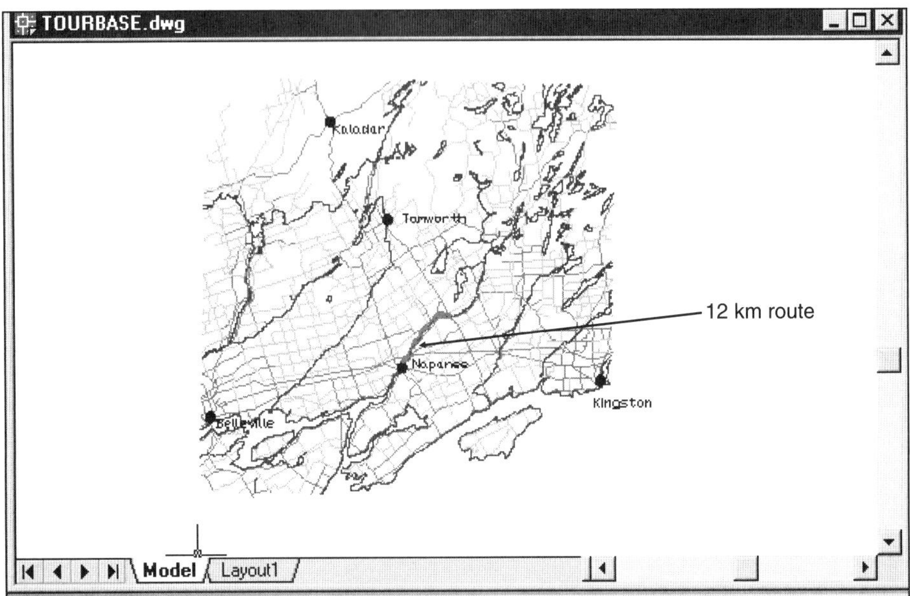

Figure 9–10

Step 9 The group now wants to locate the walking trail that is closest to the other sites of interest, but is at the same time a wilderness trail. Go to the **Map** menu and select **Query** and **Object Thematic Query**. Use the same circle for the location and select **Data**. Define the data using the **Walktrail** table and the **Walktrail** table field, then click on **OK**.

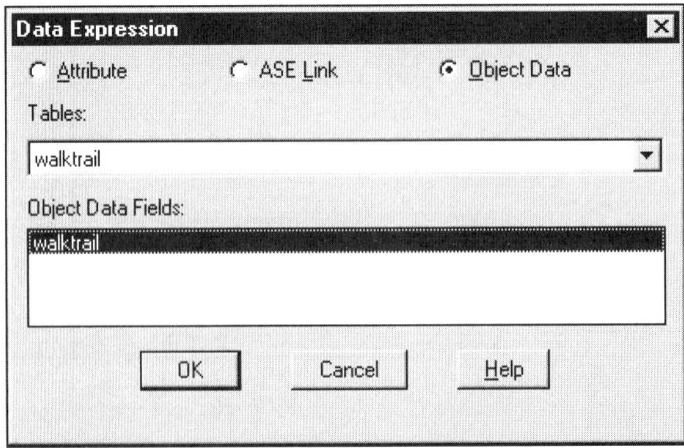

Figure 9–11

Set the Display Property to **Fill**, and click on **Define**. **Add** a Thematic Range that has a **Solid** Fill with Scale of **1** and Angle of **0**. Select a distinctive **color**, in this case **30**, and query the value **Wilderness**. Provide a description of **Wilderness**.

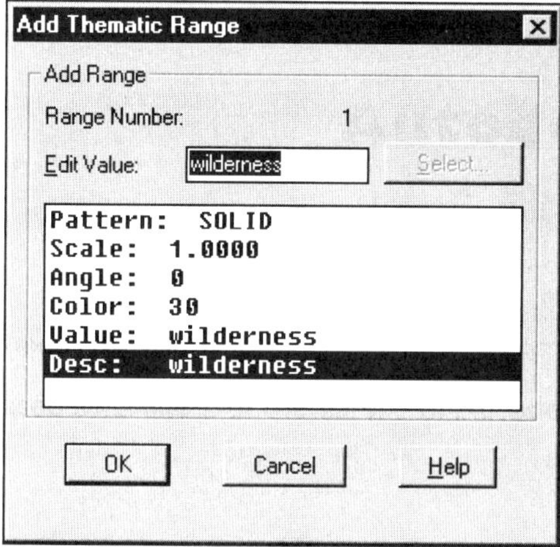

Figure 9–12

Click on **OK** then **OK** again at the next screen, then click on **Proceed**. Two trails with Wilderness classification will be filled in.

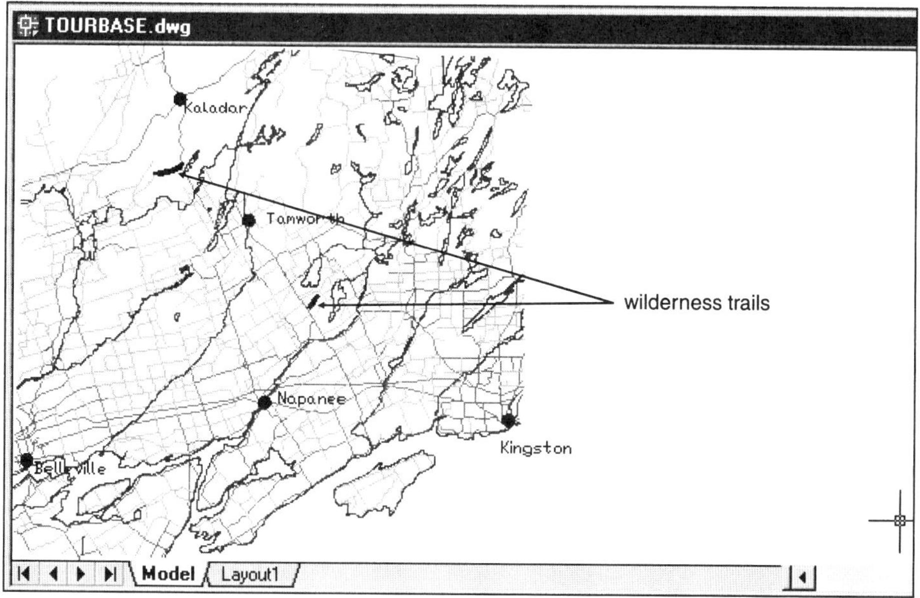

Figure 9–13

Step 10 The final step is to select the closest campground to all the points of interest. Select **Map** then **Query,** and **Object Thematic Query**. Use the same location as earlier and select **Camps** as the table and field.

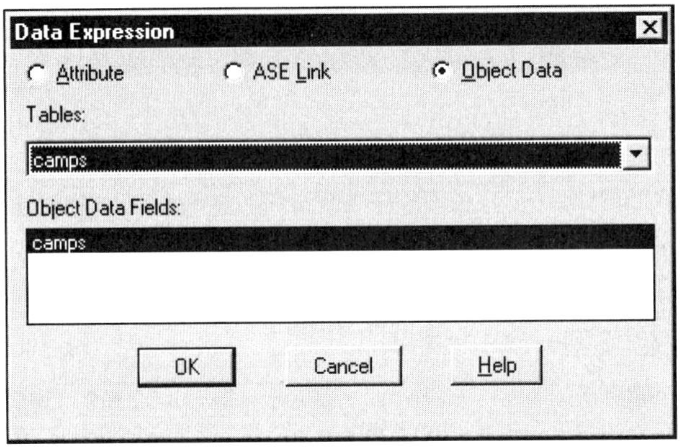

Data Expression
Attribute
ASE Link
Object Data
Tables:
camps
Object Data Fields:
camps
OK
Cancel
Help

Figure 9–14

Set the Display Property to **Fill**, then click on **Define**. **Add** a Thematic Range with a **Solid** fill pattern, a Scale of **1** and an Angle of **0**. Set the **color** to a distinctive choice, in this example, **magenta**. In the **Camps** object data there are two categories, Private and Government-owned. Enter a Value of **Private** and set the Description to **private park**.

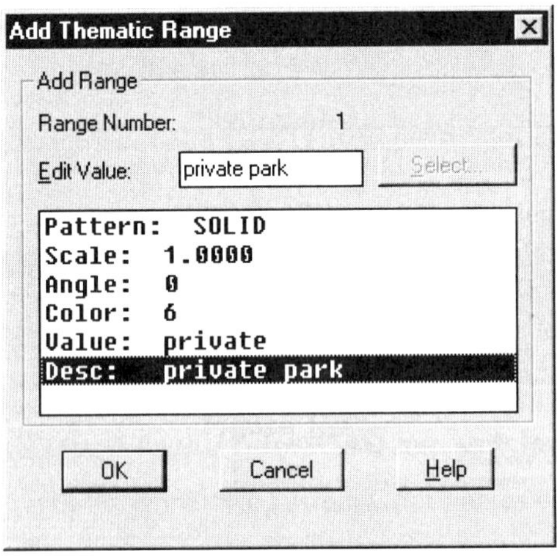

Add Thematic Range
Add Range
Range Number: 1
Edit Value: private park
Select...
Pattern: SOLID
Scale: 1.0000
Angle: 0
Color: 6
Value: private
Desc: private park
OK
Cancel
Help

Figure 9–15

Click on **OK**, click on **OK** again, then click on **Proceed**. The campground markers will appear with the Private site showing up in the color defined in the query. This process facilitates the selection of an appropriate campground in terms of proximity to the desired activities and their sites as shown by the fact that there is only one campground near the chosen sites. You may wish to zoom in on the features.

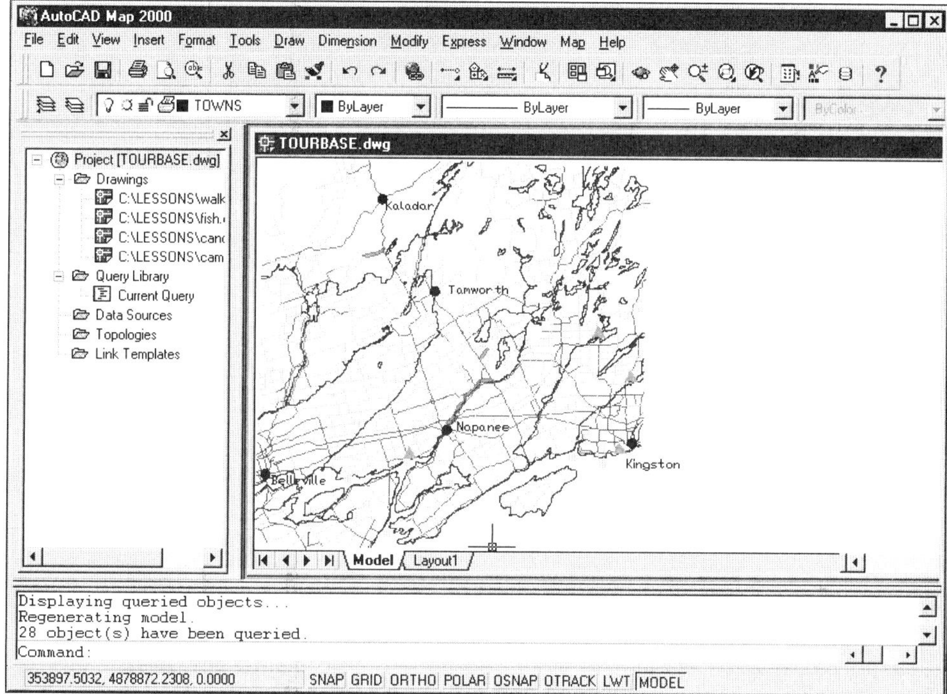

Figure 9–16

Either save the results or discard them. You can exit AutoCAD Map 2000 or continue with the tutorials.

TUTORIAL: ESTABLISHING DRAWINGS AS DATA SETS

This exercise will show you how the data base maps were prepared using the TOURIST.DWG as the base map.

Step 1 Open the **TOURIST.DWG** file. In this case, the location of the campgrounds, fishing locations, walking trails, and canoe routes were known and added to the base map one at a time, with each being placed on a separate layer (to view the layers in a drawing select the **Format** menu then **Layer**).

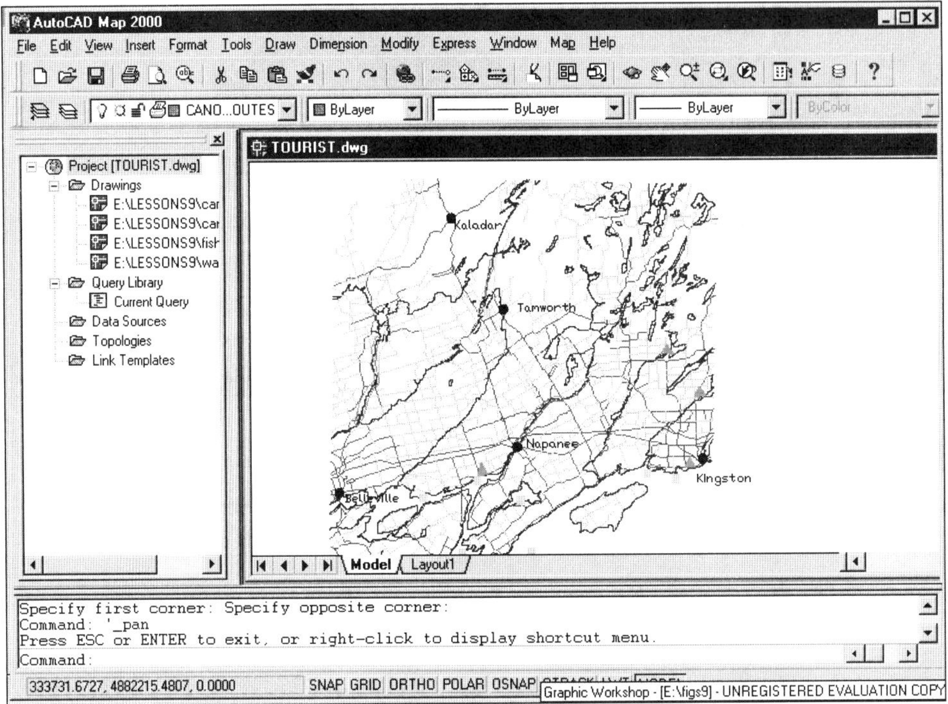

Figure 9–17

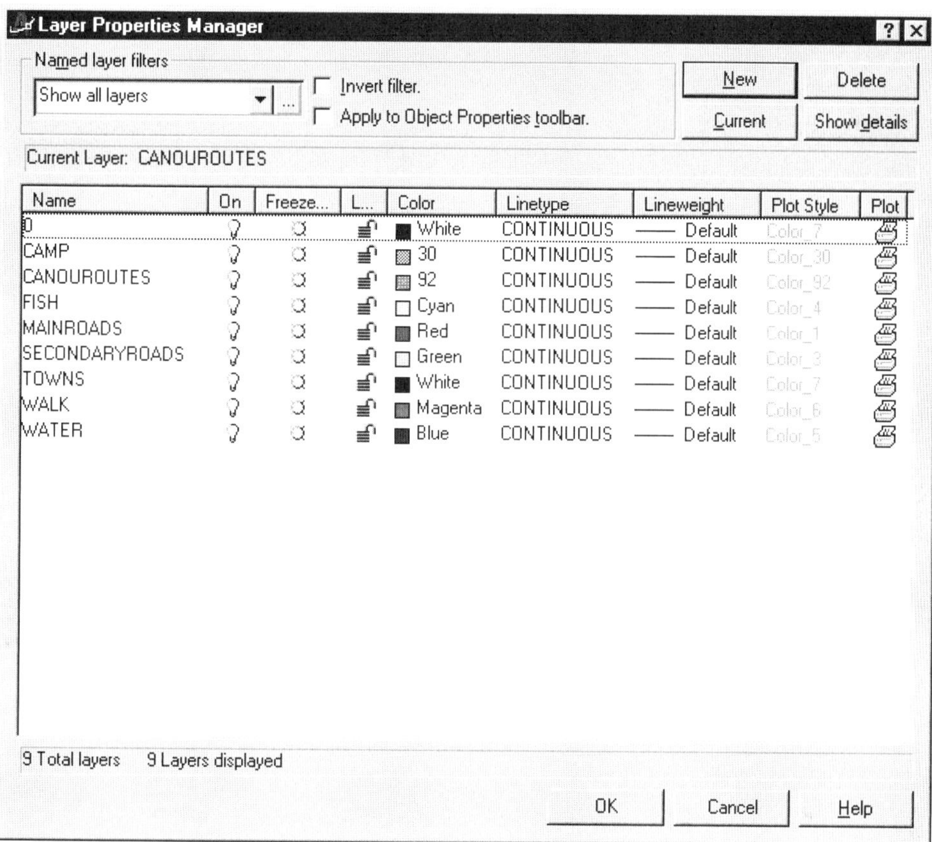

Figure 9–18

To produce a map with only campgrounds, for example, on it, select the **0** layer and make it current by clicking on the **Current** button. Turn the Camp layer **off** by clicking on the light bulb icon located to the right of the word Camp. The light bulb will turn dark. **Freeze** the camp layer by clicking on the small yellow sun shown in the same row as the name of the layer. The sun will then appear as a snowflake. **Erase** everything in the drawing. Reactivate or Thaw the Camp layer by clicking on the snowflake again, then click on the light bulb once again to turn the layer back on. Click on **OK** in the Layer Properties Manager dialog box. Only the camping sites will remain in the drawing.

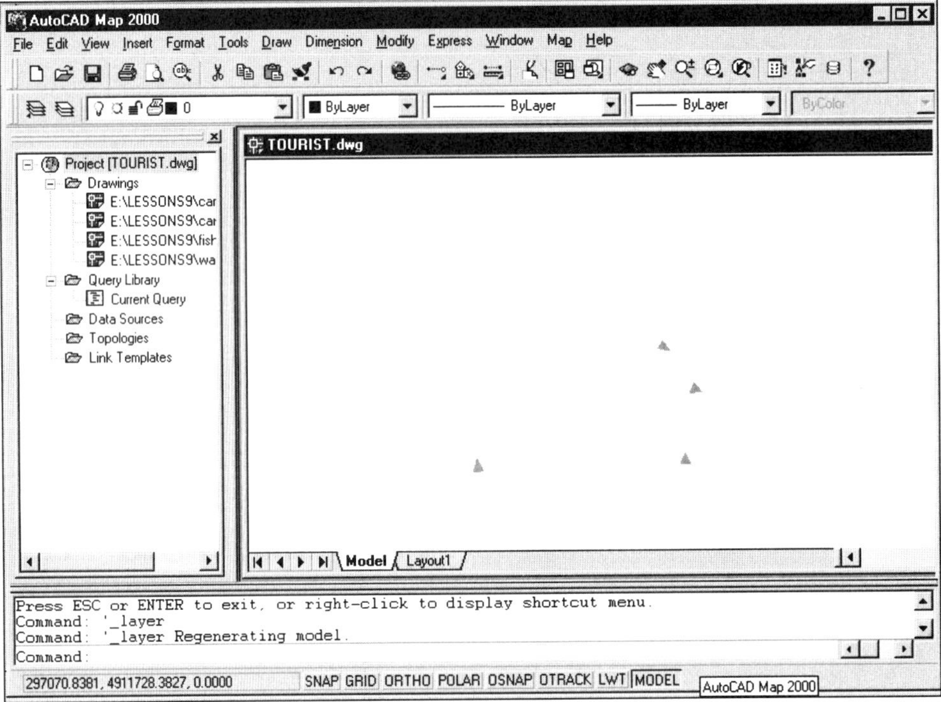

Figure 9–19

To remove the unwanted layers, set the Camps layer as the Current layer, select a layer to remove (such as Towns), and then click on **Delete**. Remove all the unnecessary layers (in this example, you would remove Water, Towns, Secondaryroads, and the Mainroads layers).

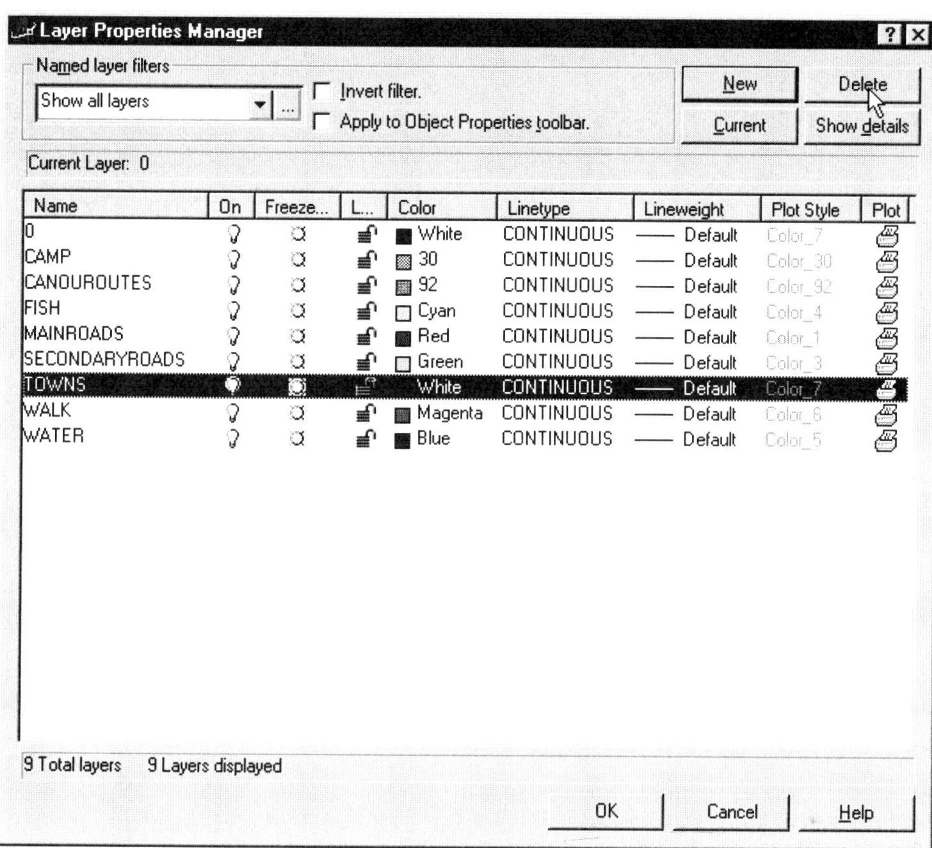

Figure 9–20

Step 2 To add data to the drawing select **Map**, then **Object Data**, and **Define Object Data** menu and create a **new table**.

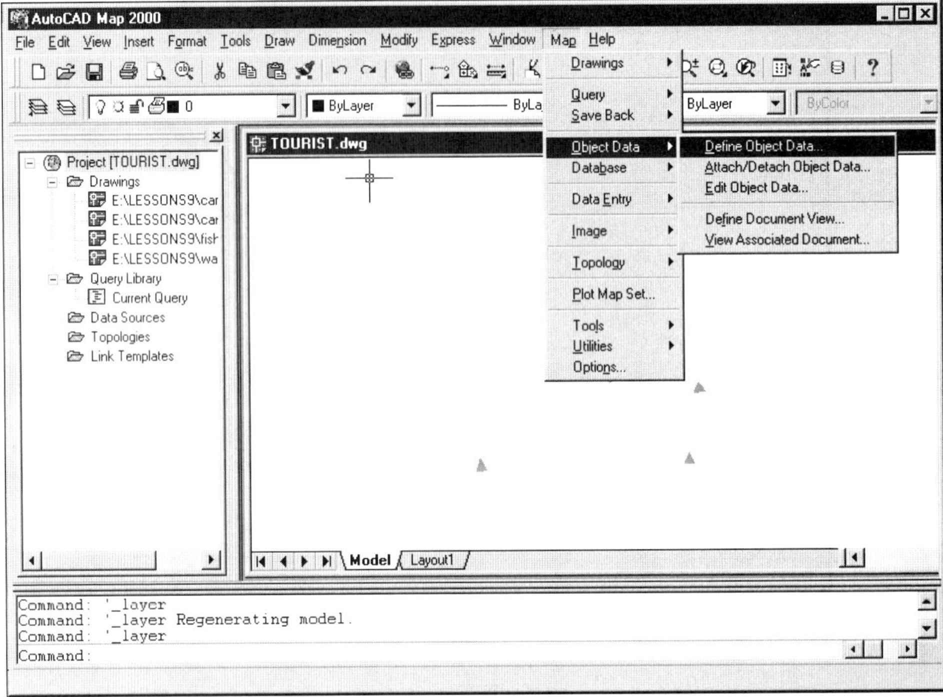

Figure 9–21

In this case, the table is called **Camps,** it is **Character** based, it has an object data field name **Camps** and a description of **Camp Type.** Click on **Add** , then **OK**, and at the **Define Object Data** screen dialog box, click on **Close.**

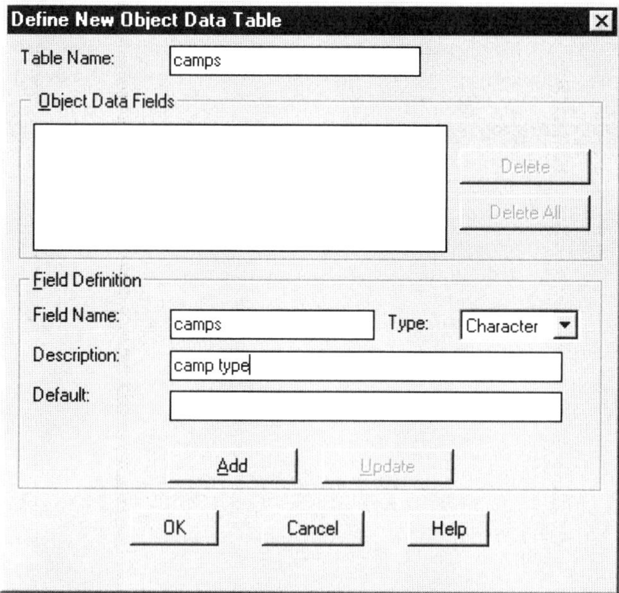

Figure 9–22

Step 3 Add data to the drawing by select **Map**, then **Object Data** and **Attach/ Detach Object Data**.

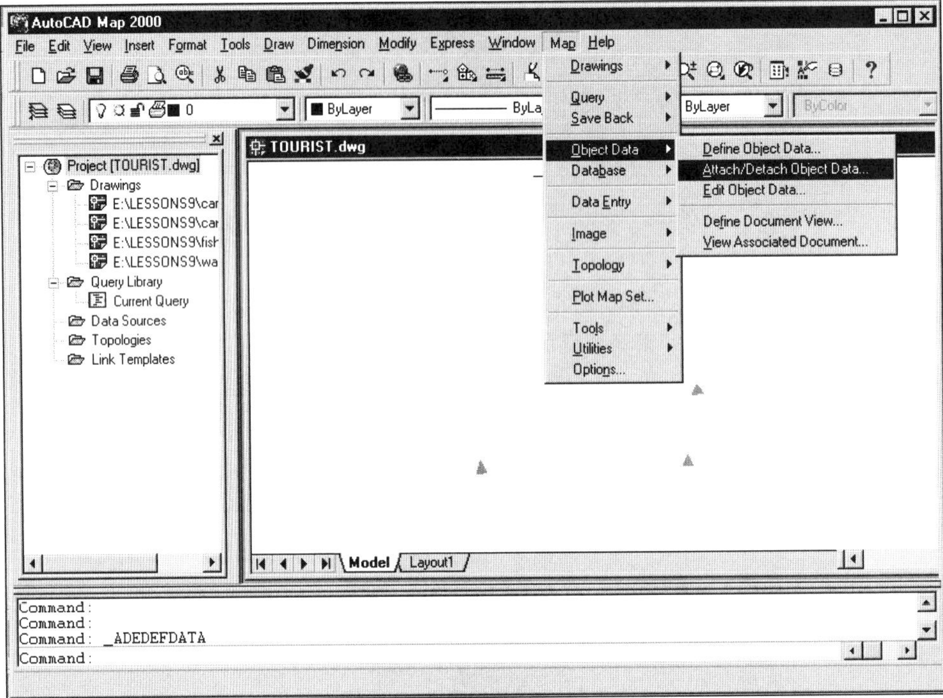

Figure 9–23

Select the table **Camps** and highlight the data field **Camp type.** In the **Value** box type in, for example **Private.**

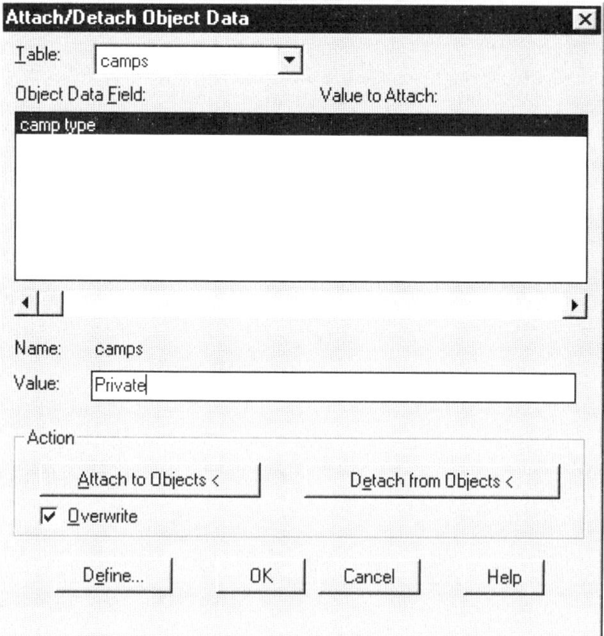

Figure 9–24

Click on the **Attach to Objects** button and **pick** the objects in the drawing that you want to attach data to. Attach the data by clicking on the object.

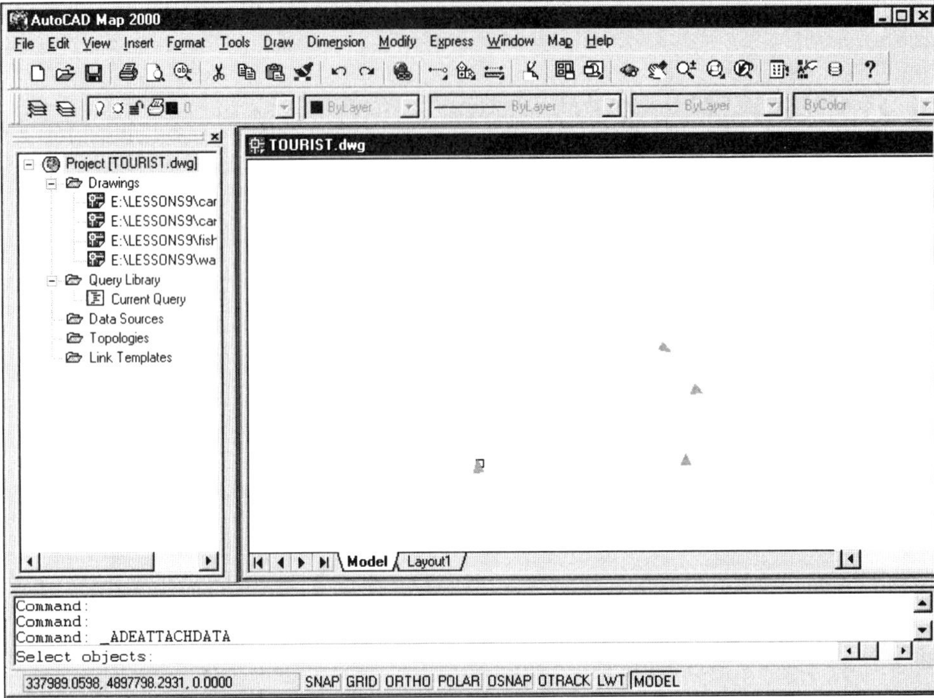

Figure 9–25

Click the right mouse button to end the Data Attach process.

Step 4 Save the drawing as **Camps**. The process of removing unwanted layers can be repeated to create the desired database of map drawings with each drawing containing only the pertinent layers. When finished, you can save the drawings and either exit AutoCAD Map 2000 or continue with the tutorials.

TOURISM EXERCISE

Part 1 Use the base map TOUREX.DWG to compile a data set of four drawings consisting of:

- Historic sites
- Museums
- Nature sites
- Accommodations

Generate a hard copy of each.

Part 2 Add object data to each drawing as follows:

- Historic sites: Fort or Home
- Museums: Regional or Thematic
- Nature sites: Geologic or Biologic
- Accommodations: Motel or Bed and Breakfast

See Figure 9–26 for a sample of how it might look when finished.

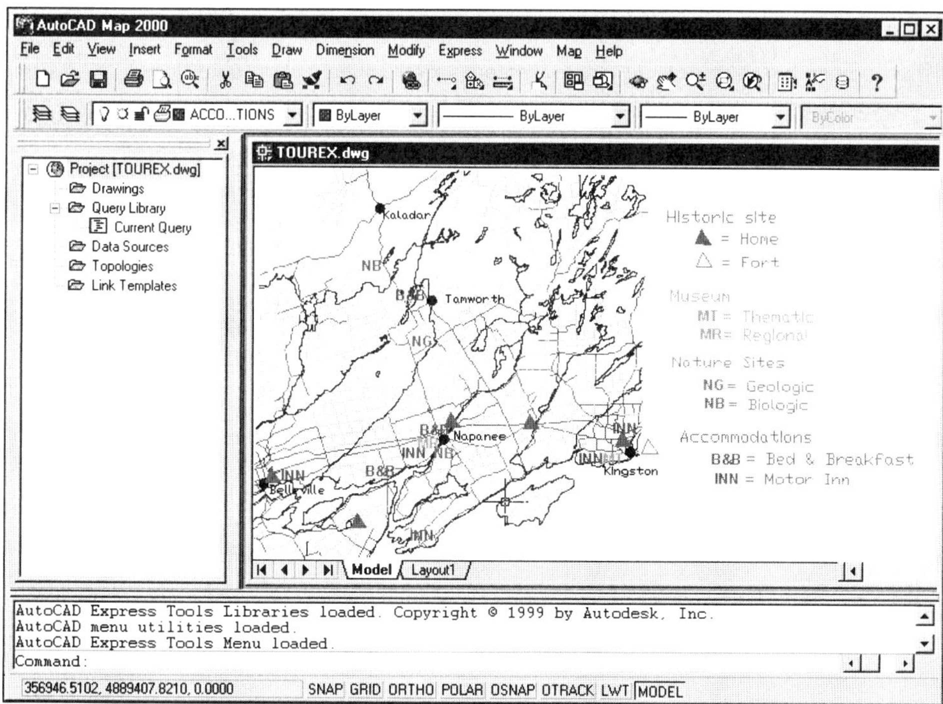

Figure 9–26 *(This textual information is also provided on TOUREX.DWG)*

Part 3 Develop a series of Object Thematic Queries that will generate a map to show the points of interest identified as the Fort, the Geologic Nature Site and the Thematic Museum (a museum devoted to a particular topic or theme). Make a hard copy of this map.

Part 4 Identify the accommodation most conveniently located for a day trip to these sites.

Part 5 Create Network Topology(see chapter 2) for the roads and generate shortest-distance route maps between the accommodation and each point of interest identified in Part 3. Generate hard copies of each route.

SUMMARY

This tutorial incorporates several of the procedures introduced earlier in the text. The provided drawings relate to a variety of tourism options and a series of queries leads the user to a conclusion regarding the most favorable location for camping. The provided drawings are used as examples for the purpose of generating drawing files for use in similar applications. The reinforcement exercise calls upon the skills introduced in the tutorial as well as techniques from earlier chapters to assist in a decision making process.

Site Selection

OBJECTIVES

In this chapter you will learn how to

- attach a second drawing (an external reference) to one already open
- utilize the Quickview feature to facilitate simple location analysis

The selection of an appropriate site for a business activity or any other type of specific land use is a common task in GIS. In this exercise, a process for combining two maps in order to have a larger spatial framework is introduced.

This exercise oversimplifies the site location issue by making an assumption that transportation costs are only a function of linear distance. No consideration is given in this case to different modes of transportation or any other factor that could impact the decision-making process.

TUTORIAL: A SIMPLE SITE SELECTION

Step 1 Start AutoCAD Map 2000 and open the supplied **CANADA** drawing.

Step 2 Go to the **Insert** menu and select **External Reference**.

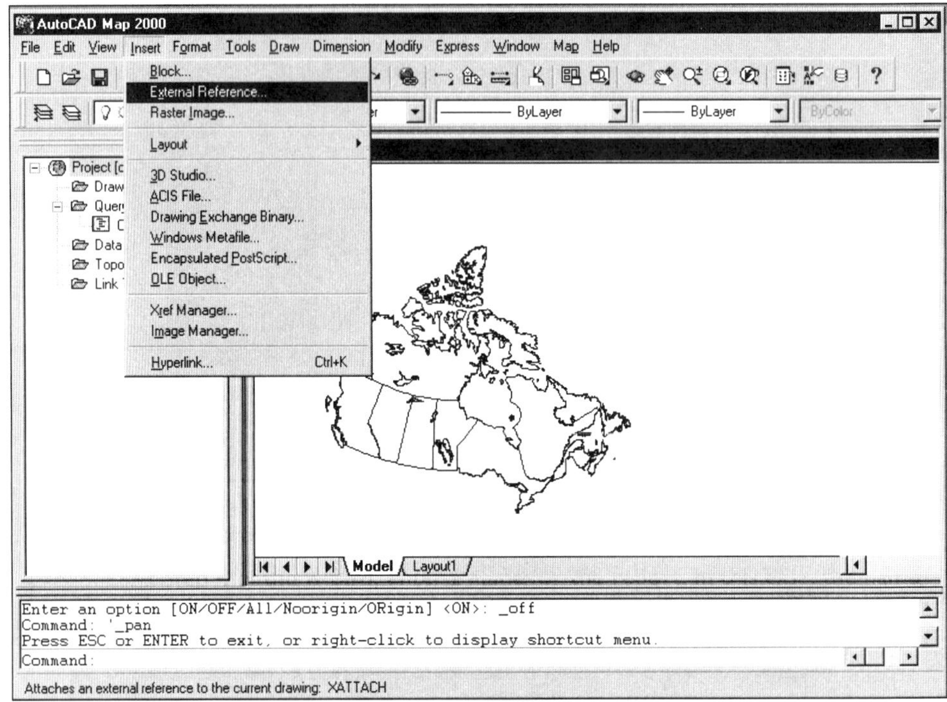

Figure 10–1

The Select Reference File dialog box is presented. Select the **US** drawing as the file to attach and click on **Open**.

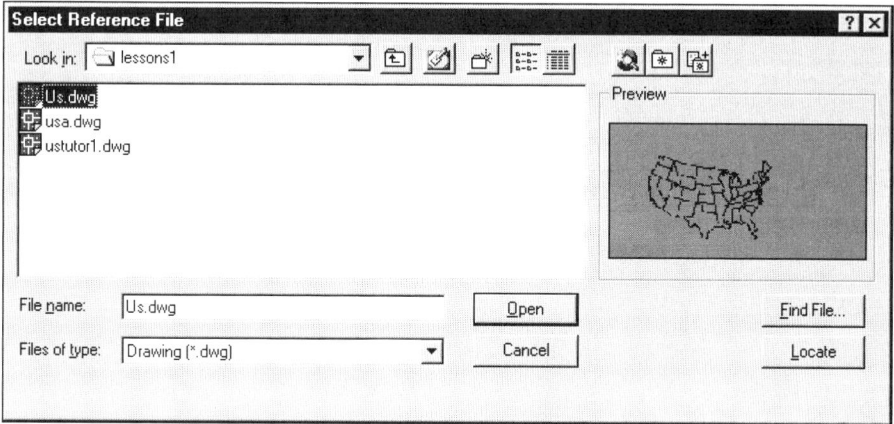

Figure 10–2

Step 3 The External Reference dialog box will appear.

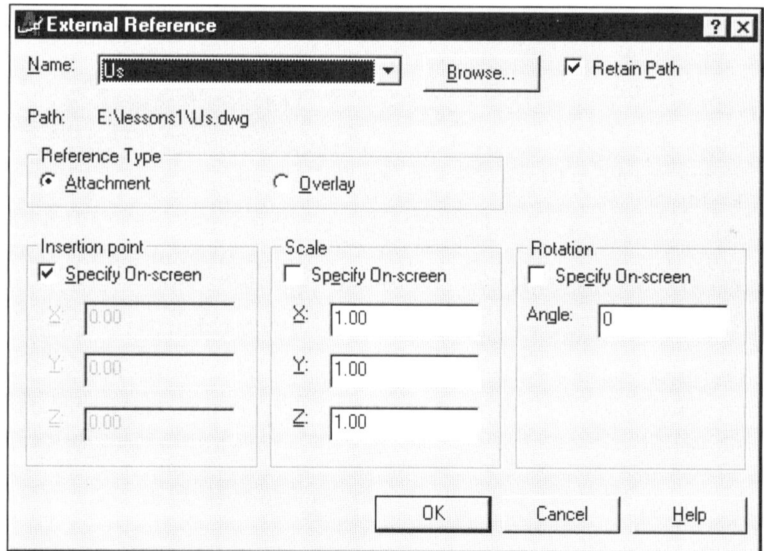

Figure 10–3

Leave the reference type as **Attachment**. **Deselect** the Specify On-screen option for both the Insertion Point and Scale. Click on **OK** to proceed. The US map will be attached to the Canada map.

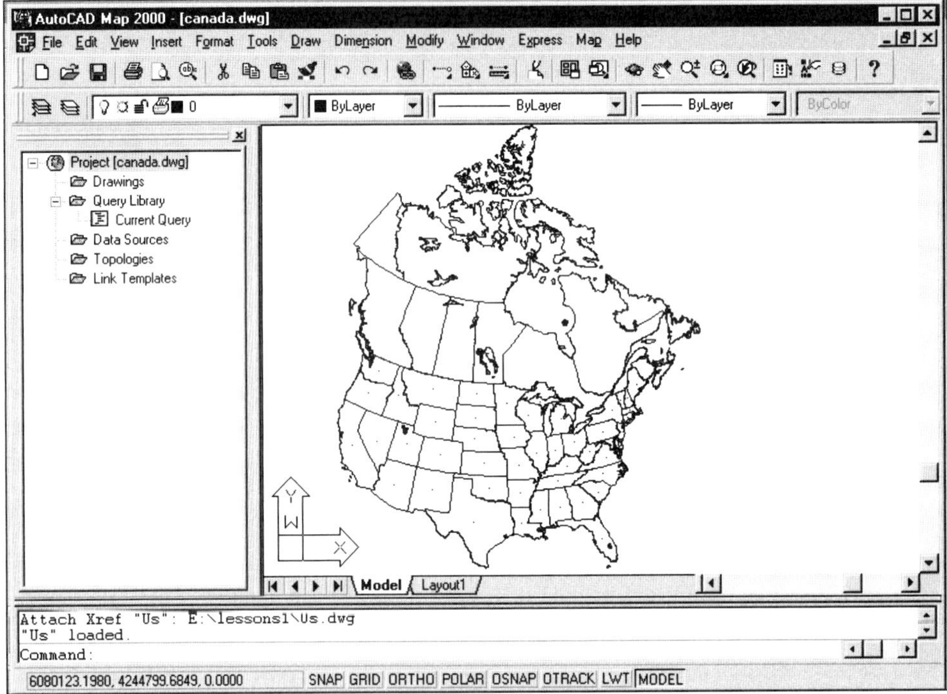

Figure 10–4

Step 4 Save this map as CANUSA. With this drawing active use the **Map** and **Define/Modify Drawing Set** menu selection to bring up the Select Drawings to Attach dialog box and **attach** the **COAL**, **IRON**, and **LIMESTONE** drawings.

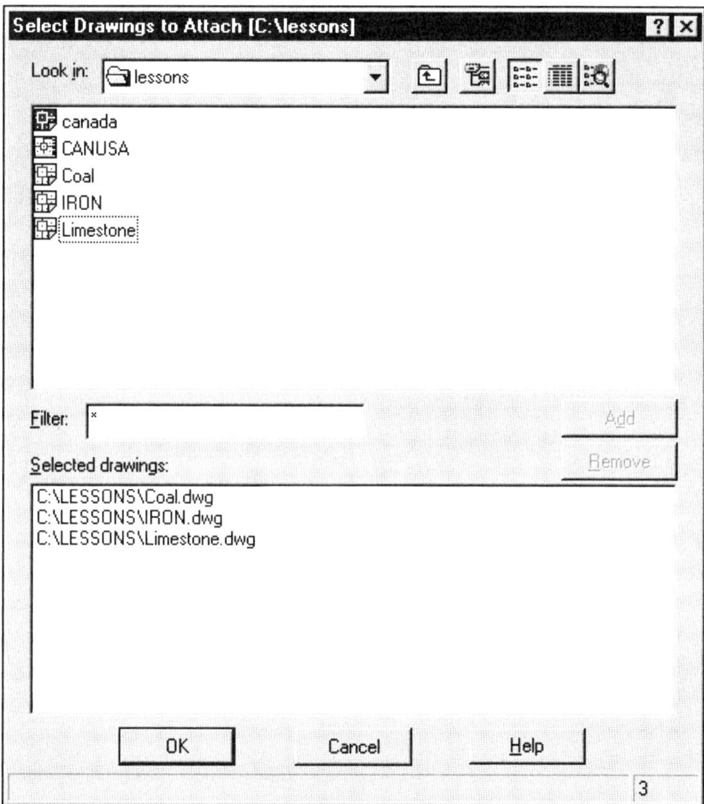

Figure 10–5

Step 5 Select **Map** then **Drawings** and, **Quick View Drawings**. The Quick View Drawings dialog box will appear. Select **Coal, Limestone** and **Iron** as the drawings to view then click on **OK.**

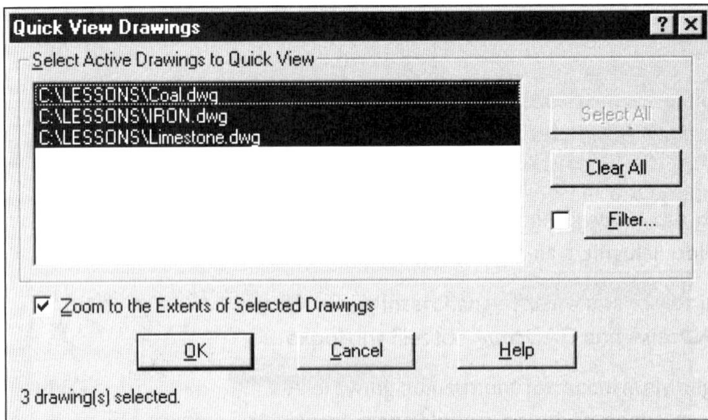

Figure 10–6

The composite map shown reveals lines of equal distance from three sources of raw materials—iron, coal, and limestone—used in steel manufacturing. Each ring, or isotim (a line joining points of equal transportation cost), represents 200 km or 125 miles. If each concentric ring represents 1 unit of cost based on transportation costs for raw materials being generally related to distance, one can determine the optimal location for a steel processing plant based entirely on the cost associated with moving the raw materials.

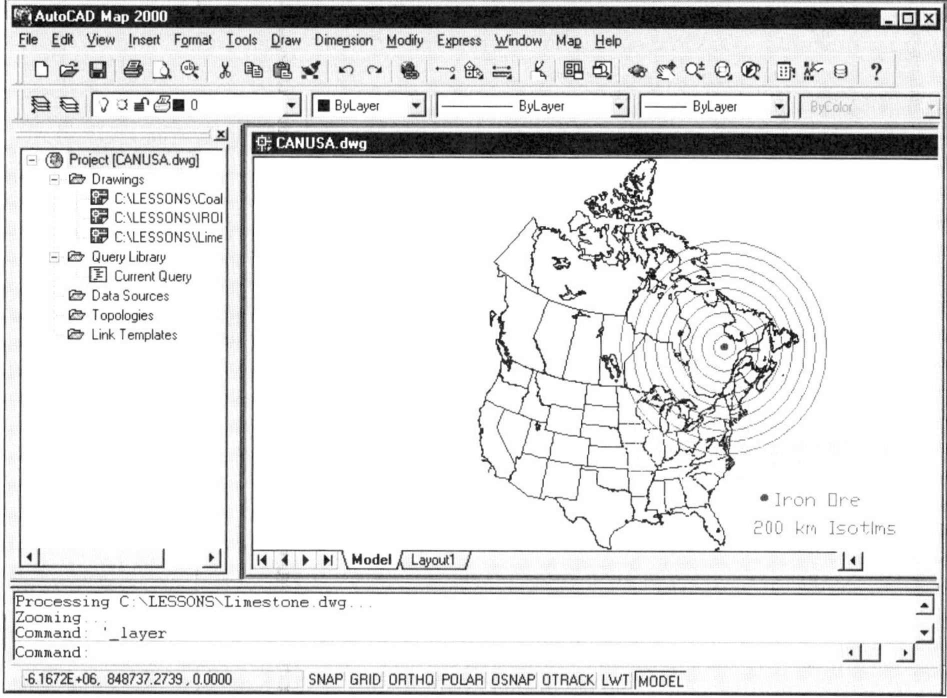

Figure 10–7

Step 6 Count rings from each material site and add up the total of all rings at intersection points of the isotim lines.

Step 7 Identify the area with the lowest total as the area to which the cost of moving the raw materials is the least.

Step 8 Draw a closed polygon around the region and hatch it. Print the map, and then compare it to known locations for steel production. The Toronto-Hamilton area on the west end of Lake Ontario should be close to the area you identified. Either exit AutoCAD Map 2000 or continue with the tutorials.

SUMMARY

The selection of the best site for a specific land use is a very common requirement for a GIS technician. In this example, factors like the location of raw materials and transportation costs were considered in order to locate the most appropriate site for a steel mill. In a real application, many more variables, such as land values and distribution routes may have to be considered. The information associated with these variables might be accessed from the base drawing as an external reference or through an attached drawing file.

Facilities Management

OBJECTIVE

After completing this chapter you will know how to

- conduct a facilities management investigation

This tutorial exemplifies the use of building layout drawings, which have simple features identified with object data or hatching to assist in the decision-making processes.

In this hypothetical case, a school is being used by a seniors group to explore the Internet. Some of the participants require wheelchair access. The drawings provided include information regarding the location of access ramps, Internet-equipped computer rooms, backbone lines for present and future links, and rooms that can be used for this purpose at the present and in the near future.

The task is to determine, through the process of accessing several drawings, the best location to be assigned to this group and to assess whether or not this location will change over the next few years based on the long-range planning of the school.

TUTORIAL: ROOM UTILIZATION

Step 1 Open the **BUILDING.DWG** file, which shows the general layout of the building.

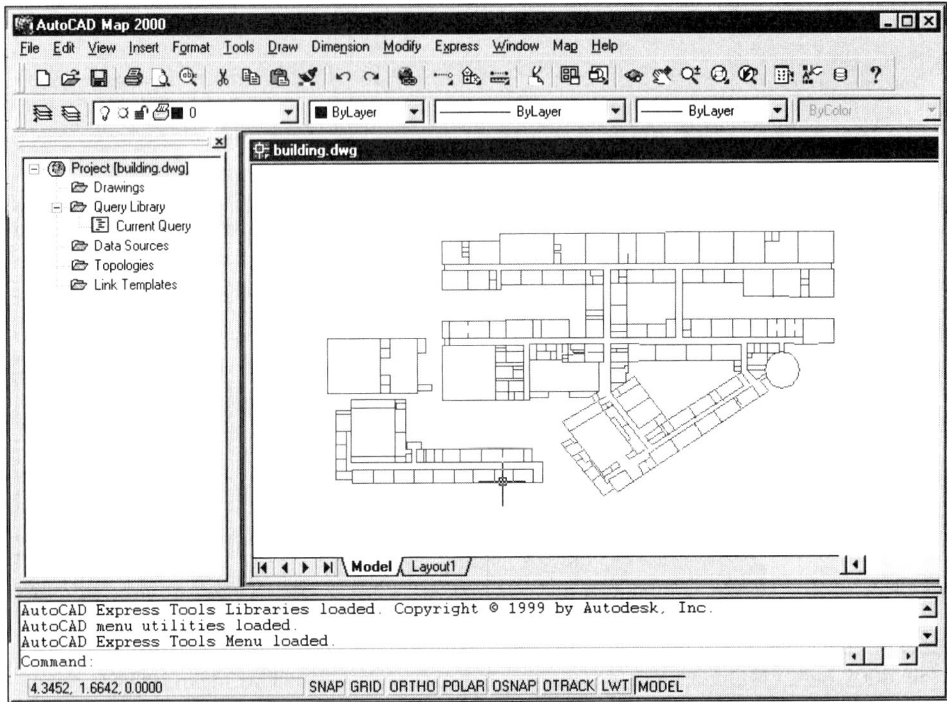

Figure 11–1

Step 2 **Attach** the **Backbone**, **Internet Ready**, and **Wheelchair** drawings by selecting **Map** then **Drawings** and **Define/Modify Drawing Set** and highlighting the drawings from the Select Drawings to Attach dialog box.

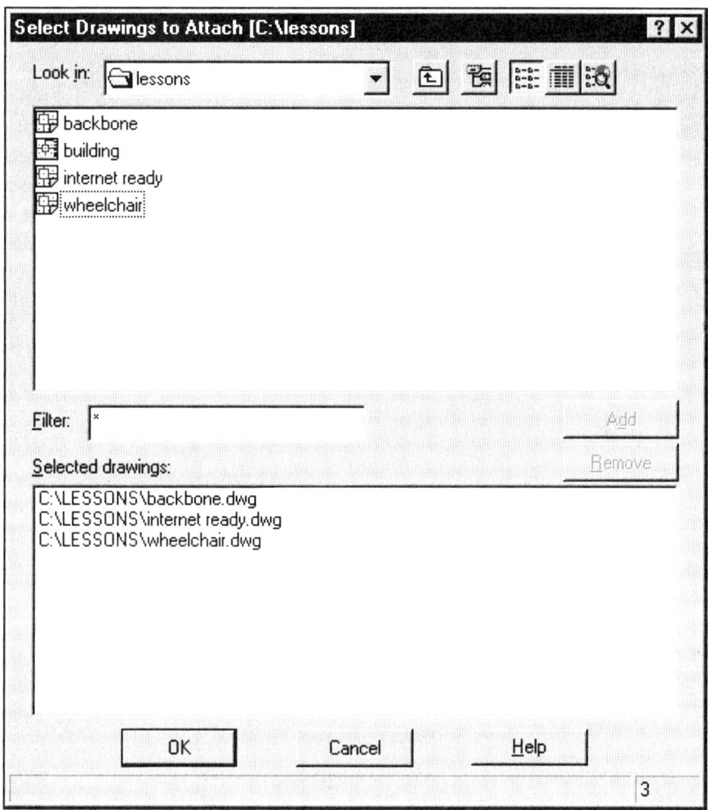

Figure 11–2

Step 3 Go to the **Map** menu and select **Query** then **Define Query**.

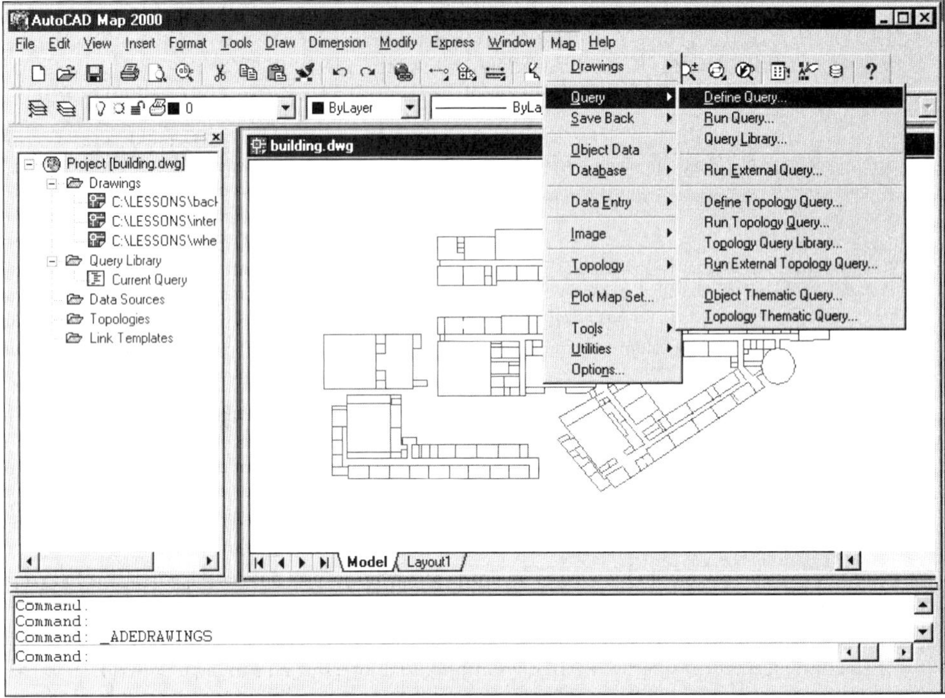

Figure 11–3

Step 4 Set the location to **Circle** and **Inside** then **Define** the entire school area.

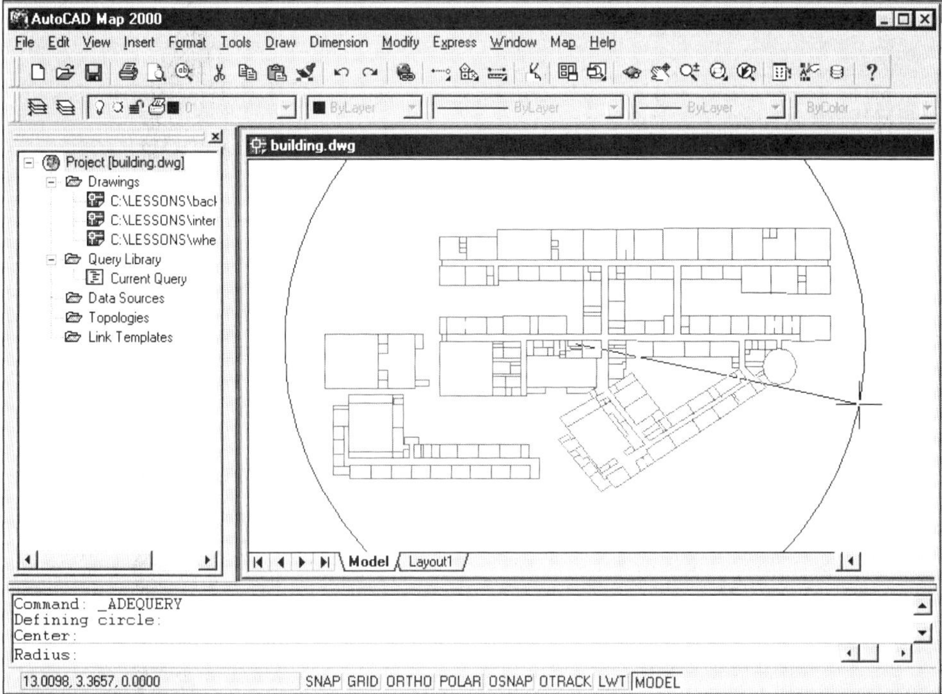

Figure 11–4

Step 5 At the Define Query dialog box select the **Property** item, and at the Property Condition dialog box select the **Layer** option.

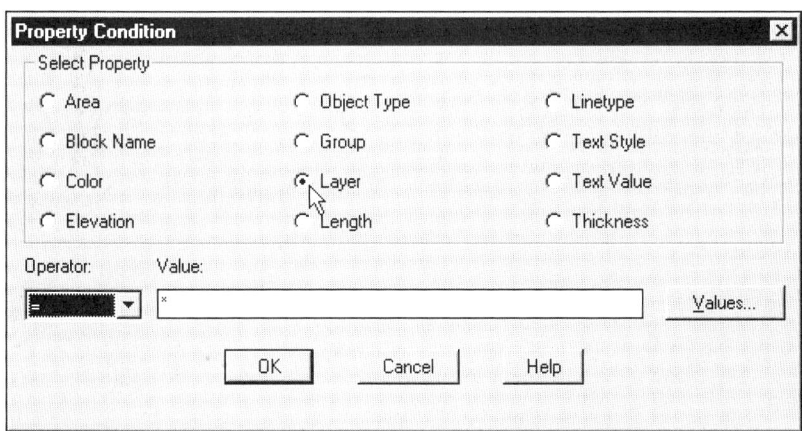

Figure 11–5

Step 6 Click on the **Values** button, highlight the **Wheelchairaccess** item, and click on **OK**. The Property Condition dialog box is presented with the **Operator = WHEELCHAIRACCESS**. Click on **OK** to accept.

Property Condition
Select Property
Area
Object Type
Linetype
Block Name
Group
Text Style
Color
Layer
Text Value
Elevation
Length
Thickness
Operator:
Value:
=
WHEELCHAIRACCESS
Values...
OK
Cancel
Help

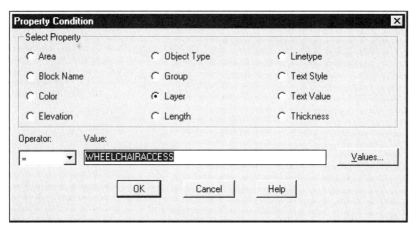

Figure 11–6

The Define Query dialog box will reappear, showing the query as it has been defined.

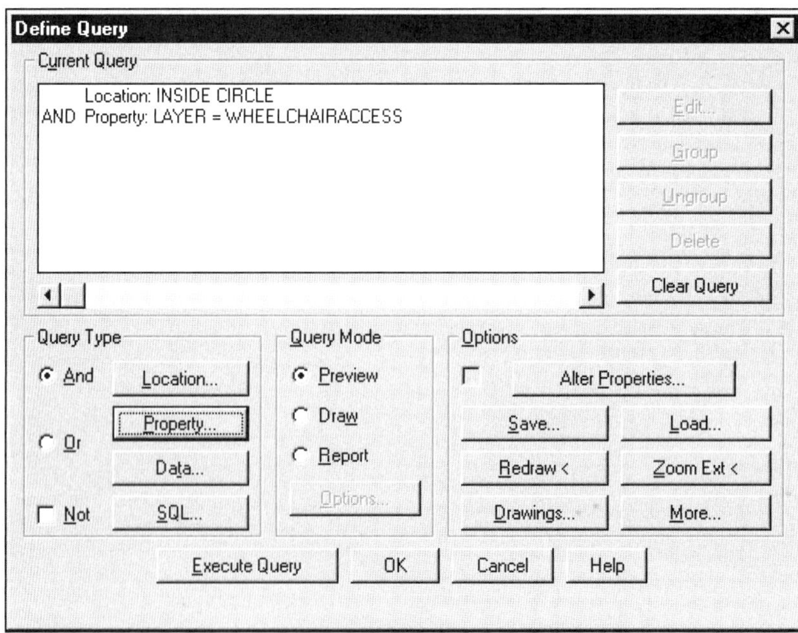

Figure 11–7

Click on the **Draw** radio button under the Query Mode section, and then click on **Execute Query**. There will be two arrows drawn on the map to indicate where wheelchair access is feasible.

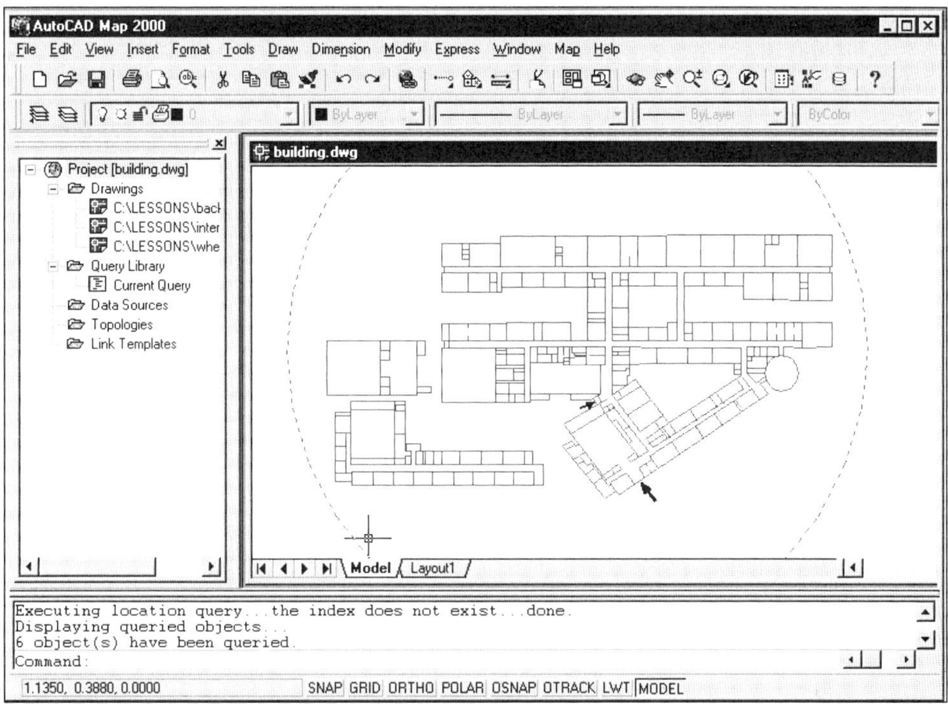

Figure 11–8

Step 7 Select **Map** then **Query**, and **Define Query** again, and at the Define Query dialog box, click on the **WHEELCHAIRACCESS** item.

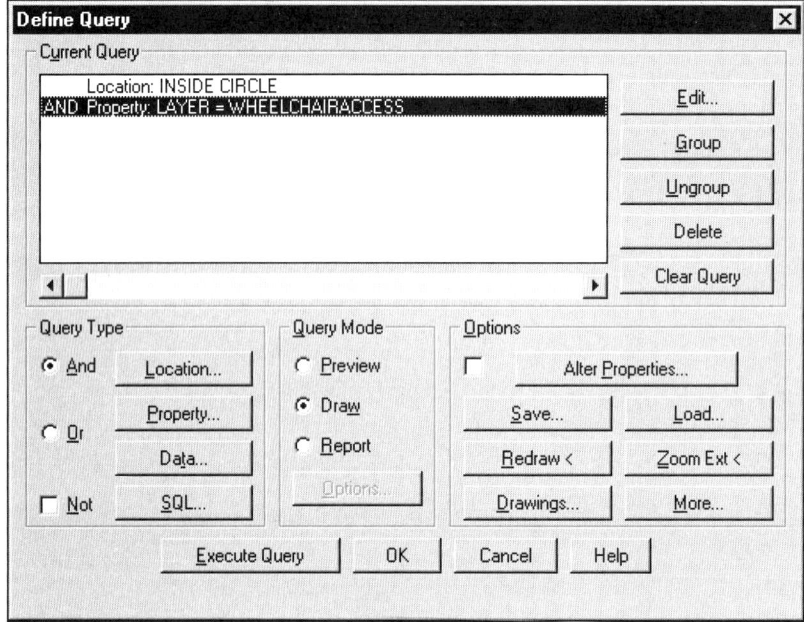

Figure 11–9

Click on **Delete** to remove this item from the query definition.

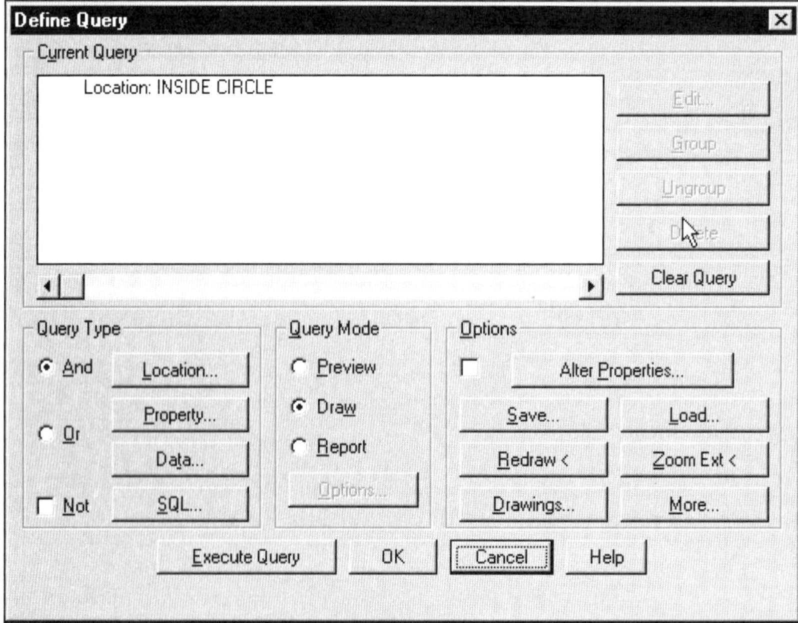

Figure 11–10

Step 8 Click on **Property** again, then on **Values**. At the next screen, select the layer **INTERNETNOW** and click on **OK**.

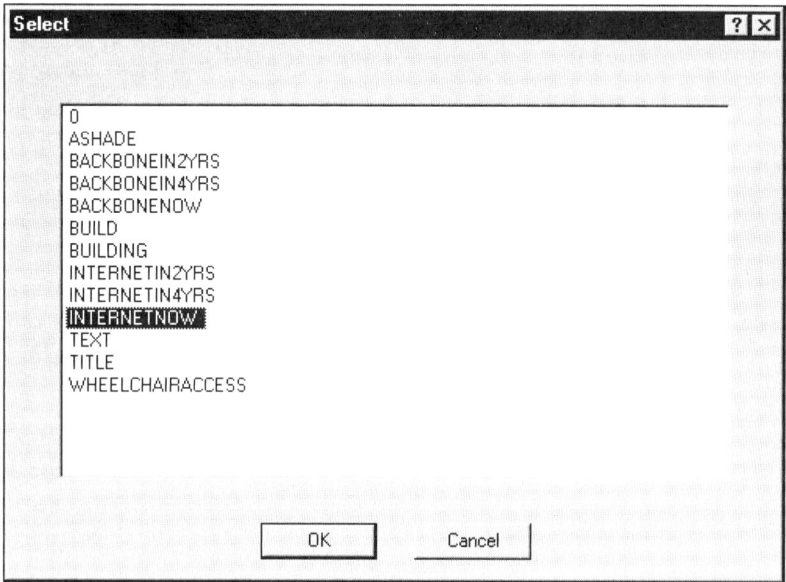

Figure 11–11

The Property Condition dialog box, with the Operator set at = and the Value established as **INTERNETNOW**, will be presented. Click on **OK**.

Step 9 The Define Query dialog box will reappear, showing the layer to be queried.

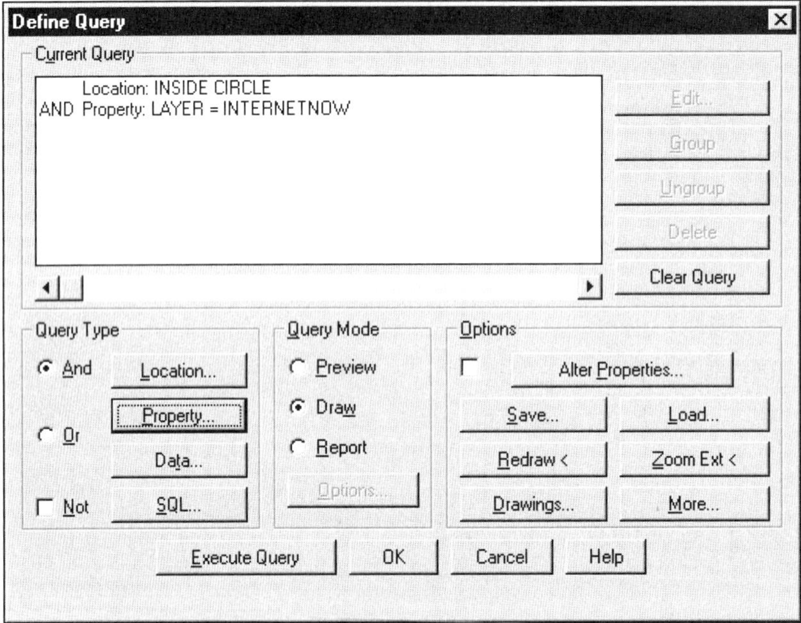

Figure 11–12

Click on the **Draw** radio button to select it, and then click on **Execute Query**. The rooms in the building with Internet access will be identified.

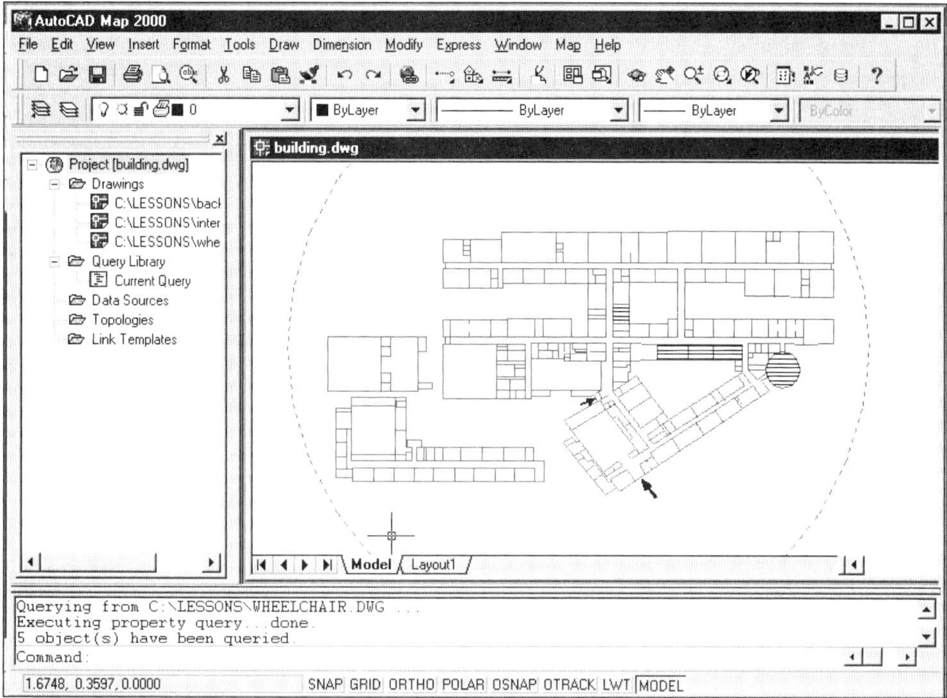

Figure 11–13

Go to the **Layer Control** icon and turn the **text** layer on to identify the Internet-equipped room that is closest to a wheelchair access point. You may wish to zoom in to read the room numbers.

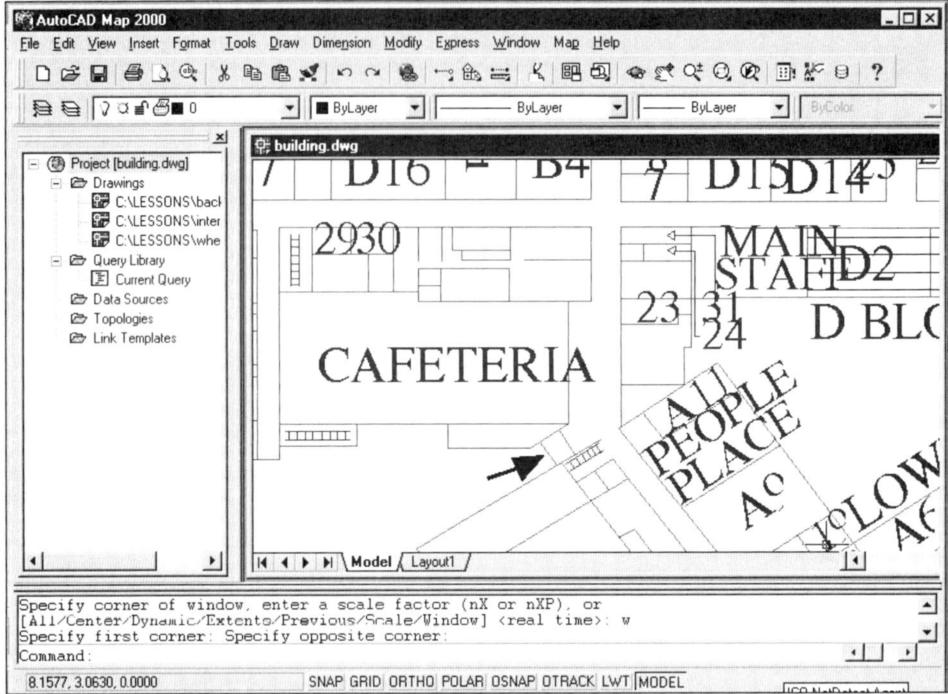

Figure 11–14

The following exercise is an enhanced version of the tutorial just completed.

FACILITIES MANAGEMENT EXERCISE

Use the drawings provided to determine which room would be closest to a wheelchair access point in two years, after upgrades are made to the building and after four years, when even more Internet accessibility is to be available.

(This exercise is an extension of the tutorial using provided files)

SUMMARY

Almost all enterprises and public agencies are in the process of creating digital versions of their old paper floor plan drawings. The availability of this type of spatial data will significantly increase the use of GIS technology in facilities management applications. The tutorial and reinforcement exercise provide examples of how such floor plan drawings, when enriched with data, can be used to provide precise solutions to complex questions.

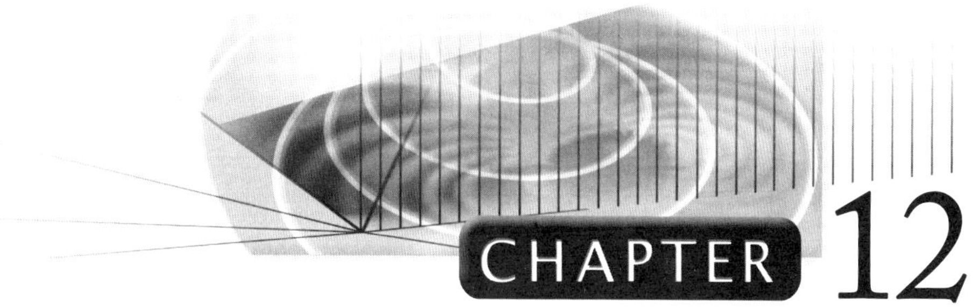

Optimal Business Delivery Routing

OBJECTIVE

After completing this exercise you will know how to

- utilize network topology in conjunction with external database queries to assist in decision making

This tutorial is designed to illustrate how you can use AutoCAD Map 2000 to select the shortest delivery route for a set of customers. Also included in this tutorial is an example of how to automatically identify which customers should be serviced based on a time-related service cycle. Dates in the form of month numbers for the most recent service are provided in an Access table that will be attached to the drawing and then queried to determine which customers require service. This table also contains data regarding the customer account balances, which can be useful in having the service agent "drop by" the delinquent account clients if they were along the route being followed.

The first part of this tutorial has you identify the shortest route to follow to provide service to four client sites identified as A, B, C, and D. Prior to starting this tutorial, you should ensure that the REGION2.DWG, and REGIONBASE.DWG files have been copied into a subdirectory of AutoCAD Map 2000.

The CUSTOMER1.MDB file was added to the Samples directory of OFFICE, identified by the path statement:

C:\progra~1\micros~2\office\samples\customer1.mdb

TUTORIAL: ROUTE DETERMINATION

Step 1 Start AutoCAD Map 2000 and open the **REGION2.DWG** file. This drawing reveals client locations and identifies each by number. This file has undergone a Drawing Cleanup as well as having a Network Topology, identified as Alroads, created. Zoom in on the central portion of the map to enable better visibility of the client sites A, B, C, and D.

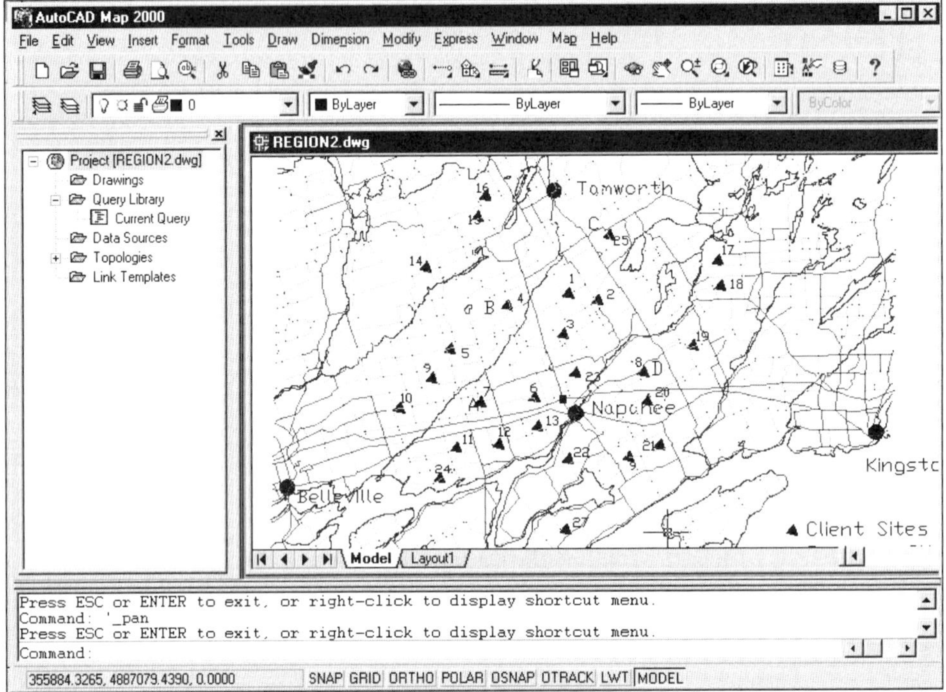

Figure 12–1

Step 2 Go to the **Map** menu and select **Topology** and **Administration**.

Step 3 At the Administration dialog box click on **Load** and choose **Alroads** as the topology to load from the Load Topology dialog box.

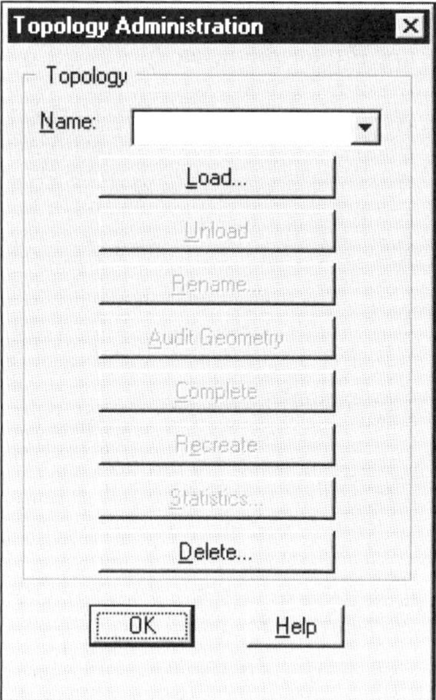

Figure 12–2

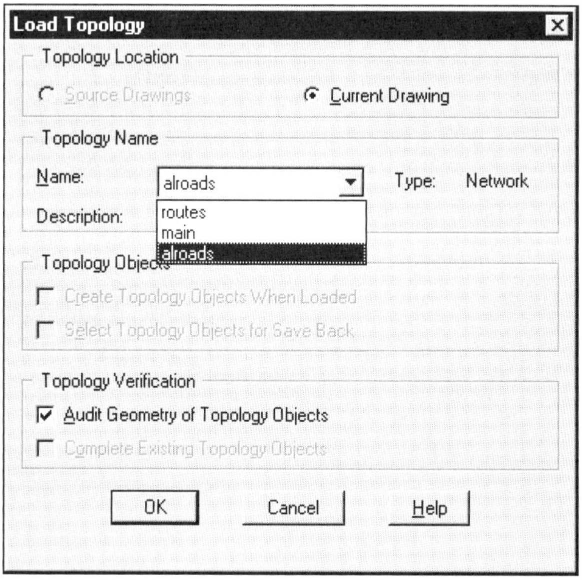

Figure 12–3

Click on **OK**. The topology will be verified, and a message informing you that the topology files are intact will appear.

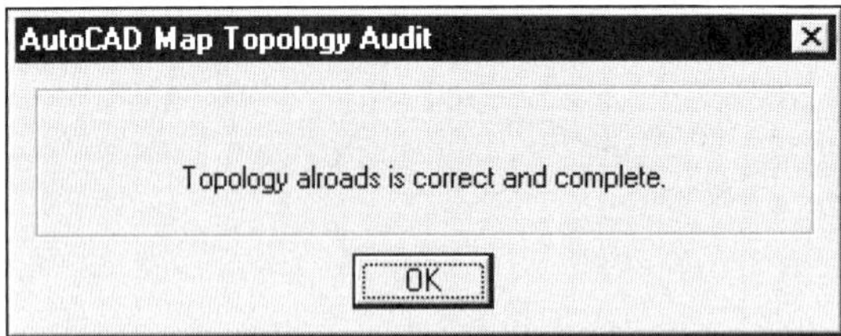

Figure 12–4

Click on **OK** and at the Topology Administration dialog box click on **OK** again.

Step 4 You will now conduct a Path Trace analysis between the points that include the business site and each of the four clients then back to the business site. Go to the **Map** menu and select **Topology** then **Path Trace.**

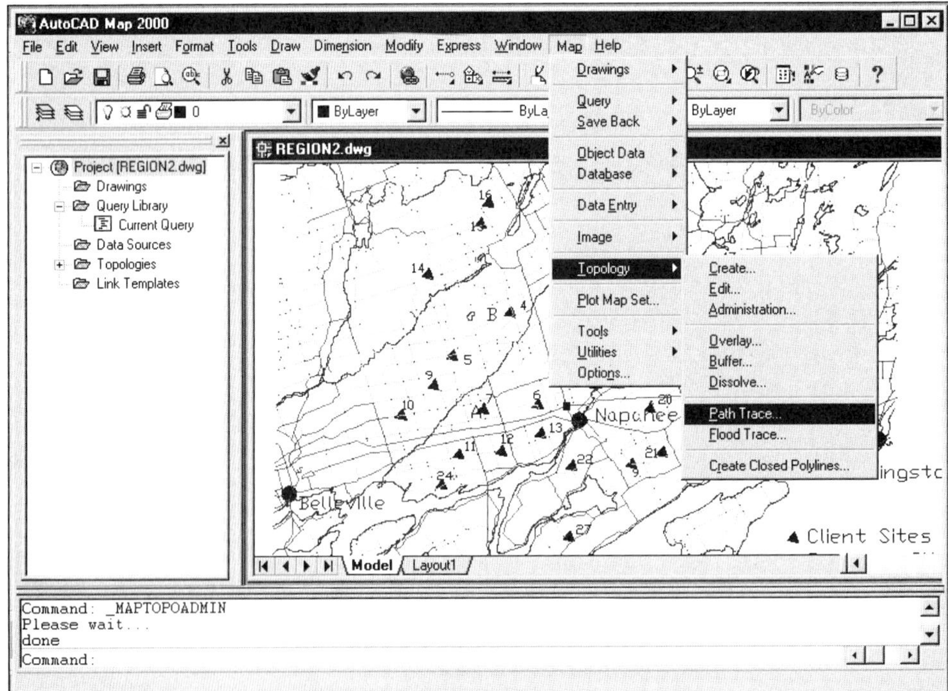

Figure 12–5

The Shortest Path Trace dialog box will appear showing the Alroads topology as the active topology.

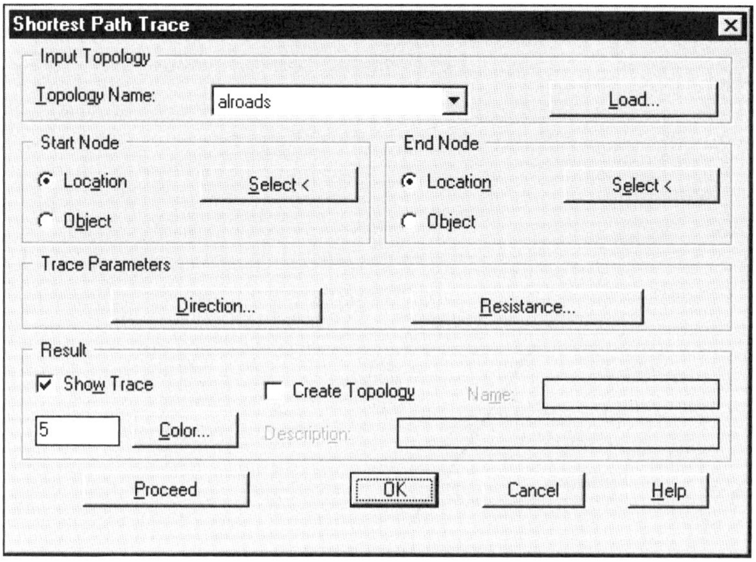

Figure 12–6

Click on the **Select** button under **Start Node** and highlight the Business Site. A small marker will appear on the map indicating the location of the start node. Click the right mouse button to indicate that you have finished selecting objects. Click on the **Select** button under **End Node**, then highlight the client site identified by A. Once again, a marker will appear on the map verifying your selection. Click the right mouse button or press the ENTER key to indicate that you have finished selecting the objects and to return to the Shortest Path Trace dialog box. Select the color you want to have the route displayed in, then click on **Proceed**. The shortest route will be presented.

Step 5 Repeat the Shortest Path Trace in Step 4 above, from A to B, B to C, C to D, and D to the business site to obtain the shortest route for the trip. The route line in Figure 12–7 below has been enhanced for presentation purposes.

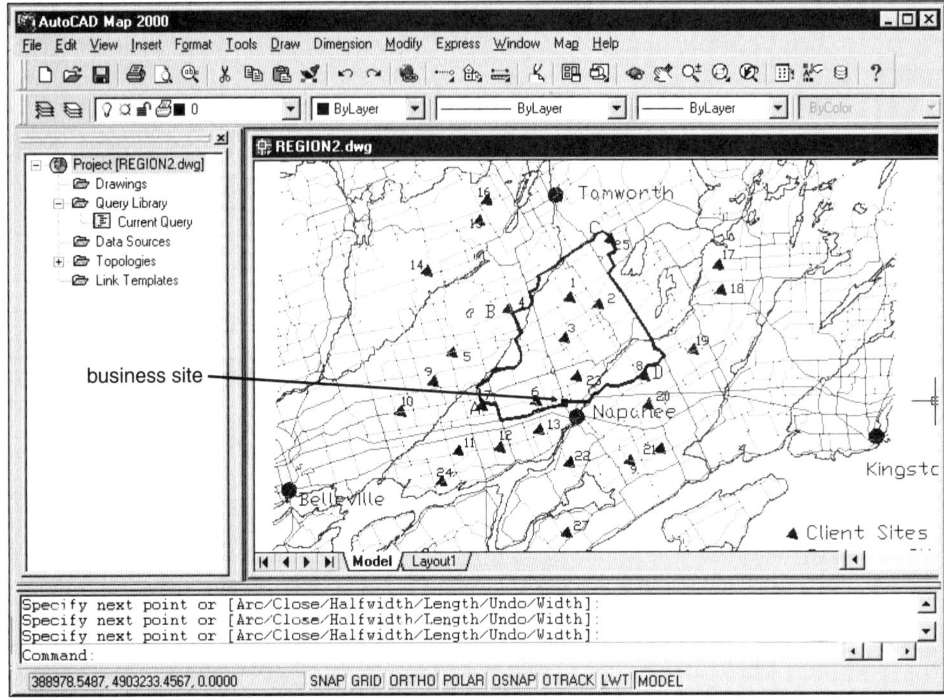

Figure 12–7

Exit AutoCAD Map 2000. There is no need to save the drawing changes.

Step 6 The next component of the tutorial is to determine if any clients along the route have an outstanding balance on their account. The data for the business is stored in Access database format. The file is accessible as an External Database file. An earlier tutorial established the ODBC drivers for Access. For this tutorial you need to set the location of the new file for the ODBC drivers. Activate the **Control Panel** and select **ODBC**. From the list of Data Sources select **ODBC_Access** and click on **Configure**.

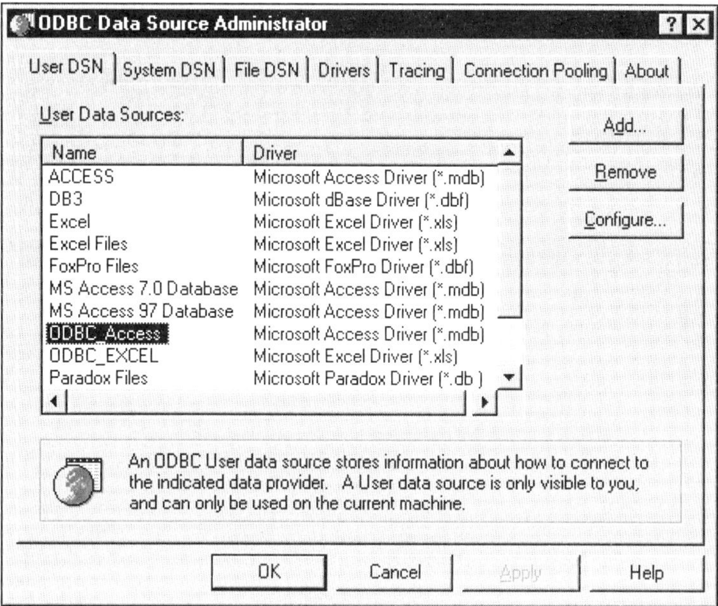

Figure 12–8

Use the **Select** feature to provide a path to the **CUSTOMER1.MDB** file.

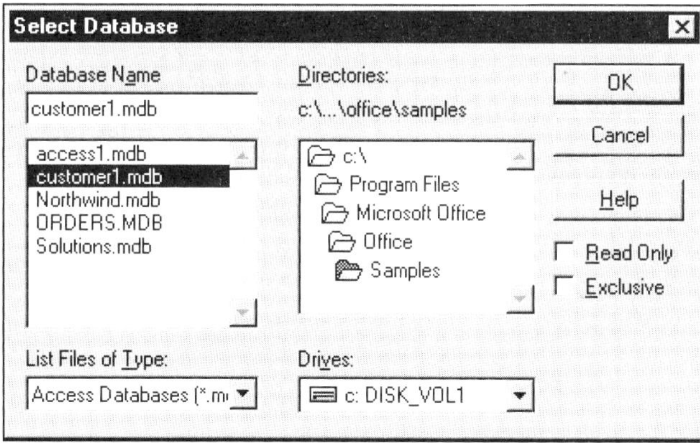

Figure 12–9

Click on **OK** to have the path registered to the database file.

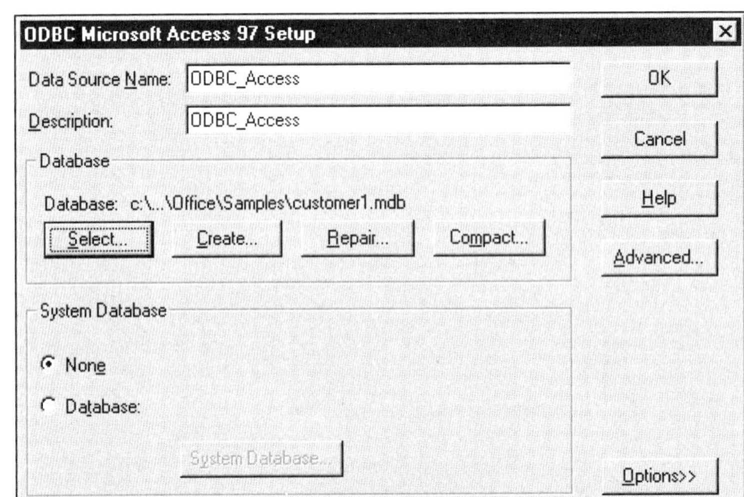

Figure 12–10

Click on **OK** again to return to the ODBC Data Source Administrator dialog box and on **OK** once more . Close the Control Panel.

Step 7 Start AutoCAD Map 2000 and open the **REGION2.DWG** file. Use the **Map** menu and select **Database,** then **Data Sources,** and **Configure**.

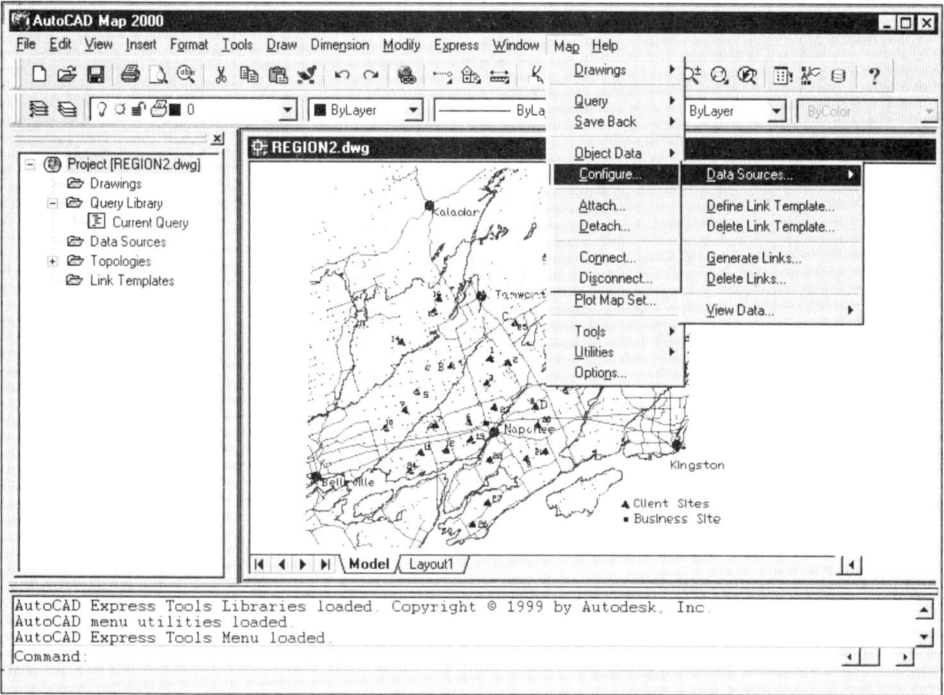

Figure 12–11

In the Data Source Name text box of the **Configure Data Source** dialog box, type in **ODBC_Access** (if ODBC_Access is listed, then select it from the list) and click on **OK**. The Data Link Properties dialog box will appear. In section 1 of the **Connection** tab use the pull-down menu to select **ODBC_Access** from the list provided.

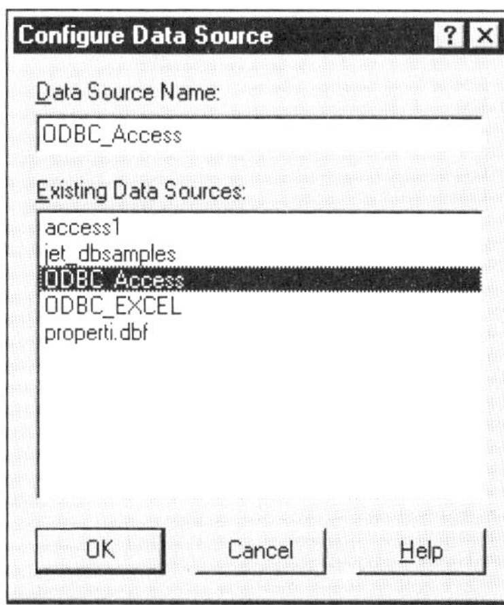

Figure 12–12

If you are working on a network and require passwords and usernames to log into the data provide this information in section 2. In section 3 of the Data Link Properties dialog box on the Connection tab, use the pull down-menu to select the database and its location.

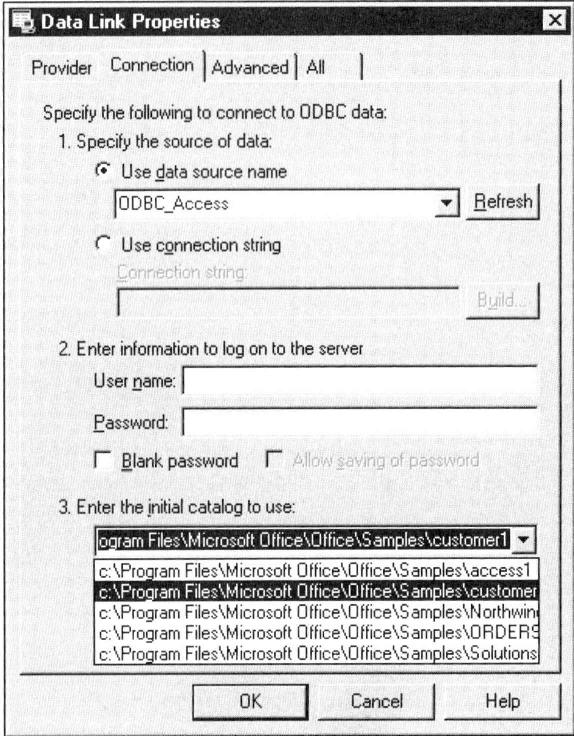

Figure 12–13

To verify that AutoCAD Map 2000 will be able to attach this data to the drawing select the **Test Connection** button. A message will indicate the successful outcome of the test.

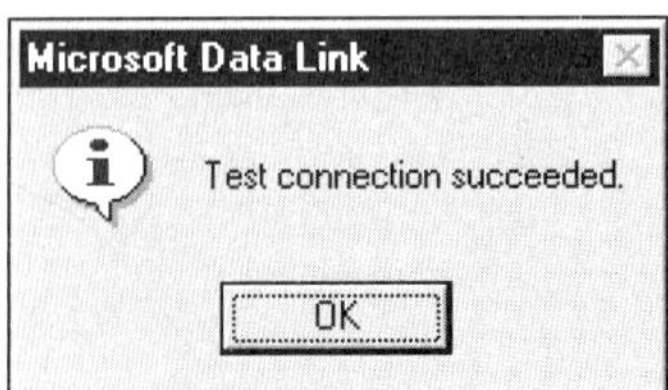

Figure 12–14

Click on **OK** to close the Data Link Properties dialog box.

Step 8 Go to the **Map** menu and select **Database** then, **Data Sources,** and **Attach**.

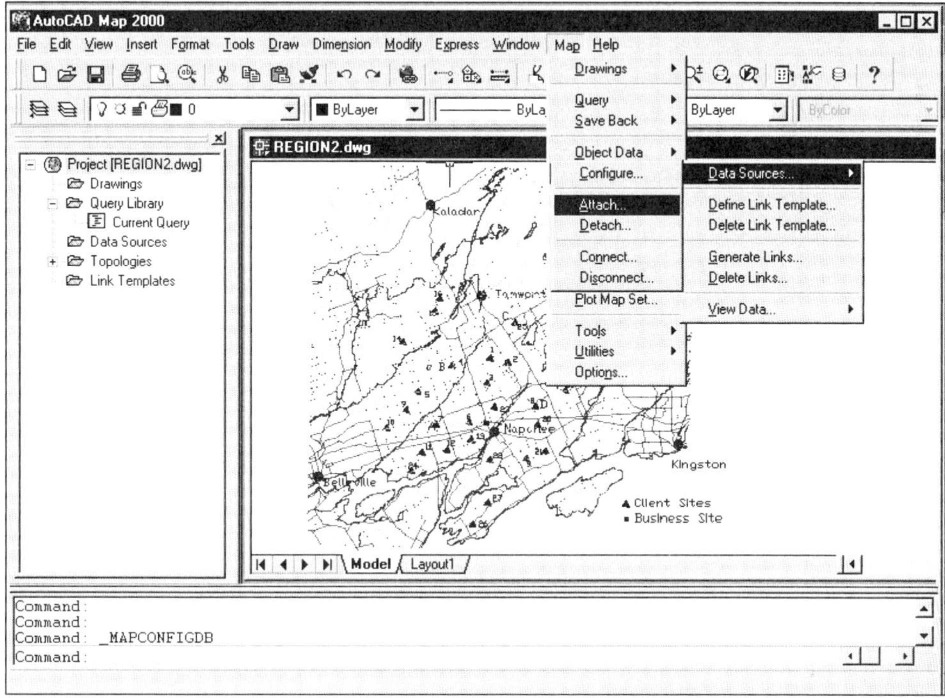

Figure 12–15

The Attach Data Source dialog box is presented. From the list of options select the **ODBC_Access.udl** item. Click on **Open**.

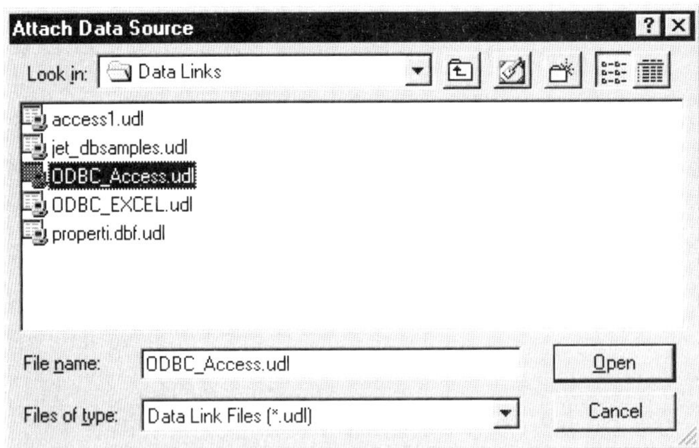

Figure 12–16

The attachment of the database will be indicated in the Work Session information area of the AutoCAD Map 2000 screen.

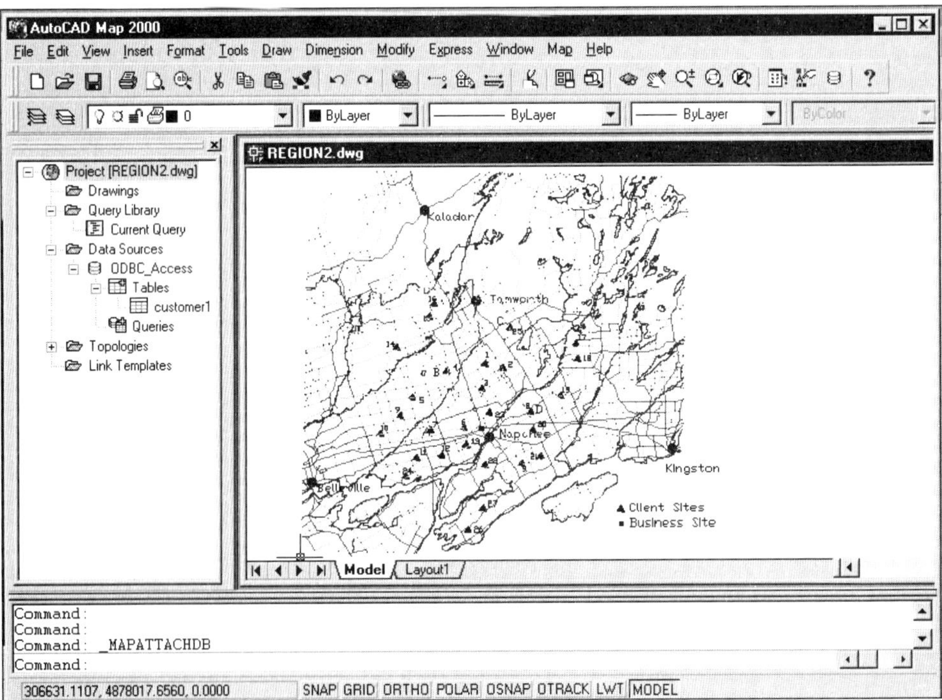

Figure 12–17

Step 9 To view the data in the database from within AutoCAD Map 2000, go to the **Map** menu and select **Database** then **View Data** and **View Table**.

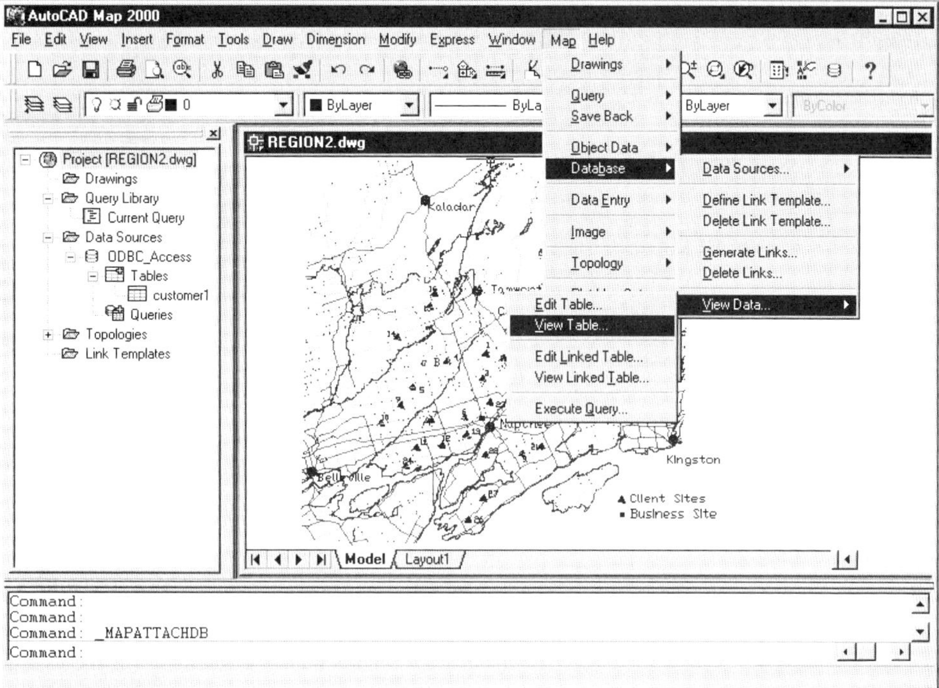

Figure 12–18

The Browse Database Table dialog box will permit you to select the table you want to view. In this case, select **ODBC_Access** as the database and customer1 as the table.

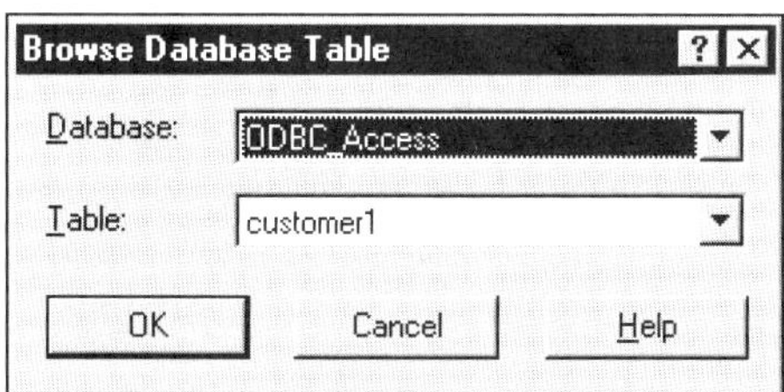

Figure 12–19

Click on **OK** and the table will appear.

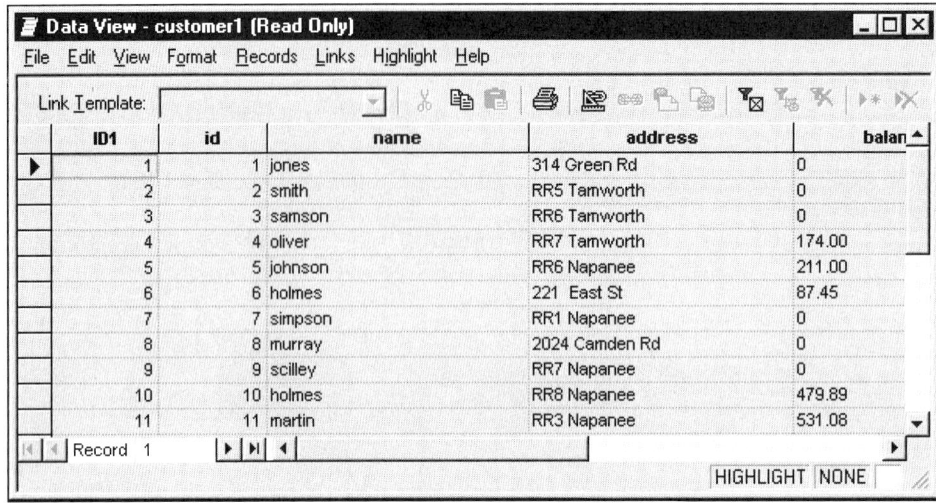

Data View - customer1 (Read Only)

ID1	id	name	address	balan
1	1	jones	314 Green Rd	0
2	2	smith	RR5 Tamworth	0
3	3	samson	RR6 Tamworth	0
4	4	oliver	RR7 Tamworth	174.00
5	5	johnson	RR6 Napanee	211.00
6	6	holmes	221 East St	87.45
7	7	simpson	RR1 Napanee	0
8	8	murray	2024 Camden Rd	0
9	9	scilley	RR7 Napanee	0
10	10	holmes	RR8 Napanee	479.89
11	11	martin	RR3 Napanee	531.08

Figure 12–20

Step 10 To establish a query go to the **Tools** menu and select **dbConnect**.

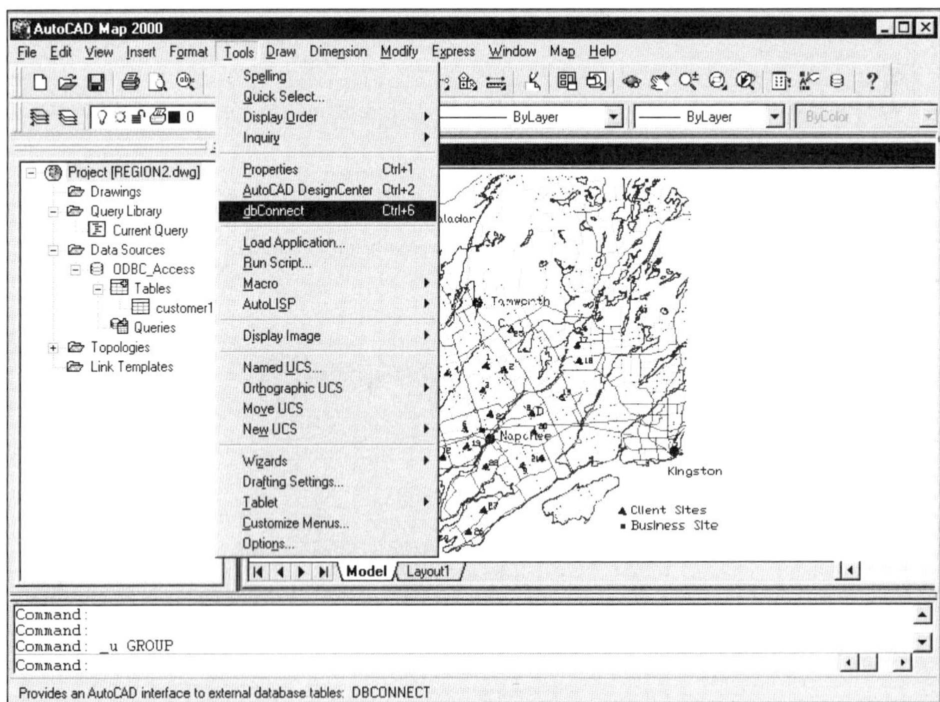

Figure 12–21

This will introduce a new menu item, dbConnect to the main Map screen as well as a second project information section to the screen.

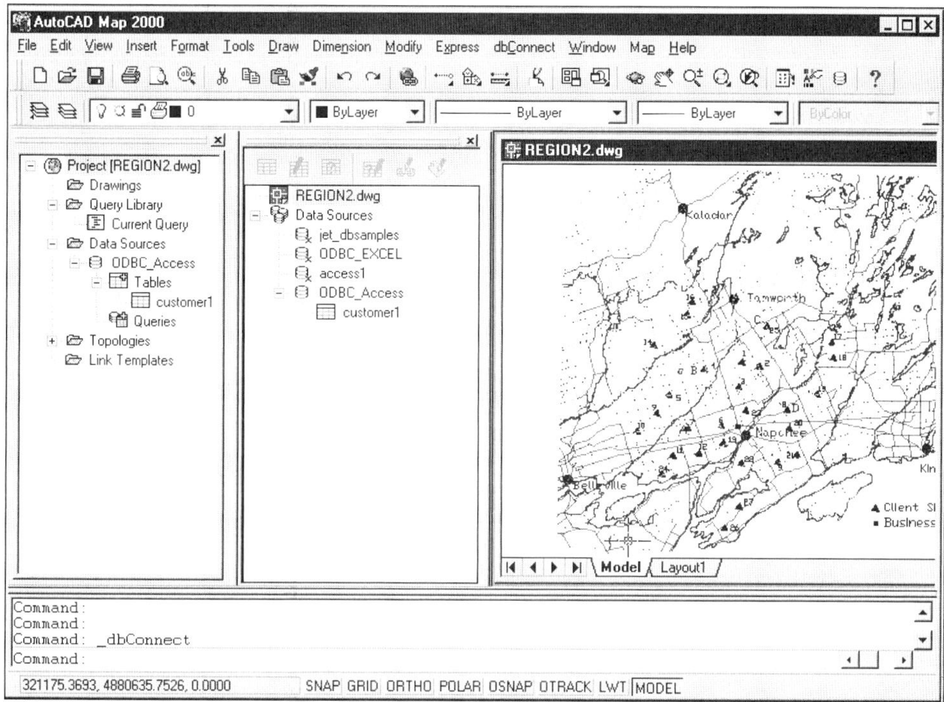

Figure 12–22

Step 11 To run a query on the data, you first have to build a link to the database. This is referred to as a Link Template. Go to the **Map** menu and select **Database,** and **Define Link Template**.

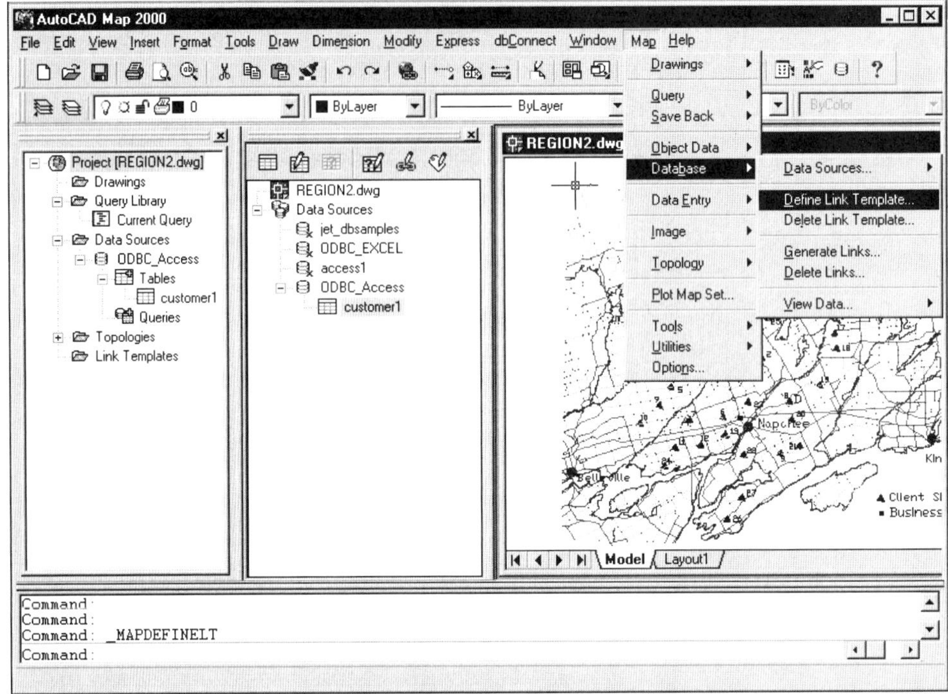

Figure 12–23

The Define Link Template dialog box will appear. Type in a link name, in this case **Link1,** and select a key field. In this example, the **ID1** column was chosen. Click on **OK**.

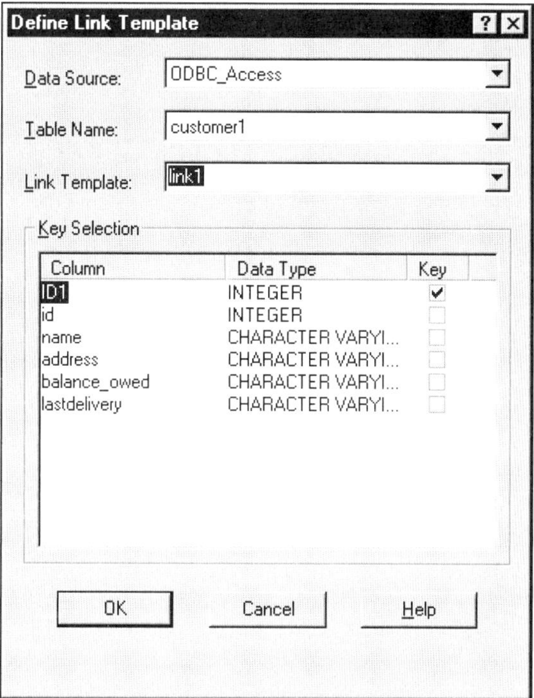

Figure 12–24

Now the Link Template is used to generate the links between the data table and the features in the drawing. Go to the **Map** menu and select **Database,** and **Generate Links**.

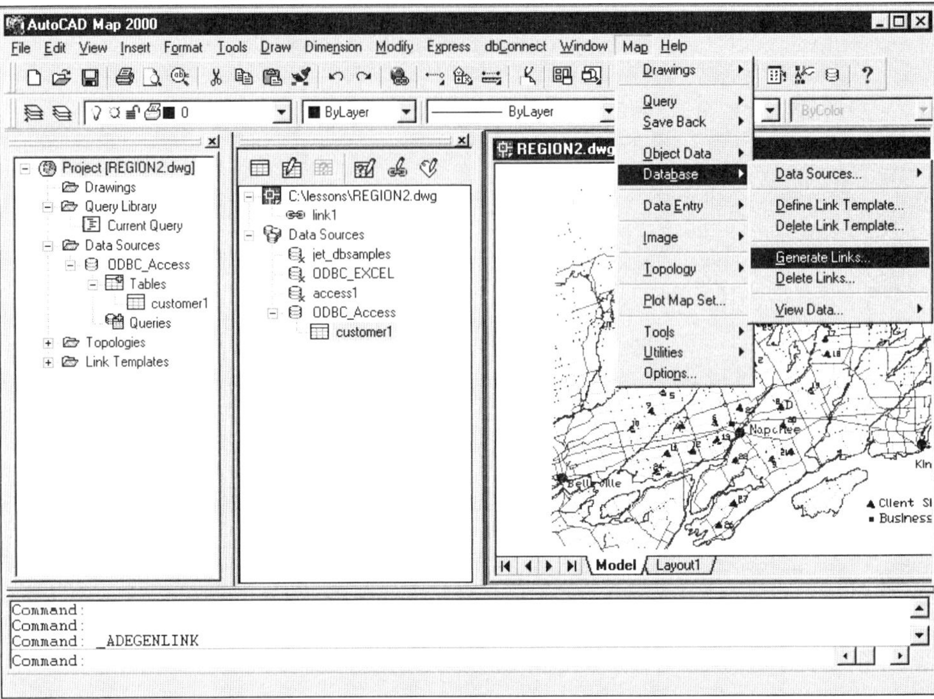

Figure 12–25

In the Generate Data Links dialog box, you set the Linkage Type. In this instance, the linkage will be made through the identifying text (numbers) of the customers. Select the **Text** option for the Linkage Type. In the Data Links section, choose the **Create Database Links** option by selecting the radio button, and use the **Link1** template. In the Database Validation section, select the radio button for the **Links Must Exist** option. Click on **OK**.

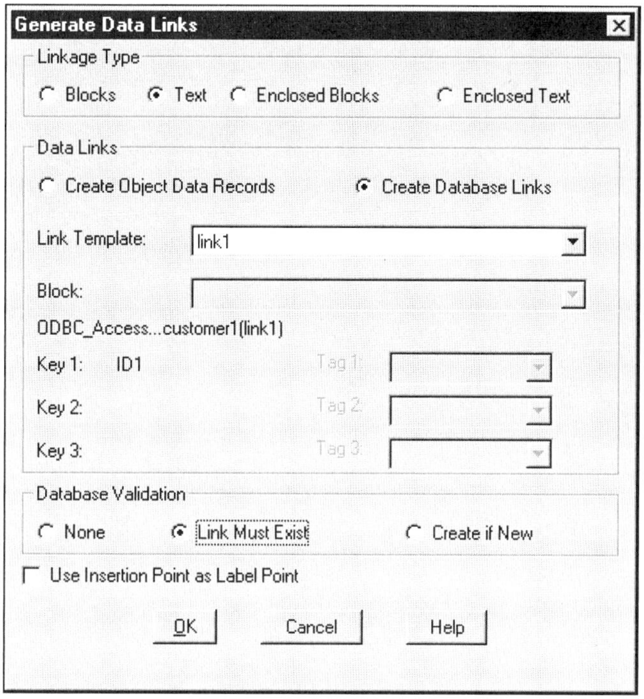

Figure 12–26

In the command prompt area of the AutoCAD Map 2000 screen you will be prompted to select the objects for which the links are to be created. Choose the **All** option and press ENTER. Links are created for the objects. Some text objects not in the database table will be rejected.

Step 12 To run a query, go to the dbConnect menu and select Queries, and New **Query on an External DatabaseTable.**

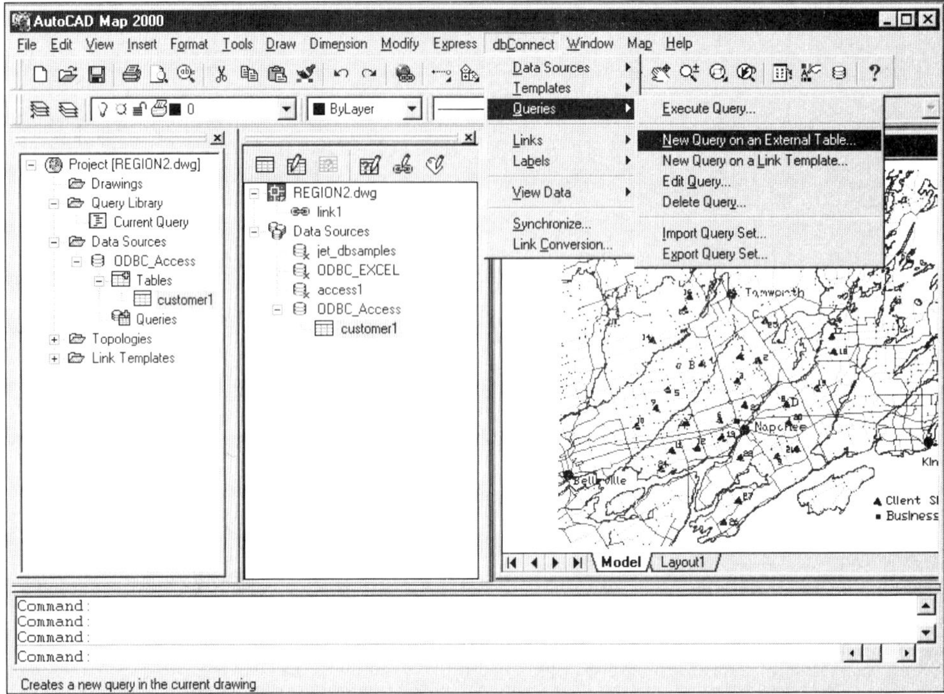

Figure 12–27

In the Select Data Object dialog box select the **customer1b** table as the table to use for the query and click on **Continue**.

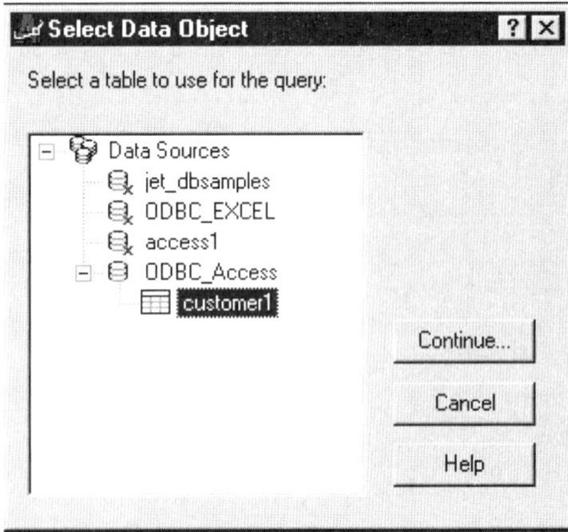

Figure 12–28

In the New Query dialog box ensure that a name is given to the query. In this case the name was **customer1Query1**. Click on **Continue**.

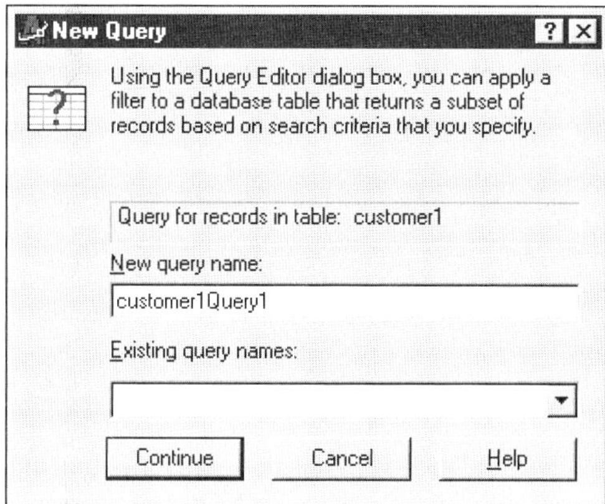

Figure 12–29

The Query Editor dialog box is presented. To search for customers that have outstanding account balances, for example, conduct a query as follows. In the Field category select the **balance_owed** item, and for the operator choose **>Greater than** and for a value type in **b**. At the bottom of the screen, select both checkboxes for **Indicate Records in Data View** and **Indicate Records in Drawing**.

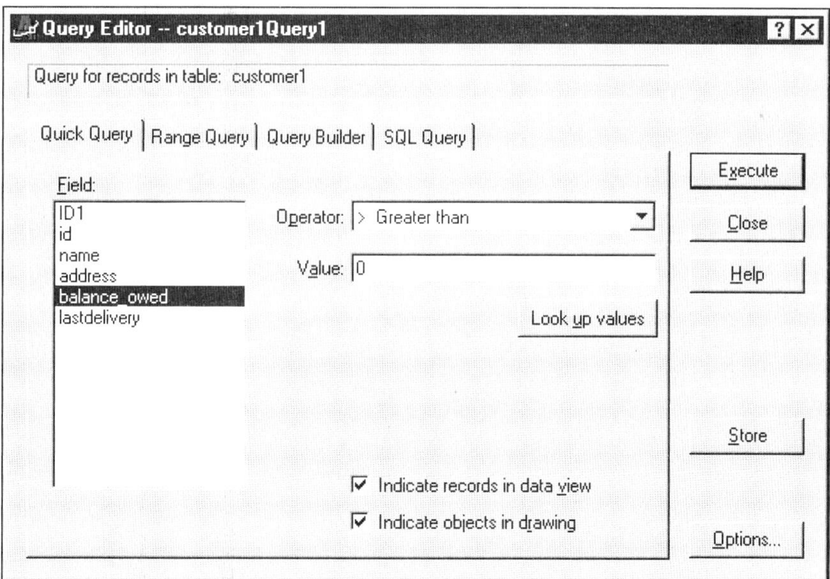

Figure 12–30

Click on **Execute** and the results of the query will be presented in tabular form as well as on the map.

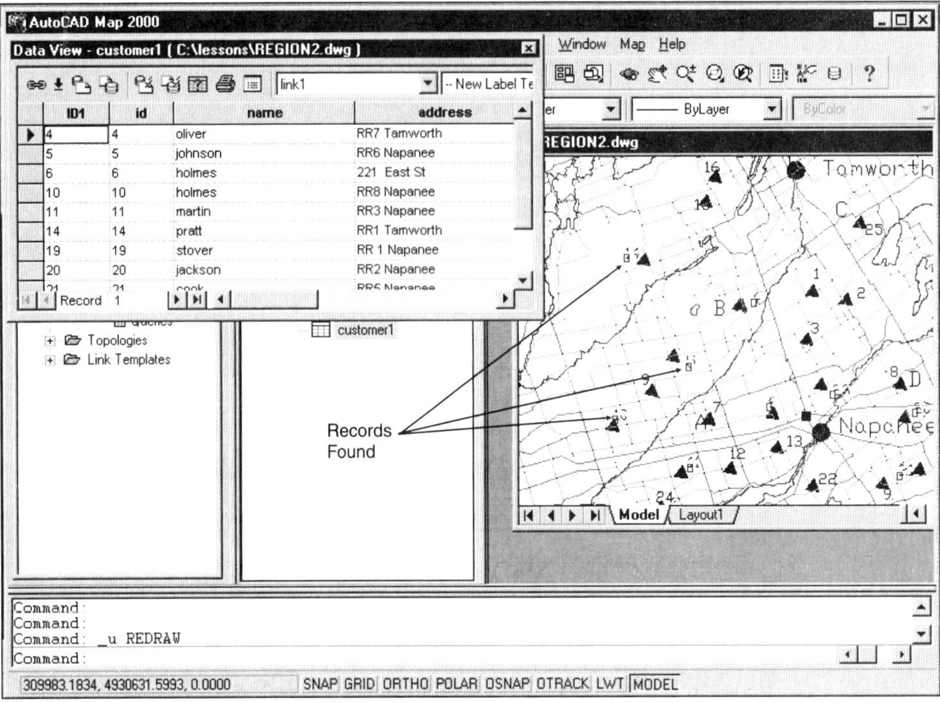

Figure 12–31

Step 13 To view customers in need of service or a delivery, another query could be conducted. A query can also be initiated by using the **New Query** icon in the dbConnect work session area.

Figure 12–32

You can use the same name, customer1Query1, if you wish. Click on **Continue**.

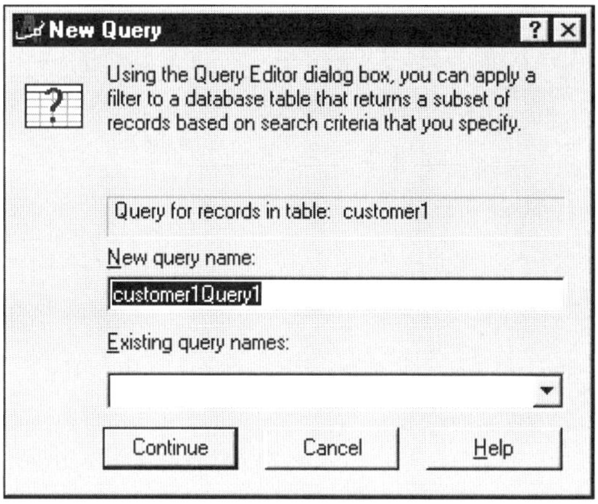

New Query

Using the Query Editor dialog box, you can apply a filter to a database table that returns a subset of records based on search criteria that you specify.

Query for records in table: customer1

New query name:

customer1Query1

Existing query names:

Continue Cancel Help

Figure 12–33

In the Query Editor select the **SQL Query** tab. For the Fields, select **lastdelivery**, then click on **Add**.

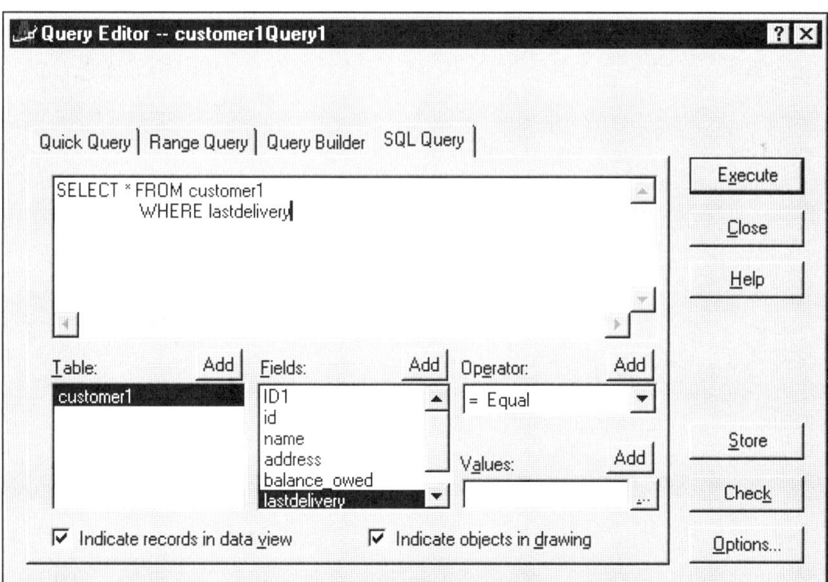

Figure 12–34

For the Operator select **<Less than**, then click on **Add**.

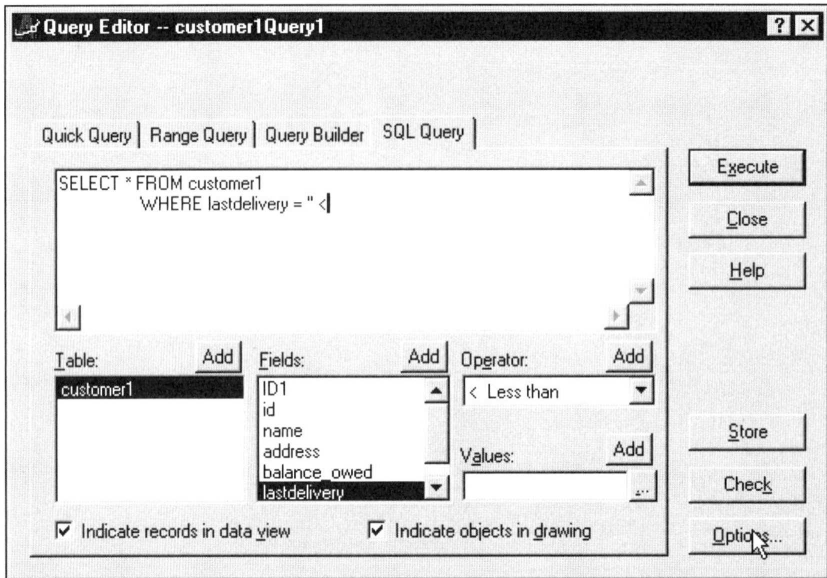

Figure 12–35

For the Values, type in **9** then click on **Add**. With both data display options selected **Execute the query.**

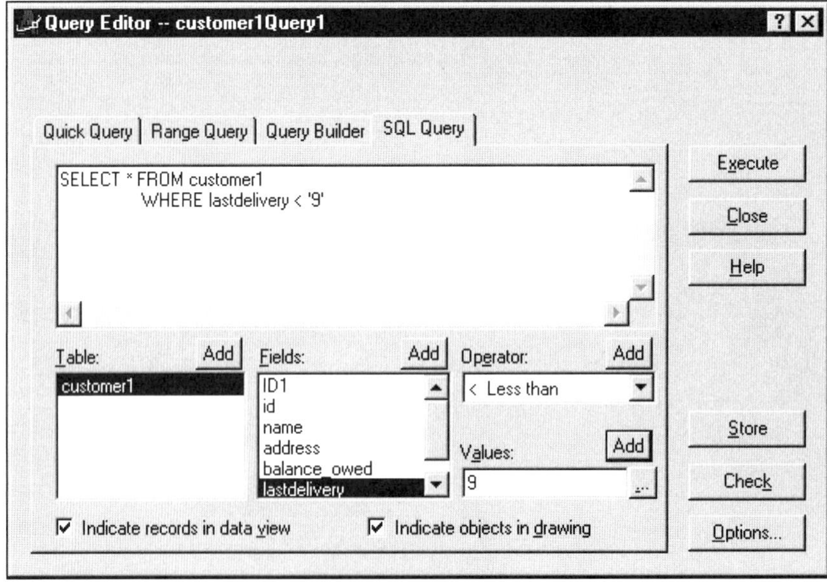

Figure 12–36

The customers who have not received a delivery since month 9 are indicated in a table and on the map.

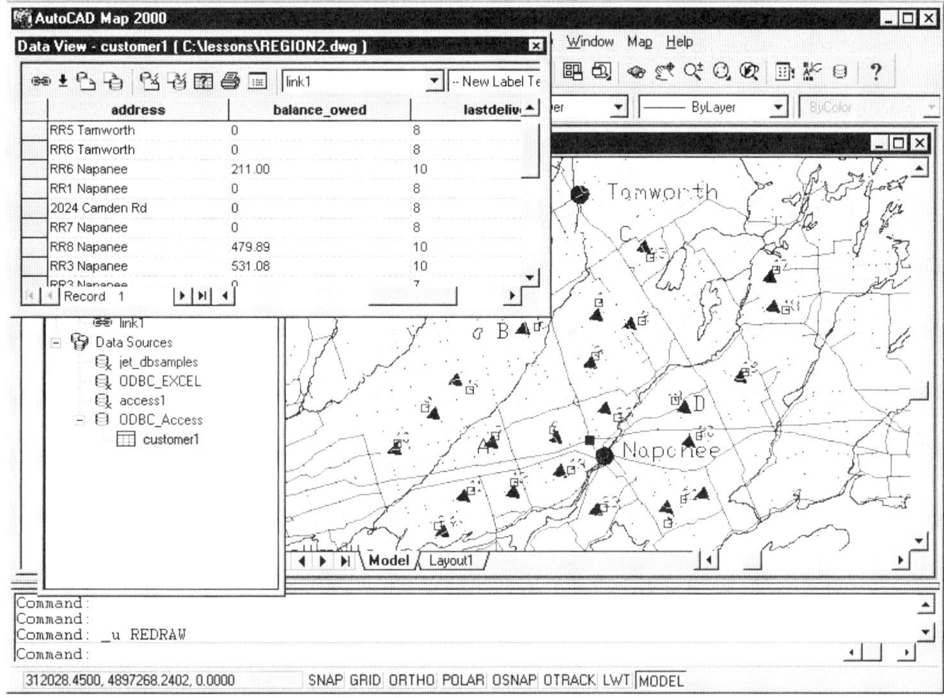

Figure 12–37

It is also important to note that data can be edited from inside AutoCAD Map 2000. To edit data go to the **Map** menu and select **Database,** then **View Data,** and **Edit Table**.

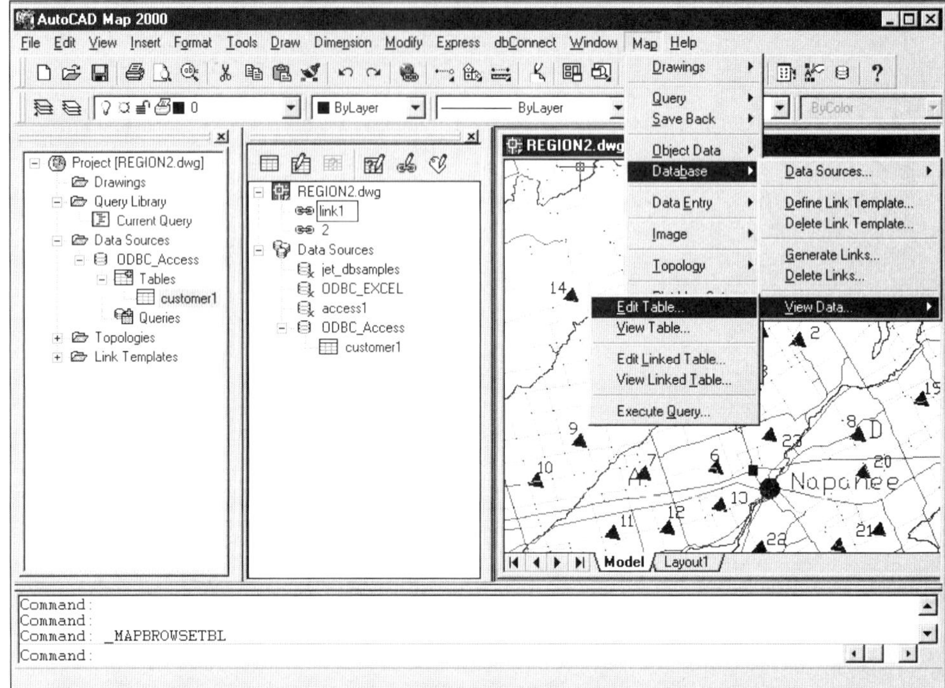

Figure 12–38

Select the **customer1** table at the **Browse Database Table** dialog box, and the table will appear in an editable format.

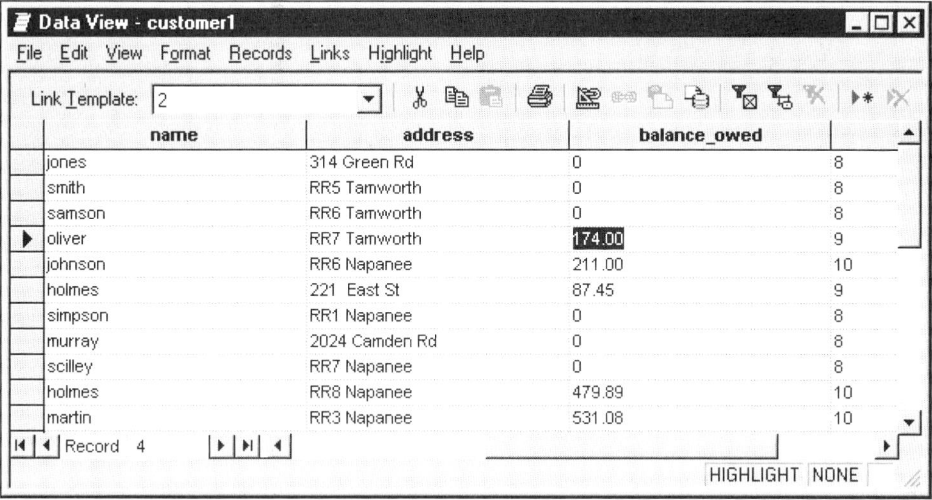

	name	address	balance_owed	
	jones	314 Green Rd	0	8
	smith	RR5 Tamworth	0	8
	samson	RR6 Tamworth	0	8
▶	oliver	RR7 Tamworth	174.00	9
	johnson	RR6 Napanee	211.00	10
	holmes	221 East St	87.45	9
	simpson	RR1 Napanee	0	8
	murray	2024 Camden Rd	0	8
	scilley	RR7 Napanee	0	8
	holmes	RR8 Napanee	479.89	10
	martin	RR3 Napanee	531.08	10

Figure 12–39

Save the drawing as **REGION3**, then close the file. Either exit from AutoCAD Map 2000 or continue with the tutorials.

SUMMARY

It is common practice for businesses of all sizes to maintain their business records in a database. This tutorial illustrates how such business data can be accessed easily and combined with other daily business practices to save money and improve business efficiency. In this case, a regular delivery schedule is enhanced by including stops at customers whose accounts are in arrears. This chapter incorporates many of the topics introduced in the text including topology, external databases, shortest path traces and SQL queries.

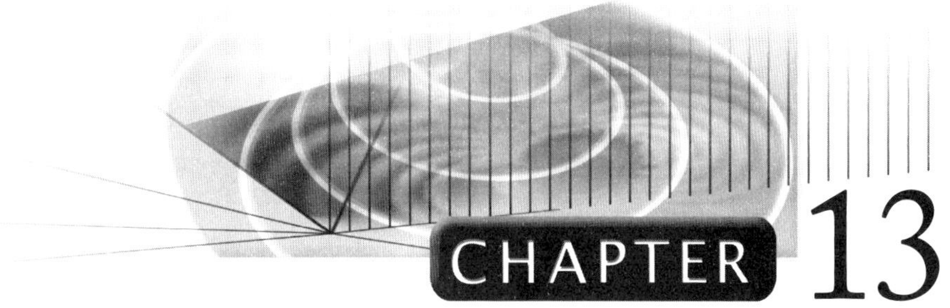

Architectural Restoration

OBJECTIVES

After completing this exercise you will know how to

- use raster image drawing units and real-world dimensions to assist in determining approximate dimensions of object in the images
- use raster image dimensions to prepare cost estimates

TUTORIAL: USING RASTER IMAGES FOR COST ESTIMATES

This exercise utilizes the raster image handling capability of AutoCAD Map 2000 to assist in an appraisal for restoration work to be done on an older building.

The cost per square foot of repointing a brick wall is given as $10 for this tutorial. In this case the RFP (request for proposal) for the architectural restoration of the library can be prepared without erecting scaffolding and conducting a labor-intensive measurement process. Instead, a photograph of the wall in question in conjunction with a linear measurement of the building's width will suffice.

The photograph could be taken with a digital camera or a conventional one with the processed image scanned and saved. In this case, the image that will be used was scanned and saved as LIBRARY.PCX, and the linear measurement found the building to be 60 feet wide. Prior to starting this tutorial, ensure that this file has been copied into your computer.

Step 1 Start AutoCAD Map 2000. Go to the **Insert** menu and select **Raster Image**.

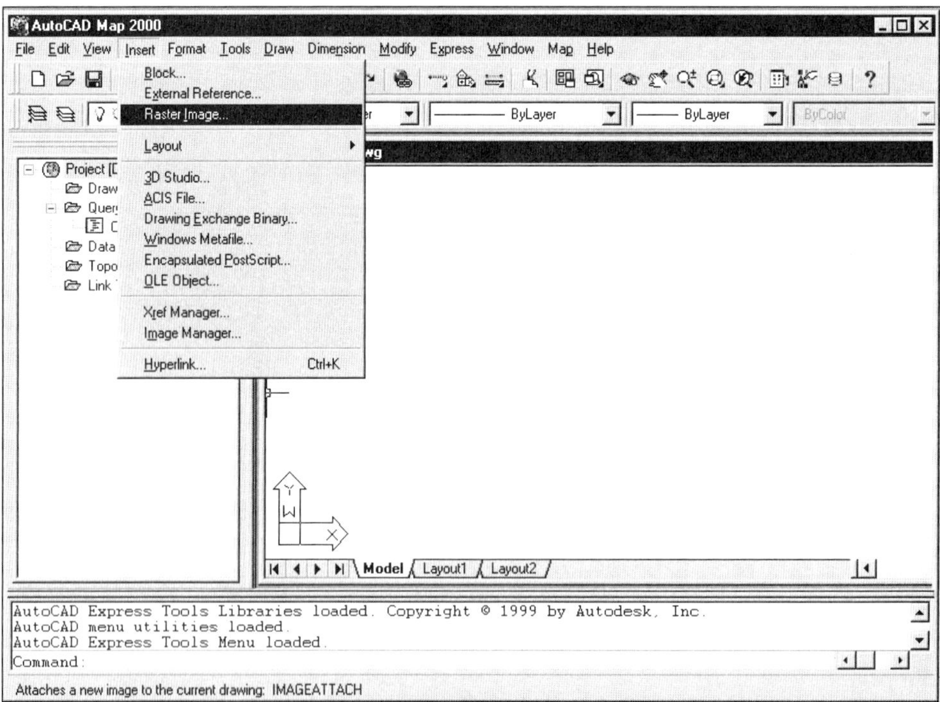

Figure 13–1

The Select Image File dialog box will appear. Select the **LIBRARY.PCX**, file then click on **Open**. For the Image Parameters, set the **Insertion Point** at **(0,0),** the **Scale** value at **10**, and leave the **Rotation Angle** at **0**. Notice in the details section of the dialog box that the image is 600 pixels wide and 414 pixels high. Click on **OK**, and the raster will be placed on the screen.

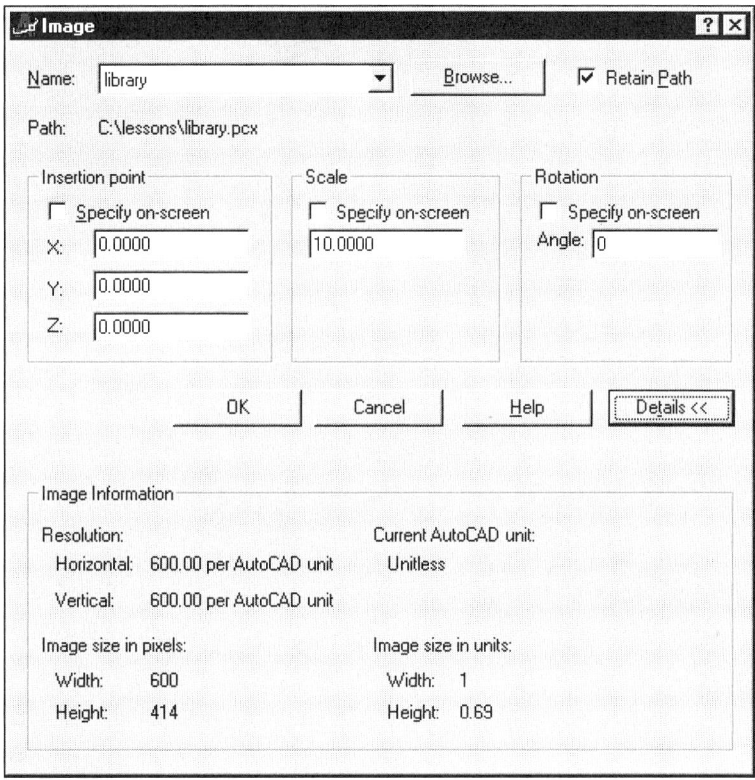

Figure 13–2

Step 2 Go to the **Tools** menu and select **Inquiry** then **Distance**.

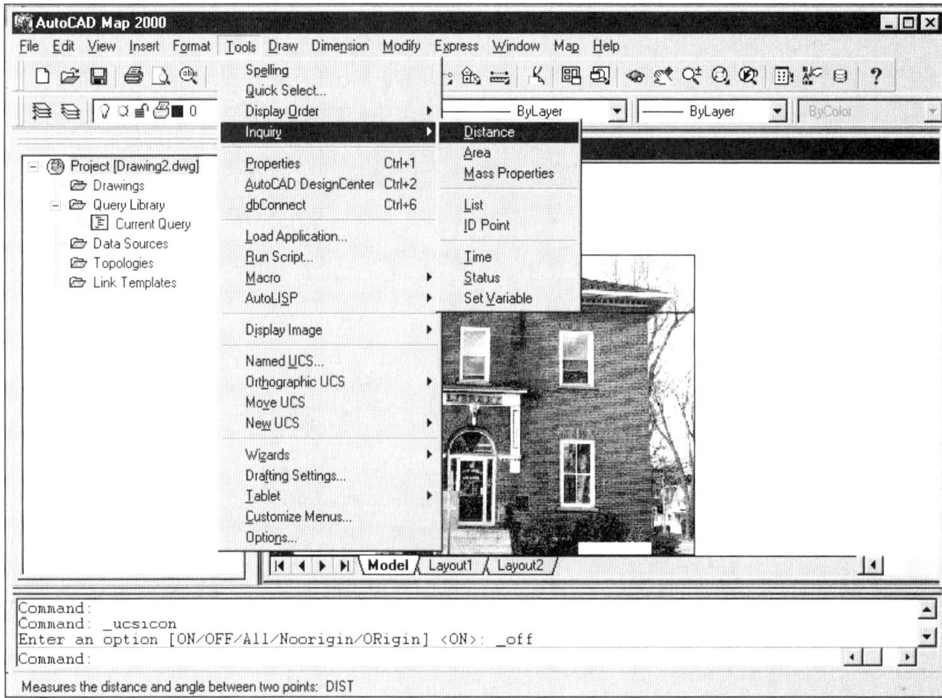

Figure 13–3

Click on one front edge of the building, and then click again on the other front edge of the building. Measure the width of the building as parallel as possible to the horizontal plane of the building. A reasonably good location to do this is just under the roof overhang.

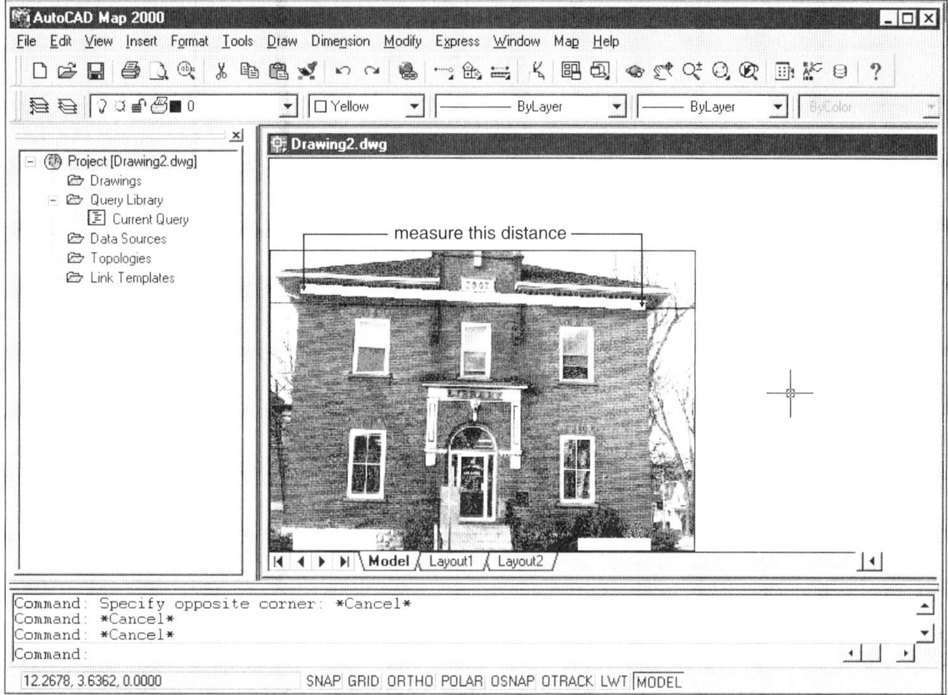

Figure 13–4

Step 3 The width of the building will be provided in drawing units in the command prompt area of the screen. The distance in this case is 8.1866 units. For estimating purposes, this can be rounded to 8.2 units.

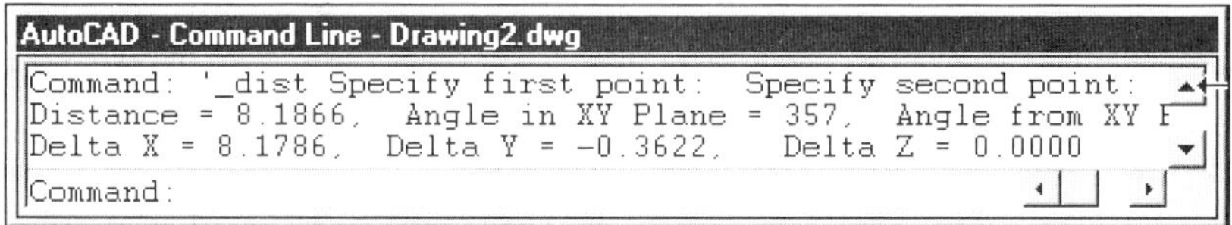

Figure 13–5

The building is known to be 60 feet wide, based on direct measurement. When divided by the number of drawing units, it leads to the conclusion that each drawing unit can represent 7.3 feet. This linear measurement estimate will be used to calculate area measurements for use in the proposal. One square drawing unit will represent 7.3 feet times 7.3 feet, equaling approximately 53.3 square feet.

Step 4 The proposal to be developed calls for repointing the front wall area that extends from the sill bottom of the three second-story windows to the foundation. Visualize a rectangular area on the front of the building.

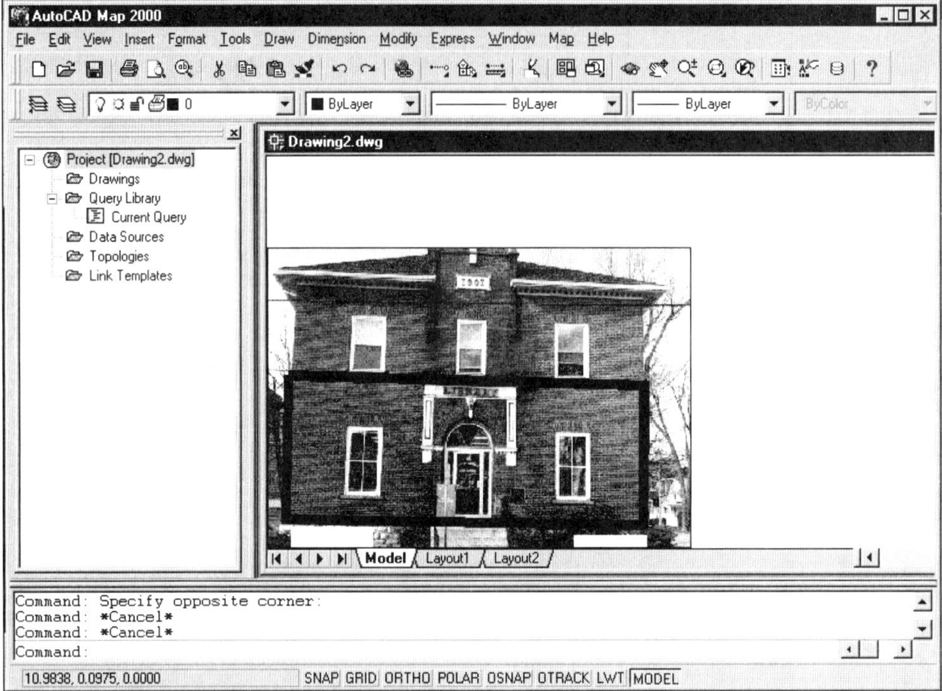

Figure 13–6

Step 5 Go to the **Tools** menu and select **Inquiry** then **Area**. Pick on the four corners of the region. The area and perimeter values will be provided in the command prompt box.

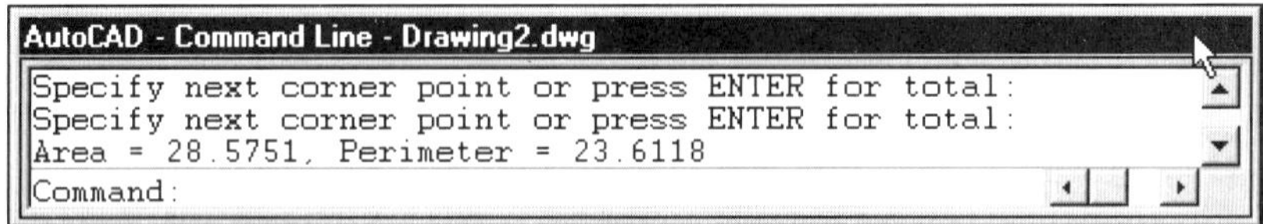

Figure 13–7

In this case, the area value can be rounded to 28.5 square units. Remember, we are using the conversion factor of 1 square drawing unit equaling 53.3 square feet. The total frontage of the building is 28.5 drawing units times 53.3 square feet, equaling approximately 1,519 square feet.

This however, represents the area of the entire wall, and allowances must be made for the windows and doorway.

Step 6 Use the **Tools** menu and select **Inquiry** then **Area** to determine the area for the two lower windows and the main door. The area of each window is approximately 1.4 drawing units, totaling 2.8 square drawing units. Because each such unit represents 53.3 square feet, the windows cover 2.8 times 53.3 square feet, equaling approximately 149 square feet. This must be subtracted from the total frontage of 1,519 square feet, yielding 1,370 square feet as the result.

Step 7 Use the same procedure to determine the area of the entrance. This process is complicated somewhat by the presence of an object in front of the left side of the door and by the semicircular portion on the entrance. It is safe to assume the door is rectangular even though the bottom left side is blocked from view. To gather the data area on the semicircular portion, pick on several points along the perimeter of the entire entrance. This area is approximately 2.3 square units, equaling approximately 123 square feet. When this is subtracted from the remaining 1,370 square feet, 1,247 square feet of frontage is left. Finally, the entrance overhang and supports must be measured and factored out in order to arrive at a final value for surface area.

Step 8 Use the **Tools** then **Inquiry**, and **Area** selections to determine the area of the overhang and supports. The process generates a total of approximately 2 square drawing units, which translates into a rounded-off value of 107 square feet. The final calculation, removing this area from the previous total, is 1,247 minus 107, resulting in 1,140 square feet.

The cost estimate can be based on $10 per square foot can be established as 1,140 times 10, equaling $11,400.00. (The figures have been rounded off throughout this exercise, and your results may vary.)

Either exit AutoCAD Map 2000 or continue with the tutorials.

SUMMARY

The raster file LIBRARY.pcx is provided, along with some of the physical dimensions of the building, for the purpose of calculating an amount of surface area which in turn is used to prepare a quote for a "Request for a Proposal." With the ready availability of digital cameras this approach for preparing costing estimates can replace the time consuming process of physical measurement. This approach could be applied to a variety of similar situations. One example would be to utilize a land use photograph as the base drawing for calculating an area to be sprayed for insect infestation.

Suggested Projects

OBJECTIVES

- After viewing the topics listed you will have greater insight into the types of projects that can be completed using AutoCAD Map 2000.

Topic	Focus
Bus routes	Maximizing population served while minimizing time and distance of routes
Wildlife areas	Location of protected areas in relation to the location of endangered species
Tourism promotion	Pamphlet mapping out tourist sites within given driving times
Wedding maps	Route maps between ceremony and reception
Chamber of Commerce	Local business promotion maps
Cross-curricular map	Map locations for settings of novels, battles, sports
Site management	Parking lot efficiency maximization
Land use	Update zoning models
Schools	Maps revealing floor plans, first-aid stations, fire exits, room utilization
Orienteering	Course layout
Fishing	Hot spots, boat launches, fuel sites, hazard areas
Crime	Type versus unemployment rate
Earning power	Income versus education levels
Social dependency	Unemployment versus education level
Social friction	Crime rates versus immigration rates
Forestry	Forest depletion by cutting versus fire
Disease	Distribution of various illnesses by state
Agriculture	Pesticide and herbicide application versus yield

Topic	Focus
Meals on Wheels	Delivery maps
United Way	Canvassing district maps
Neighborhood Watch	Neighborhood activity maps
Block Parents	Maps revealing "safe" houses
Municipal	Street repair projections
Natural disasters	Flood areas, tornado touchdowns, earthquakes, hurricanes
Health care	Population density versus hospital bed availability
Sports	Athletes by state of origin
Disaster relief	Emergency evacuation plan
Religion	Church attendance by region and denomination
Global warming	Average global temperature change
CFC impact	Ozone layer fluctuations over time
Nutrition	Daily caloric intake levels
Government	Trends in international governments over time
Geology	Bedrock type versus mineral deposition
Ecology	Endangered species by nation over time
Multinationals	Trends of employment
Pollution	Optimization of landfill site selection
Culture	The geography of language
Politics	Voter registration versus outcomes
Policing	Community-based crime patterns
Wildlife management	Species kills versus hunting permits

SUMMARY

The list of general topics is not, by any means, exhaustive. In fact, each topic listed could have numerous related topics and interpretations. It is important to note that a major reason for the rapid growth of GIS technology is the flexibility and adaptability the technology offers to countless decision making scenarios.

Related Technologies

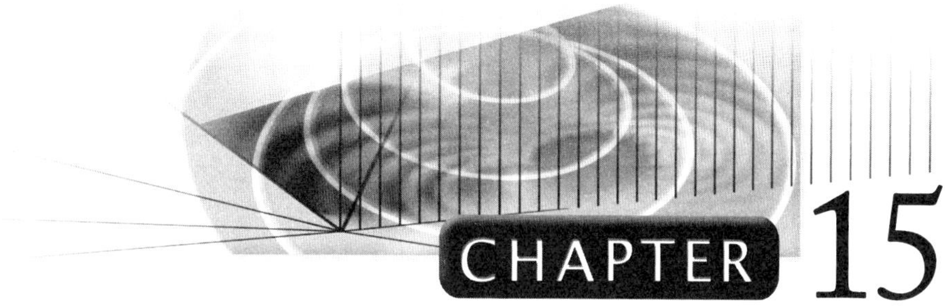

Autodesk Mapguide Viewer and the Internet

OBJECTIVES

After completing this chapter you will know how to

- download and install Autodesk MapGuide Viewer software

This tutorial guides the user through the process of obtaining the program and introduces some of the features available in it.

Autodesk has developed a suite of software products designed to facilitate a variety of GIS applications. These products include the Autodesk MapGuide products and Autodesk World. The Autodesk MapGuide products consist of three separate components: Autodesk MapGuide Viewer, Server, and Author. These programs are world class Internet information delivery tools and warrant their own dedicated documentation. Autodesk MapGuide Viewer is, as the title implies, a map viewing tool designed to allow the user to access AutoCAD Map 2000 prepared drawings that are delivered to the Internet via Autodesk MapGuide Author and Server.

TUTORIAL: INSTALLING/USING AUTODESK MAPGUIDE VIEWER

Step 1 Connect to the Web using your traditional link and browser. Autodesk MapGuide Viewer will work with either Microsoft Internet Explorer or Netscape Navigator, but for this example Internet Explorer has been selected as the browser.

Step 2 Set your browser to go to the Autodesk MapGuide Web site by typing in the address **www.autodesk.com/products/mapguide/viewer.htm**.

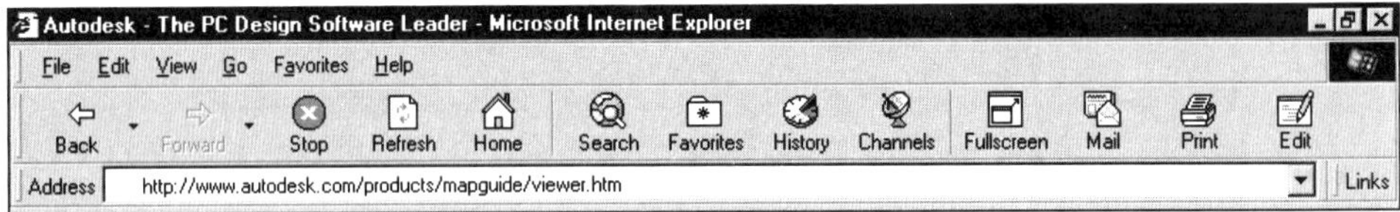

Figure 15–1

This will take you to the Autodesk MapGuide site.

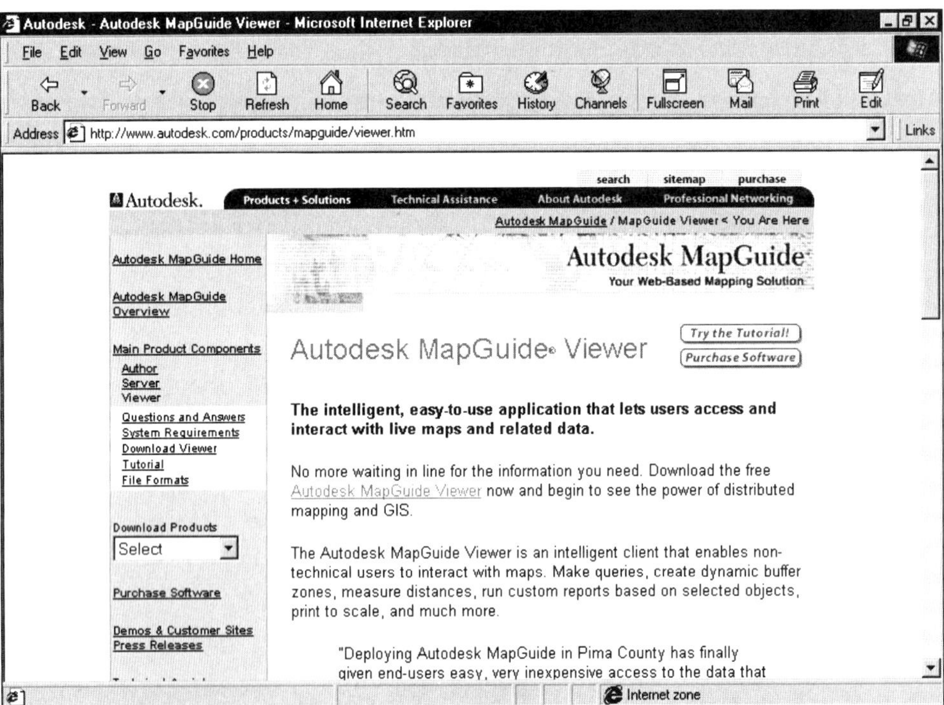

Figure 15–2

Step 3 Along the left side of the Web page you will notice a list of options that includes Download Viewer. Click on the **Download Viewer** option. This will take you to the Autodesk MapGuide Viewer Download page.

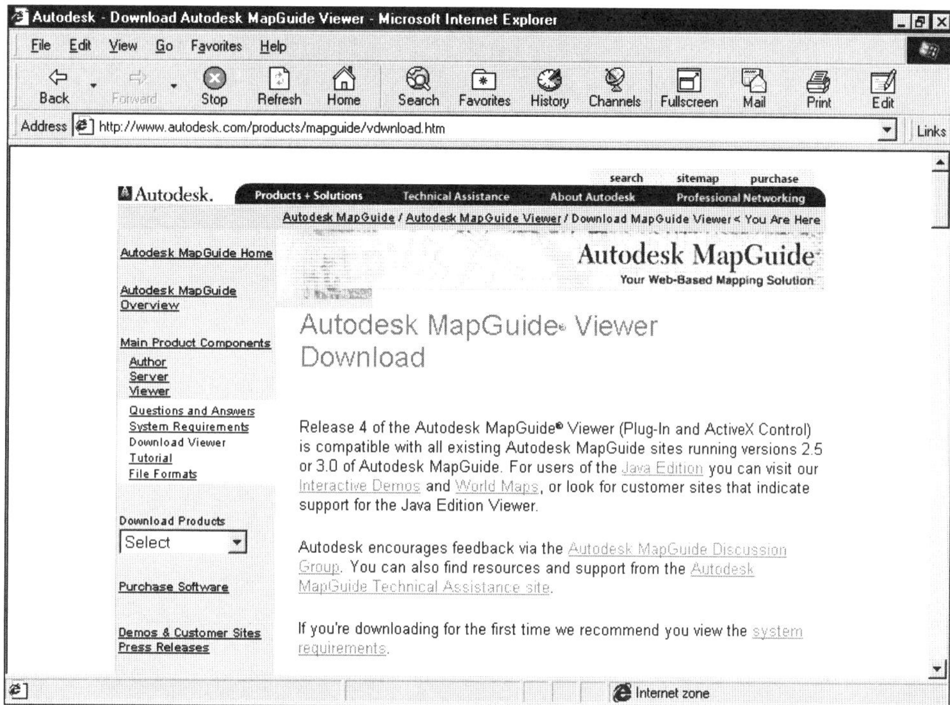

Figure 15–3

At this page you can read through the installation instructions and then **select** the **version** of the viewer you wish.

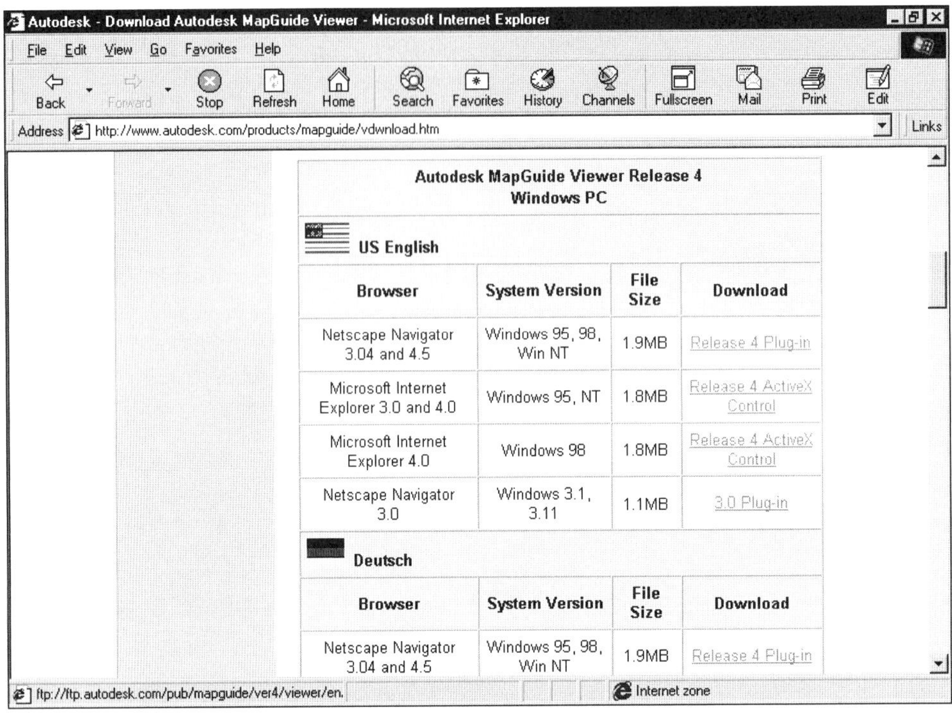

Figure 15–4

Step 4 When you determine which version you require based on your browser and operating system, **click** on it. A File Download notice will appear, providing you with the opportunity of saving the file to a disk.

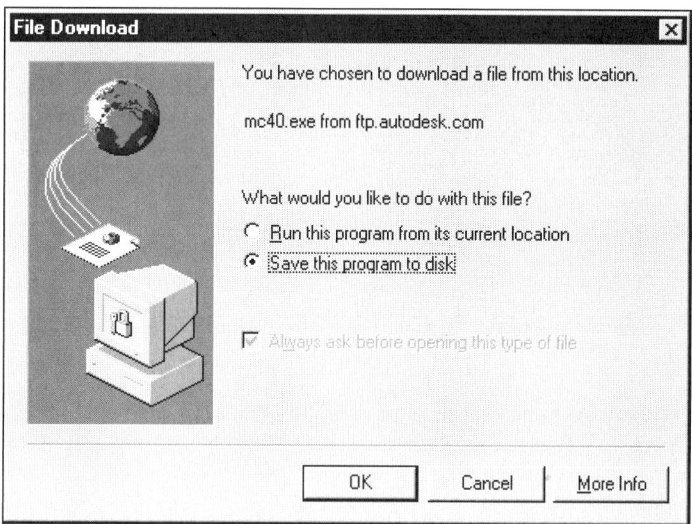

Figure 15–5

Step 5 Click on **OK** to save the file to disk and you will be prompted for a location.

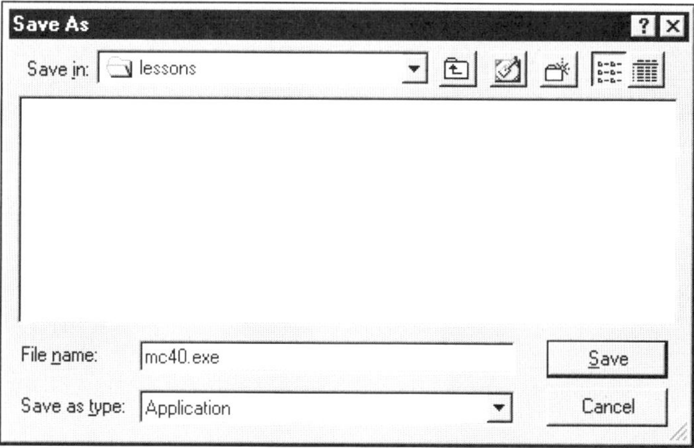

Figure 15–6

Select a directory to save it to, and click on the **Save** button.

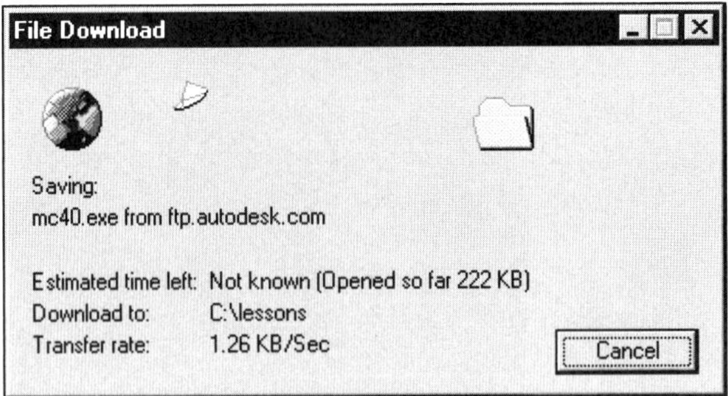

Figure 15–7

A progress report may appear on the screen to show you how much time is required for the download to be completed and this will of course, vary with connection speed. Another screen will appear to inform you that the download has been completed.

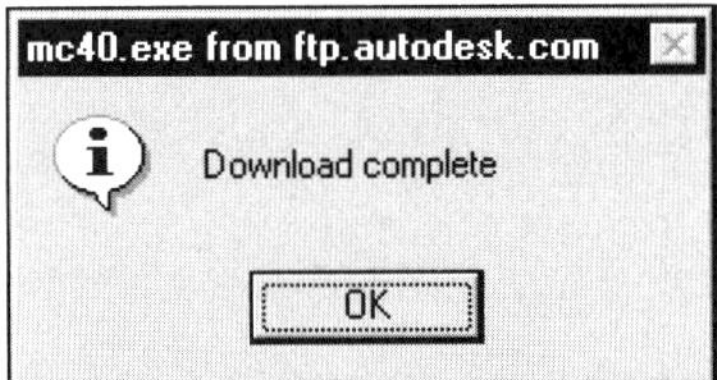

Figure 15–8

You must now close your browser and disconnect from the Internet.

Step 6 Go to the **Run** menu and select the downloaded file **MC40.EXE** (in this case).

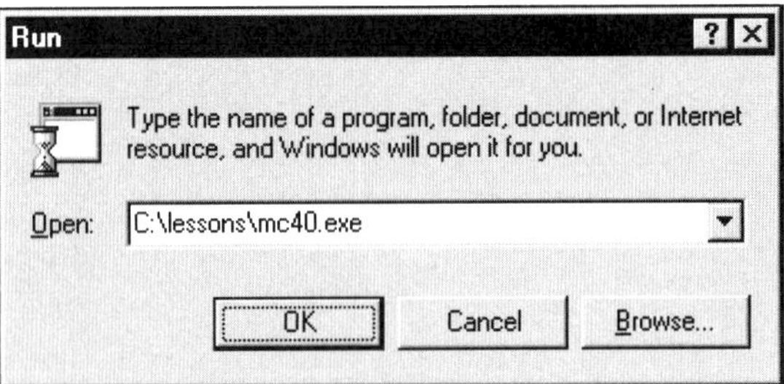

Figure 15–9

This will run the installation of the MapGuide Viewer during which you select a directory for the install.

Step 8 Reconnect to the Internet, restart your browser, and point your Autodesk MapGuide Viewer–equipped browser back to the Autodesk MapGuide Site (www.autodesk.com/products/mapguide/vdwnload.htm).

Step 9 At this page you can select demonstration sites that you can access. Spend a few minutes and explore these sites. More and more Autodesk MapGuide Server–powered sites are being developed, and a search of the Internet will provide other sites to visit. Of particular note is the real-time mapping taking place around the world and being displayed on the Internet through Autodesk MapGuide Server.

Step 10 Another good site for getting started with Autodesk MapGuide Viewer is at Gridnorth.com. Point your browser to **www.gridnorth.com/index2.htm**.

Click on the **Virtual World Map** option then select the **USA** option. A map of the continental United States will be presented. You will be able to use the mapping tools to present some very detailed maps of virtually anywhere in the country. As you zoom in, features like scale will be updated, as will the real-world dimensions that are shown on the screen bottom and more layers of information are provided. The legend items are shown along the left side of the map. This technology can even generate street-level maps if you continue zooming in. The maps are printable and therefore usable for other applications, such as generating a digital map file as discussed in Chapter 5. With a little familiarity the capabilities and information gathering power of Autodesk MapGuide Viewer will become a very welcome addition to your GIS toolkit.

SUMMARY

Internet mapping technology is still relatively new but Autodesk MapGuide Viewer is an excellent tool for acquiring insight into the capabilities of this technology. The web sites referred to in this chapter are meant to serve as starting points for using Autodesk MapGuide Viewer. There are two more components to Autodesk's Internet mapping technogolgy; MapGuide Author and MapGuide Server. These two packages allow the preparation, publishing and distribution of maps over the Internet.

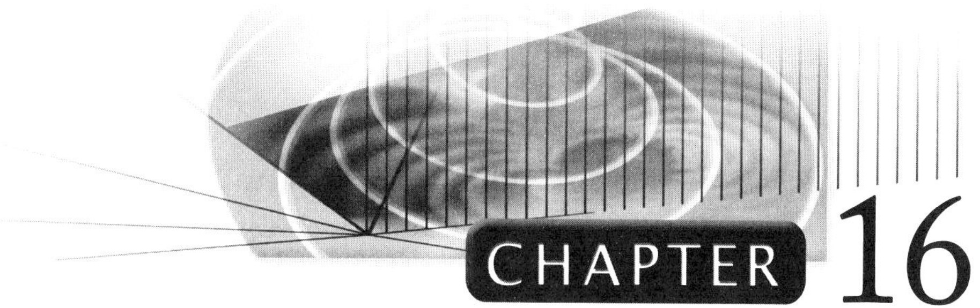

GPS and MAPS

OBJECTIVES

After completing this exercise you will know how to

- import GPS text files into AutoCAD Map 2000 as DXF fles

TUTORIAL: GPS FILES TO MAPS

GPS, or Global Positioning System technology is becoming more common as a means of collecting accurate and affordable, real-world positional data. Once this data has been collected it gains value when it can be used in conjunction with other geographic information, whether that be a map or an engineering drawing.

This exercise shows one way to transform files collected from a Magellan NAV5000 GPS unit into AutoCAD Map 2000. This technique may vary with different GPS receivers and as a function of the software provided by the manufacturers. This particular application makes use of a third-party software package called the Geographic Calculator, produced by Blue Marble Geographics.

Prior to beginning the data download, it would be prudent to create a separate directory for the files. In this example, the directory was named GPS.

Step 1 After collecting a data file with the GPS unit, connect the unit to your computer using the cable provided by the GPS manufacturer. In this exercise, the GPS unit was attached to COM2.

Step 2 Click on the **Start** button on the computer screen then select **Programs** and **Accessories**. From the Accessories activate **HyperTerminal**.

Step 3 If prompted to install a modem select **No**.

Step 4 When prompted to enter a name for the connection, type in **GPS**. Click on **OK**.

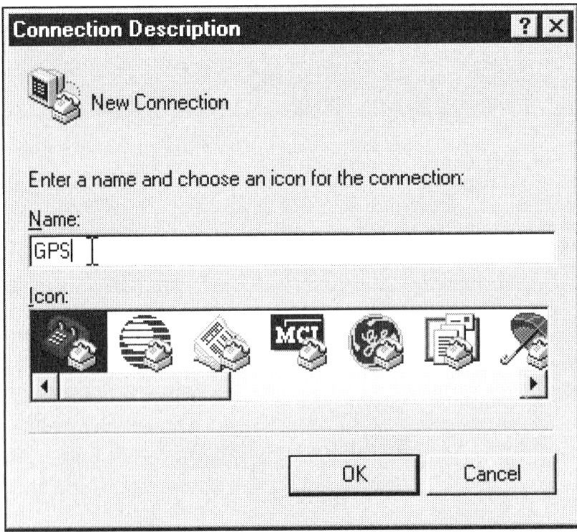

Figure 16–1

Step 5 When prompted for a phone number, set the connection to the appropriate port. In this case, COM2 was used, so the setting chosen was **Direct to Com2**. Click on **OK**.

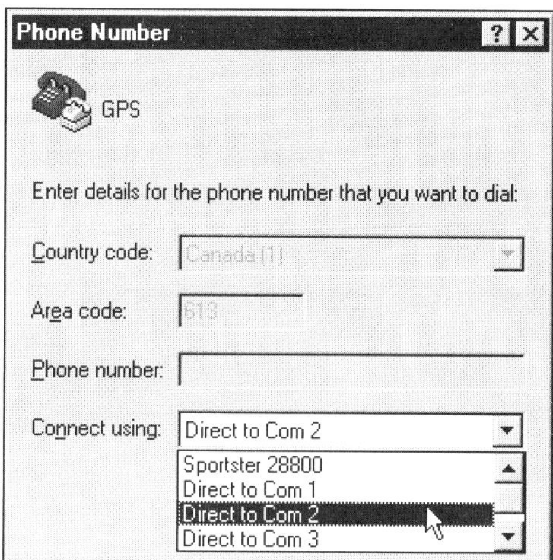

Figure 16–2

Step 6 When prompted for the COM2 properties, enter the values as described in the GPS manual. In this case it was **4800**, **8**, **None**, **1**, and **None**. Click on **OK**.

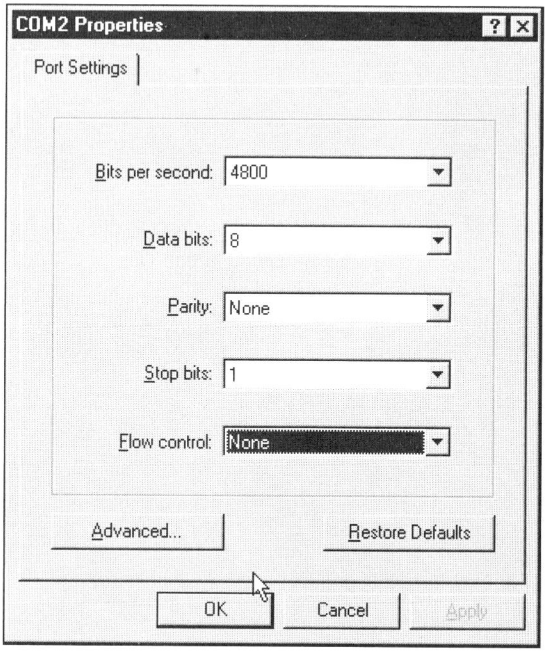

Figure 16–3

Step 7 You are now connected to the GPS unit via Hyperterminal. On your GPS unit's menu, go to **Upload Data**. In the case of this particular unit, the **Upload NMEA** option was selected.

Step 8 In HyperTerminal select **Transfer** then Capture Text then send the data from the GPS unit to the computer. The text will come in line by line and appear on the screen.

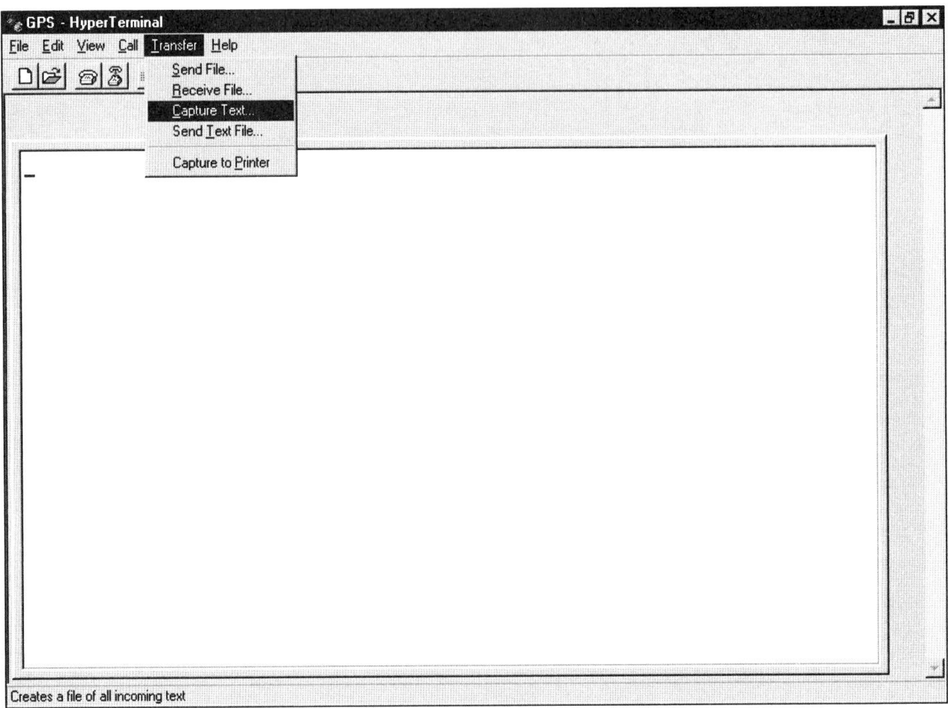

Figure 16–4

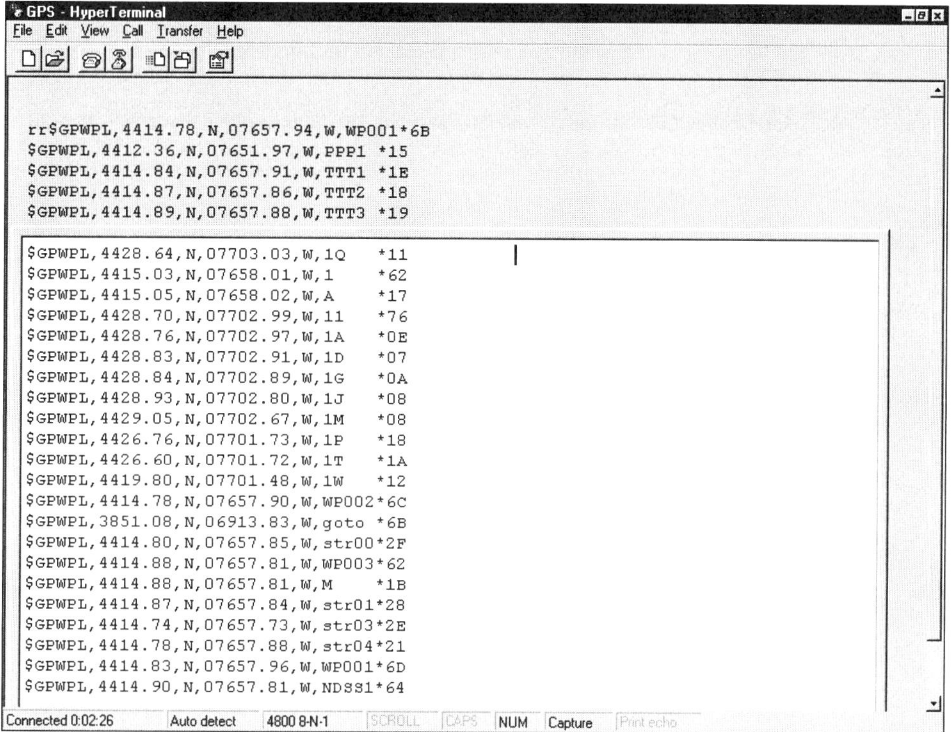

Figure 16–5

Step 9 Highlight the text file and use the clipboard function to copy the contents into Notepad, the text editor provided by the operating system and available through Start|Accessories, and save the file as a text file, such as GPS.TXT. This file contains extraneous alphanumeric characters, which must be edited out, but the numeric values for the latitude and longitude must be maintained.

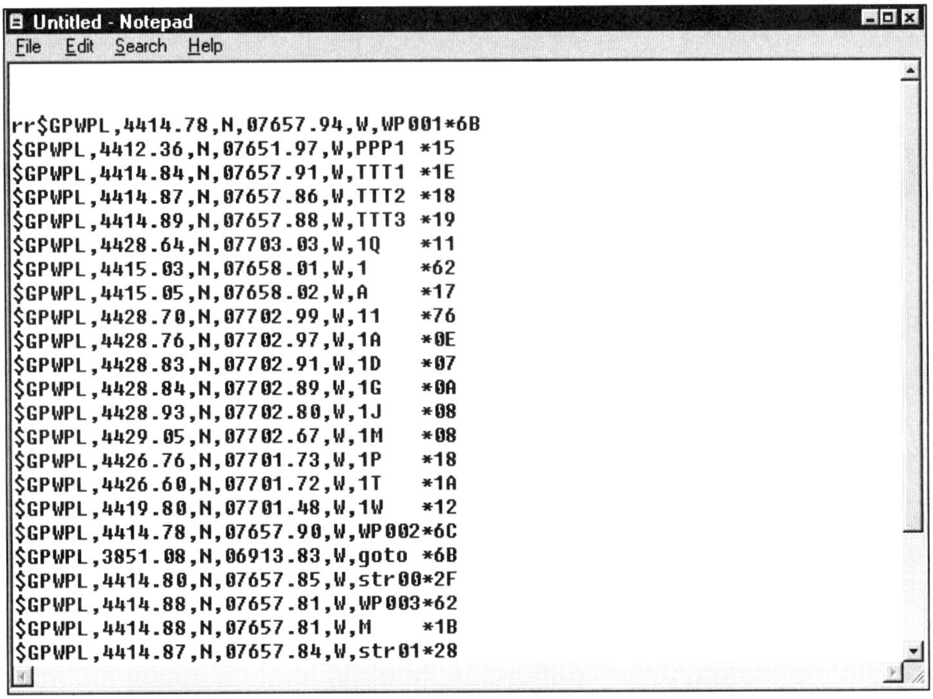

Untitled - Notepad
File Edit Search Help

```
rr$GPWPL,4414.78,N,07657.94,W,WP001*6B
$GPWPL,4412.36,N,07651.97,W,PPP1 *15
$GPWPL,4414.84,N,07657.91,W,TTT1 *1E
$GPWPL,4414.87,N,07657.86,W,TTT2 *18
$GPWPL,4414.89,N,07657.88,W,TTT3 *19
$GPWPL,4428.64,N,07703.03,W,1Q   *11
$GPWPL,4415.03,N,07658.01,W,1    *62
$GPWPL,4415.05,N,07658.02,W,A    *17
$GPWPL,4428.70,N,07702.99,W,11   *76
$GPWPL,4428.76,N,07702.97,W,1A   *0E
$GPWPL,4428.83,N,07702.91,W,1D   *07
$GPWPL,4428.84,N,07702.89,W,1G   *0A
$GPWPL,4428.93,N,07702.80,W,1J   *08
$GPWPL,4429.05,N,07702.67,W,1M   *08
$GPWPL,4426.76,N,07701.73,W,1P   *18
$GPWPL,4426.60,N,07701.72,W,1T   *1A
$GPWPL,4419.80,N,07701.48,W,1W   *12
$GPWPL,4414.78,N,07657.90,W,WP002*6C
$GPWPL,3851.08,N,06913.83,W,goto *6B
$GPWPL,4414.80,N,07657.85,W,str00*2F
$GPWPL,4414.88,N,07657.81,W,WP003*62
$GPWPL,4414.88,N,07657.81,W,M    *1B
$GPWPL,4414.87,N,07657.84,W,str01*28
```

Figure 16–6

Step 10 In Notepad, edit out these characters and add spaces between the degrees and minutes for both latitude and longitude values and save this file as **GPS2.TXT**.

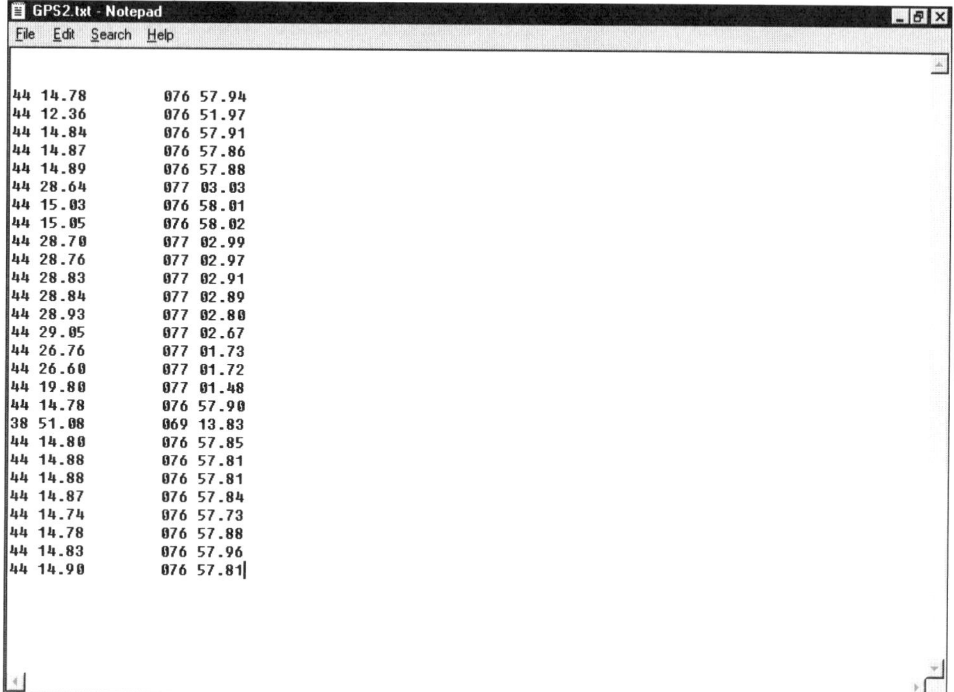

Figure 16–7

Step 11 Go to the MS-DOS Prompt or use Windows to copy or rename the text file to a points or pts file. This can be done, for example, by typing in DOS **copy GPS2.txt GPS2.pts** (leave a space between copy and GPS2.txt and a space between GPS2.txt and GPS2.pts) and then pressing ENTER.

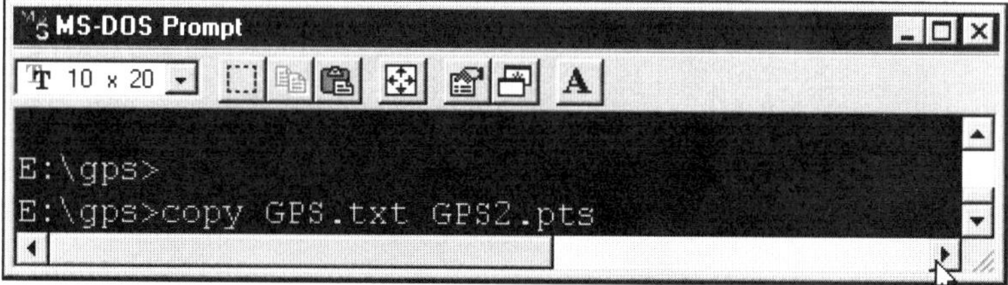

Figure 16–8

Step 12 Start the conversion program The Geographic Calculator. (Note that this program is a separate product selected for use here because it enables the particular GPS files to be converted to a DXF format. Different manufacturers of GPS receivers may provide or recommend other programs for completing the same task.)

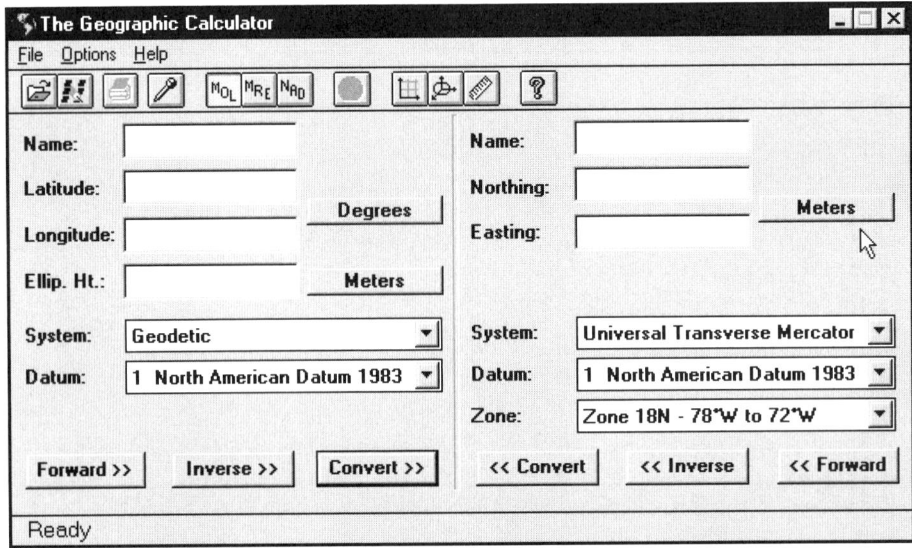

Figure 16–9

Step 13 On the left side of the dialog box you should see the system set to **Geodetic** and the datum set to North American datum 1983. On the right side, the system should be set to **Universal Transverse Mercator**, with a datum of North American Datum 1983, and the zone set to the UTM zone for the area in which the GPS unit was used, in this case UTM zone **18**. This sets up the proper coordinate system for the region in which the points were collected.

Step 14 Click once on the File Folder icon in the upper left region of the tool bar Geographic Calculator. At the Open an Input Coordinate File dialog box, select the **GPS2.PTS** file that you want to convert and click on **OK**.

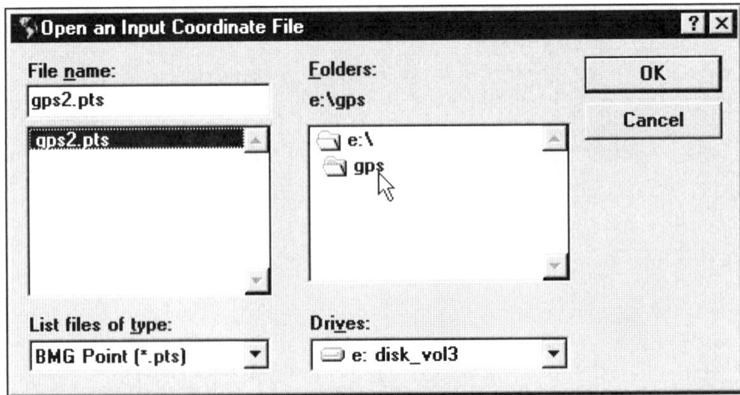

Figure 16–10

Step 15 When prompted for an Output file set the Save File as Type option to **AutoCAD DXF**.

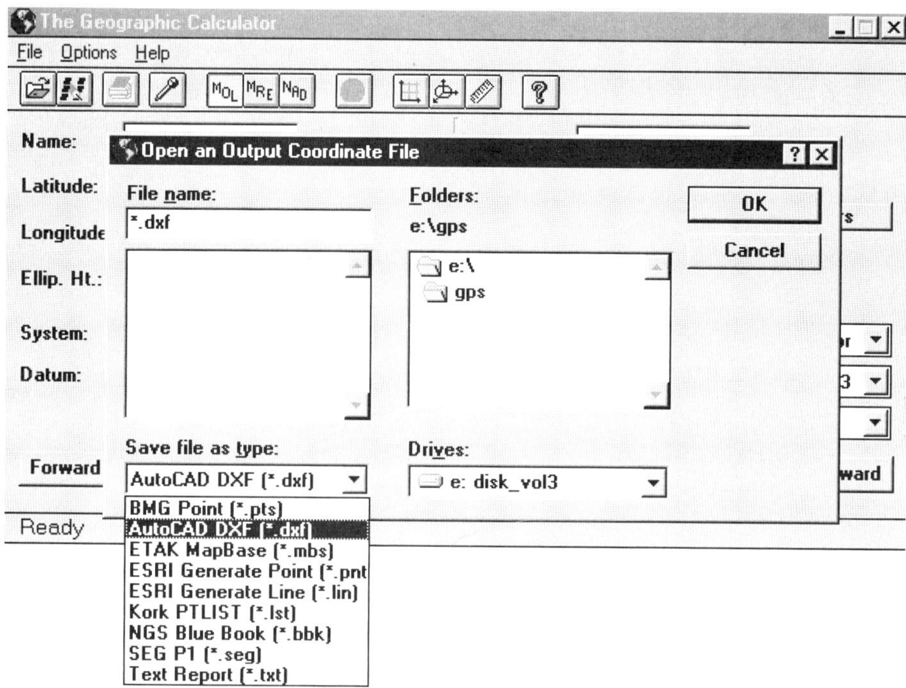

Figure 16–11

Give the new DXF file a name, in this case TEST.DXF and click on **OK**. You will be given a message stating the number of points that were processed in the conversion.

Step 16 You now have a DXF file that can be opened in AutoCAD Map 2000.

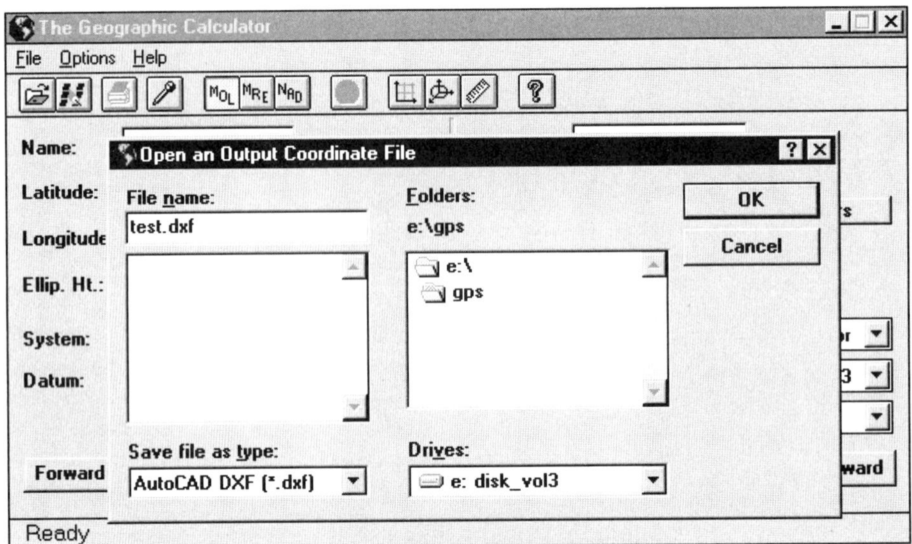

Figure 16–12

It is important to note that the GPS file will be a collection of points on the screen.

In AutoCAD Map 2000 go to **File** then **Open** and change the file type to **DXF**.

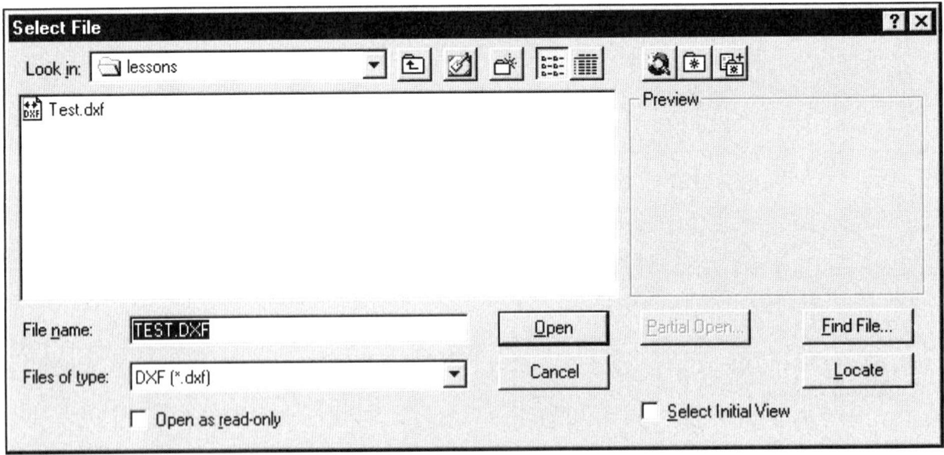

Figure 16–13

Select the **TEST.DXF** file, click on **Open**, and the points will appear.

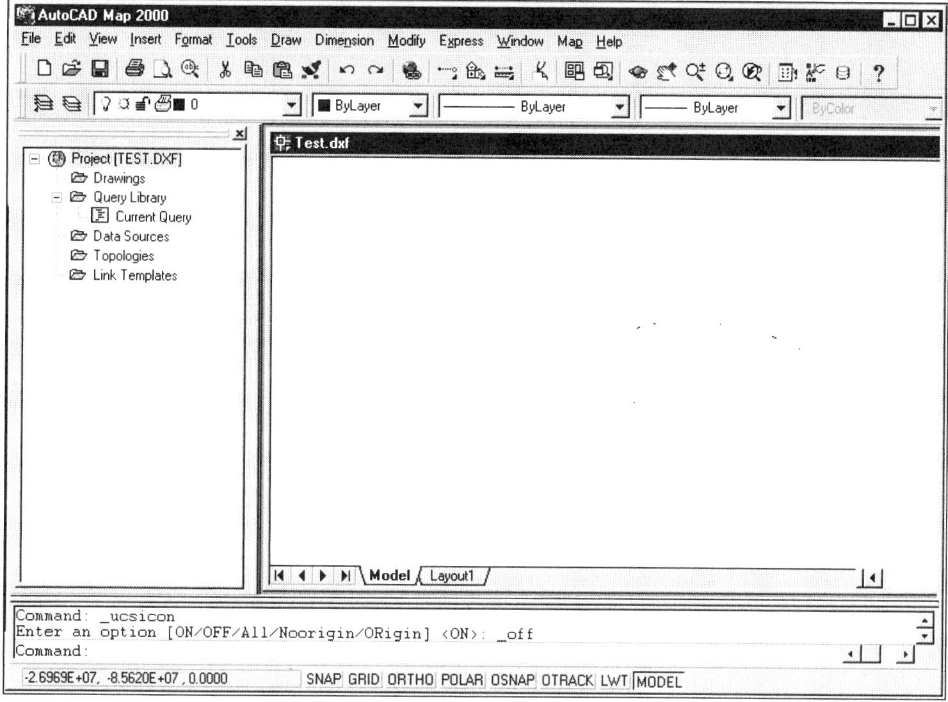

Figure 16–14

This file can be saved as a DWG file, if desired. Either continue with the tutorials or exit AutoCAD Map 2000.

SUMMARY

This chapter introduced one procedure for downloading files collected by a GPS receiver and converting them into AutoCAD Map 2000 format. Different brands of GPS receivers might have slightly different procedures. This methodology utilized a third-party software package made by Blue Marble Geographics called The Geographic Calculator 3. At the time of writing, release 4 was eminent.

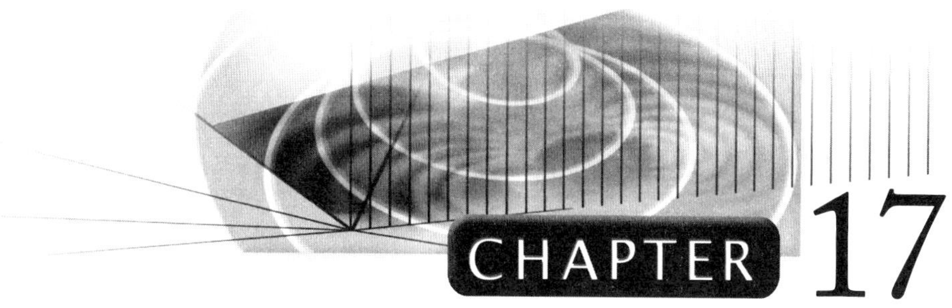

Autodesk World

OBJECTIVES

After completing this exercise you will

- have increased awareness of the nature of Autodesk World.

KEY TERMS

Geobase
Report

Autodesk World is a product from Autodesk that is designed to provide GIS functionality within a Microsoft Office environment. This program can integrate large volumes of spatial and attribute data from a variety of formats as well as compile your own data sets. The software also provides the user with the capability of running queries, generating charts, and producing maps. It is shipped with a variety of learning materials, some of which are utilized in the following tutorial.

Autodesk World is, in itself, a comprehensive GIS tool and consequently cannot be fully portrayed in one chapter. The purpose of this chapter, therefore, is to generate awareness of this powerful tool.

Prior to starting the tutorial below, ensure that Autodesk World has been installed on your computer. If you do not have this software, you can read through the tutorial to gain an appreciation for the ease with which World completes mapping procedures.

TUTORIAL: A BRIEF INTRODUCTION TO AUTODESK WORLD

Step 1 Start Autodesk World. The starting screen reveals three areas: toolbars, a Display Manager, and the Project Window.

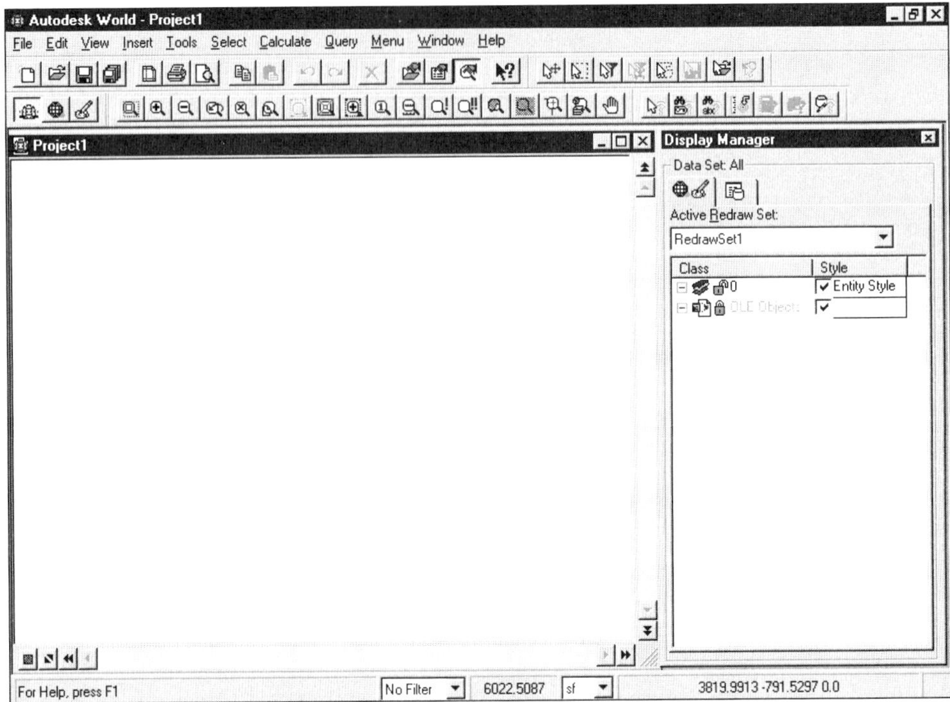

Figure 17–1

To access a stored tutorial, go the **File** menu and select **Open**. Choose the **Tutorial** item, select the **Tutorial1.APF** item, then click on **Open**. A detailed map will be drawn in the Project Window. The Display Manager lists the classes of data displayed, as well as those that are locked.

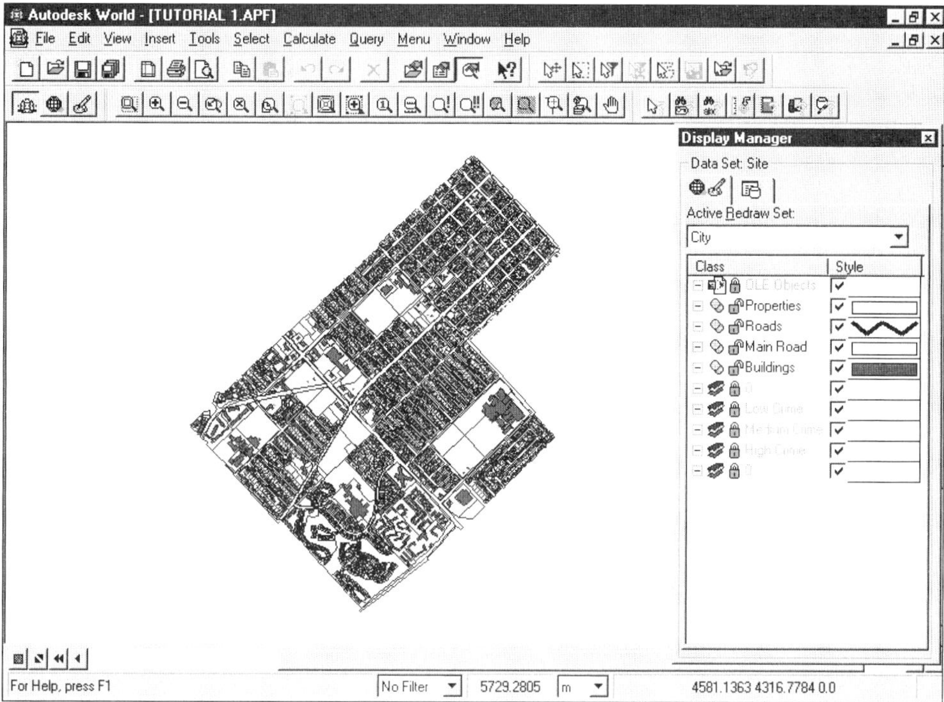

Figure 17–2

Step 2 To establish a query or Filter, go to the **Select** menu and choose **Filter**.

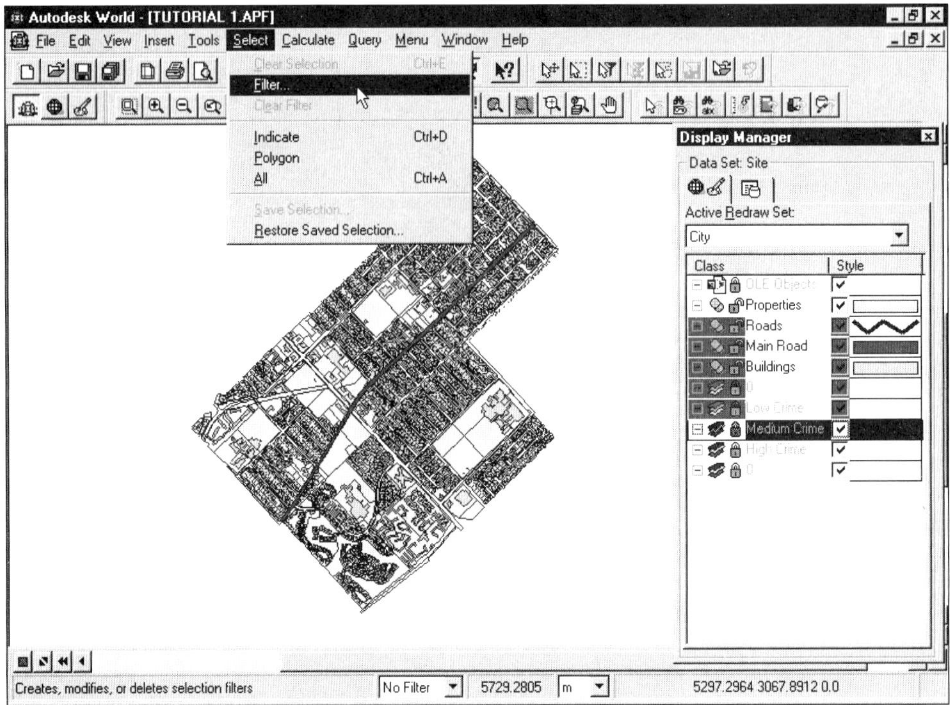

Figure 17–3

The Selection Filter dialog box will appear. Click on the active **Flashlight** icon located to the right of the text box containing the term No Filter.

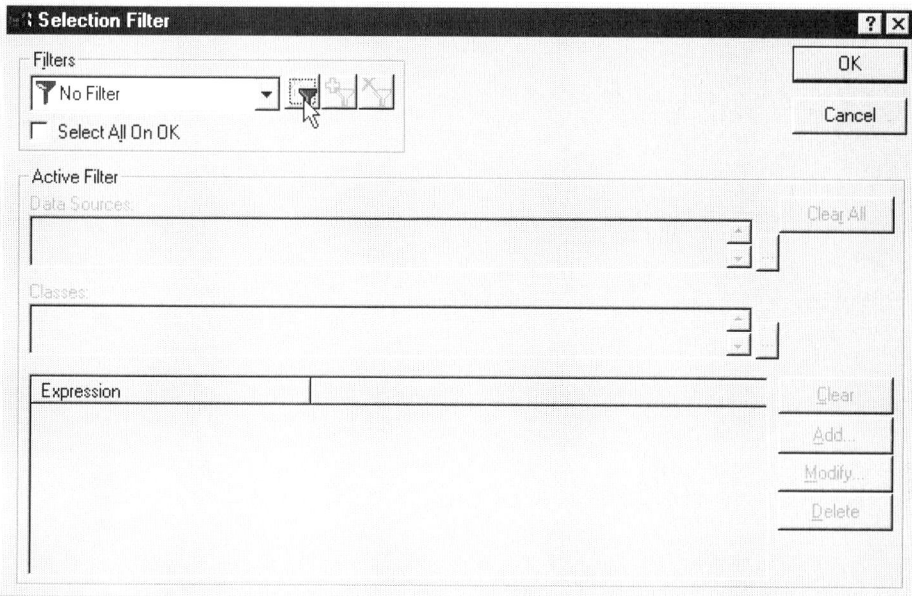

Figure 17–4

In the New Filter dialog box type the name of what you want to search for, in this case **Location** and click on **OK**.

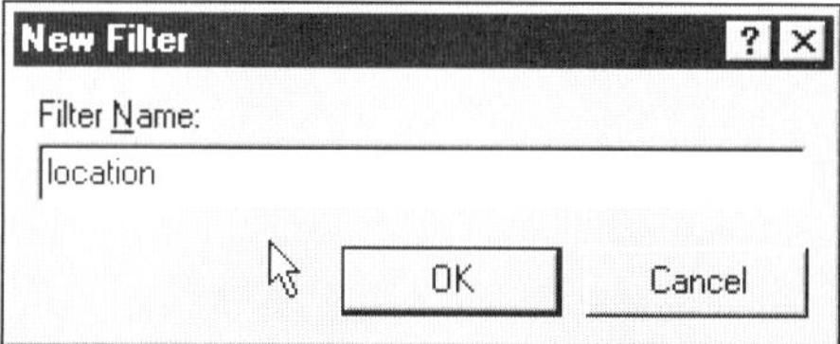

Figure 17–5

You will be returned to the Selection Filter dialog box. In the central portion, identified as Classes, click on the icon containing **three dots** (see Figure 17–6), which is located to the right of the dialog box. This will result in the display of the classes of data in the geobase.

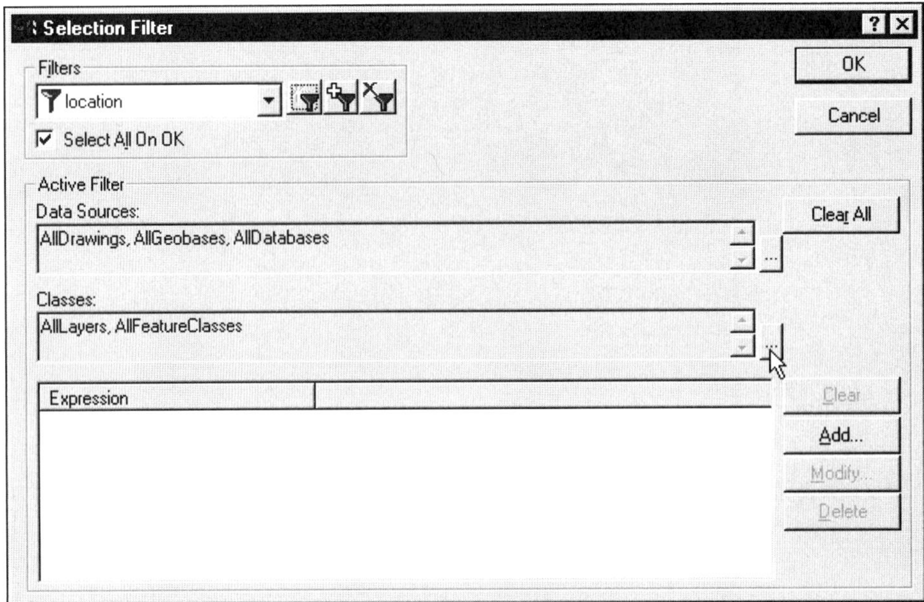

Figure 17–6

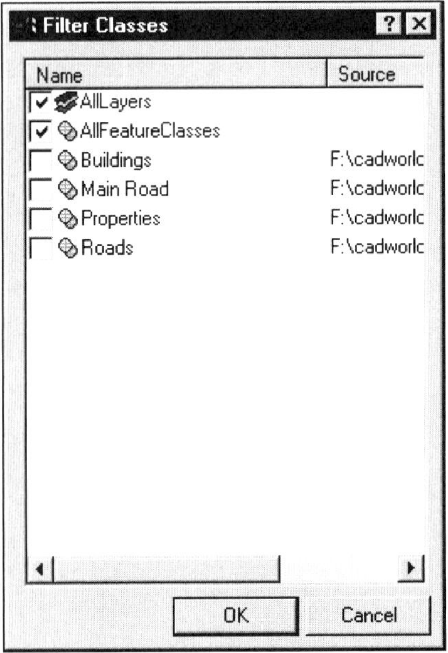

Figure 17–7

Step 3 Click on the **AllLayers** and **AllFeatureClasses** items to clear them, and click on the **Properties** item to select it.

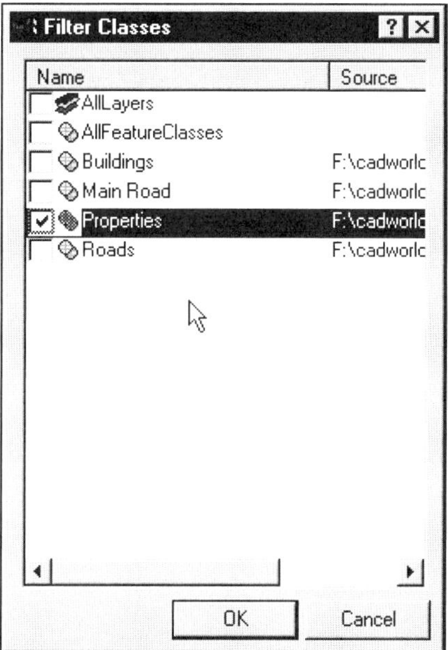

Figure 17–8

This has specified that only features in the Properties class will be available for selection. Click on **OK**. The Selection Filter dialog box returns. Click on the **Add** button located to the right of the Expression area, and the Add Expression dialog box will appear. The Graphic tab should be in the foreground.

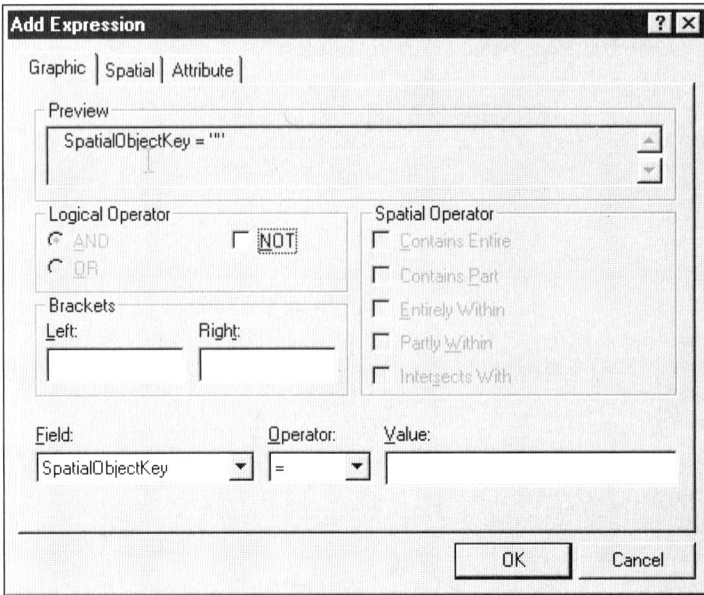

Figure 17–9

In the Field section choose SpatialObjectArea, for the operator select **>**, and for Value type **5000**.

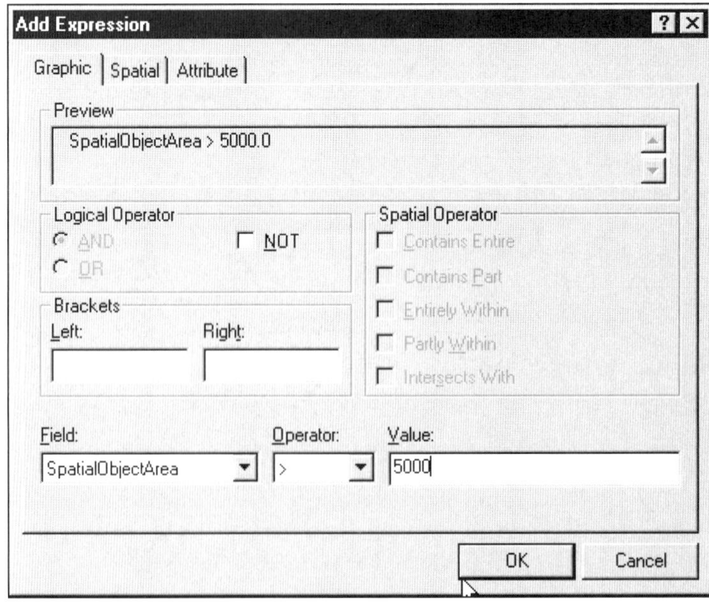

Figure 17–10

Step 4 Click on **OK** to add this expression to the Selection Filter, and then click on **OK** to run this selection filter on the drawing. The areas containing objects larger than 5,000 square meters will be highlighted.

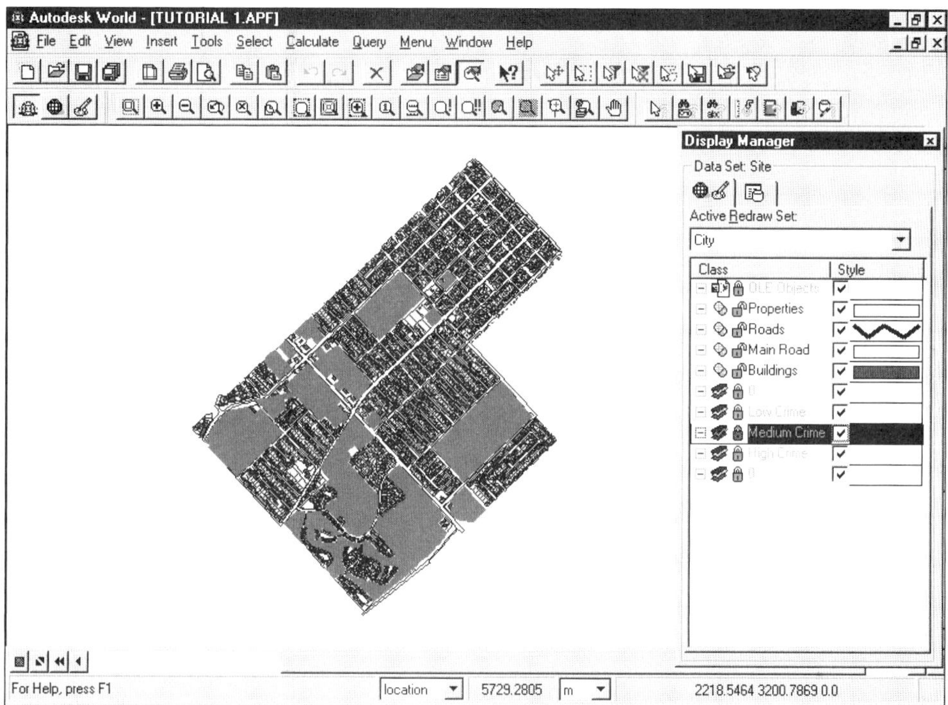

Figure 17–11

This procedure has established a query that detects properties with buildings a specified size. The tutorials provided with Autodesk World proceed to illustrate how many refinements can be made to such queries in order to enhance their utility.

Step 5 Adding annotations to a map is an automated process, assisted by a four-step wizard. As an example of how to add annotations to the map, go to the **Query** menu and select **Annotate**.

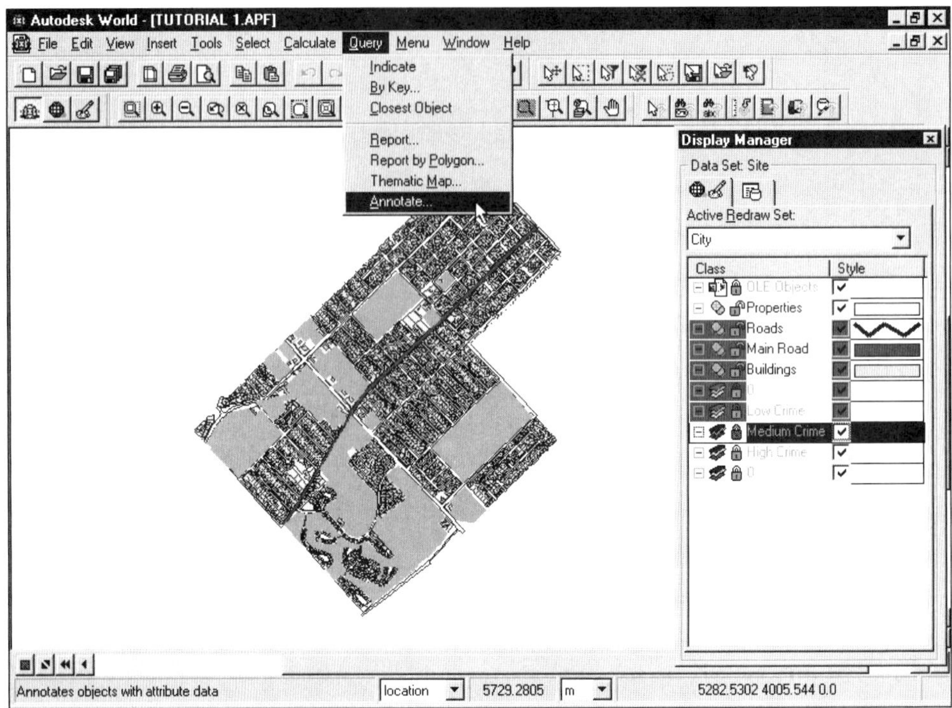

Figure 17–12

The Annotation Wizard will start. Select the **Selected Spatial Objects** option in the first screen. Click on **Next**.

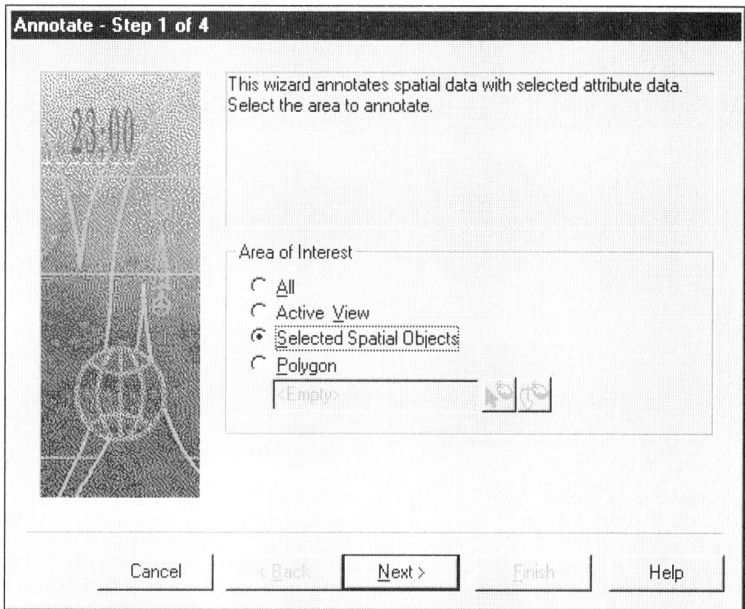

Figure 17–13

The second screen prompts you to select the features to label. Activate the **Properties** item and clear the **Buildings** item. Click on **Next**.

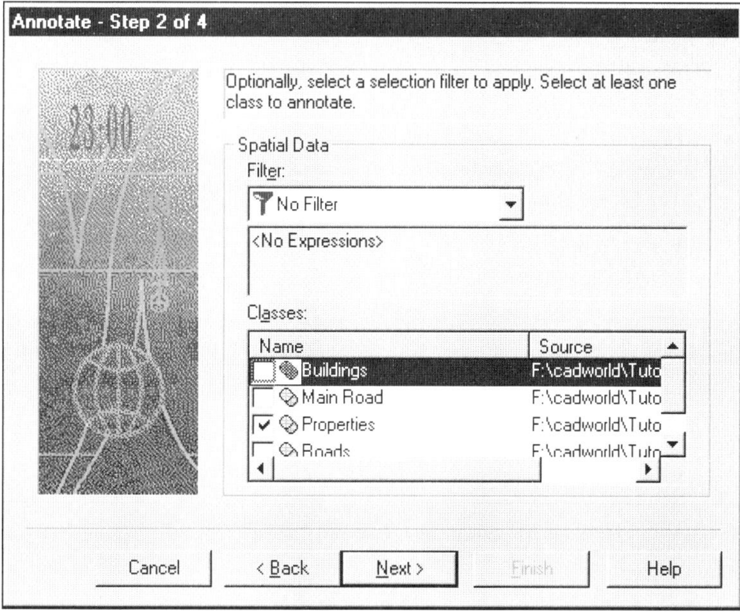

Figure 17–14

At Step 3 of the wizard, click on the **Property Number** item, then click on **Next**.

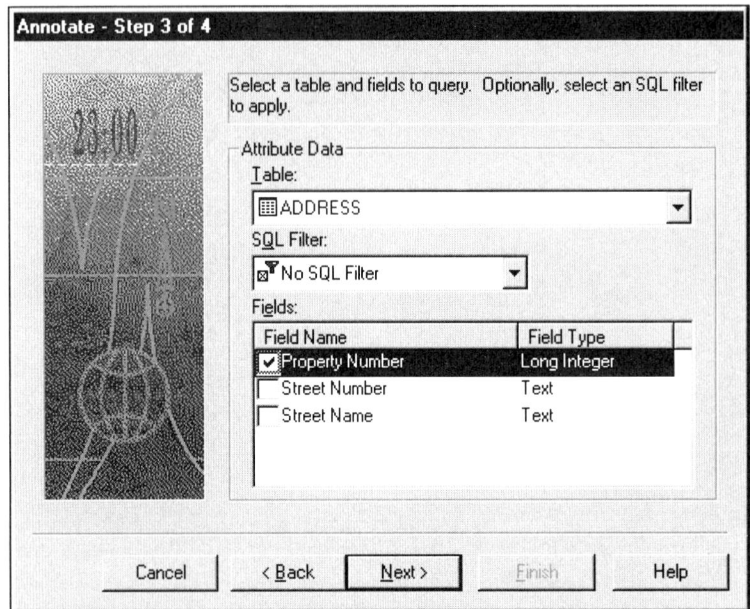

Figure 17–15

At Step 4 of the wizard, set the Annotation Style to **Annotation Text** and deselect the **Key** item. Click on **Finish** and the annotation text will appear on the map.

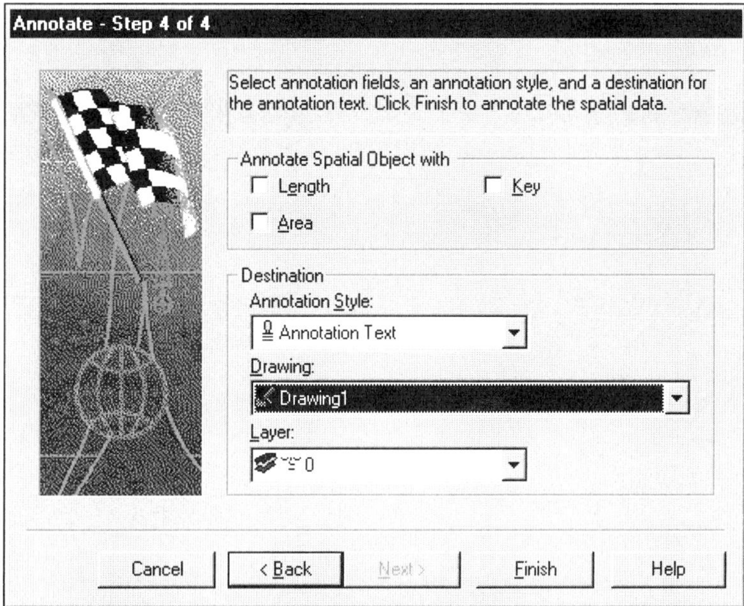

Figure 17–16

Step 6 Autodesk World also facilitates the generation of reports based on your data. To activate the Report Wizard go to the **Query** menu and select **Report**.

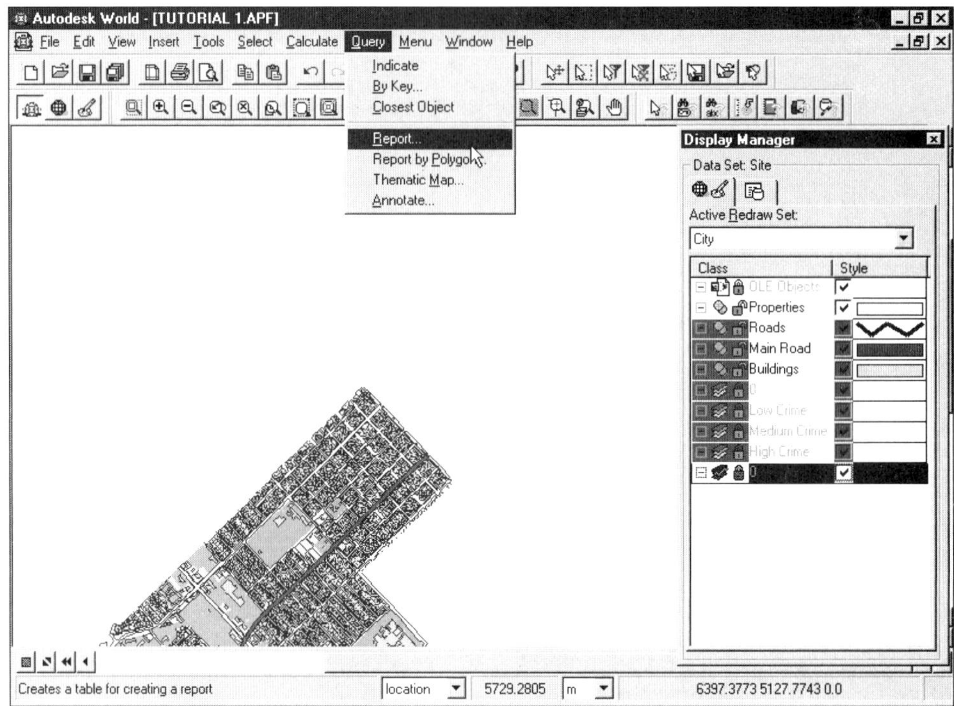

Figure 17–17

Step 7 Select the **Area of Interest** at the first screen in the wizard, in this sample **Selected Spatial Objects**. Click on **Next**.

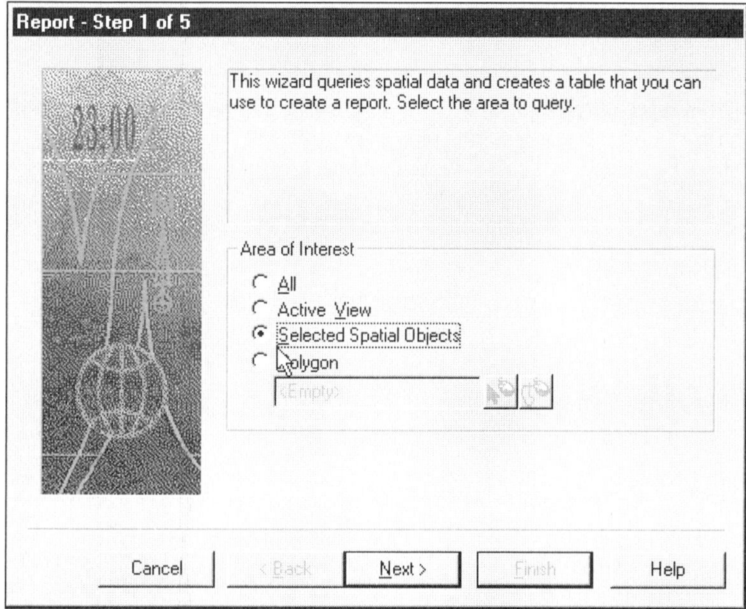

Figure 17–18

Select the **Classes** on which the report is based, in this example Properties. Click on **Next**.

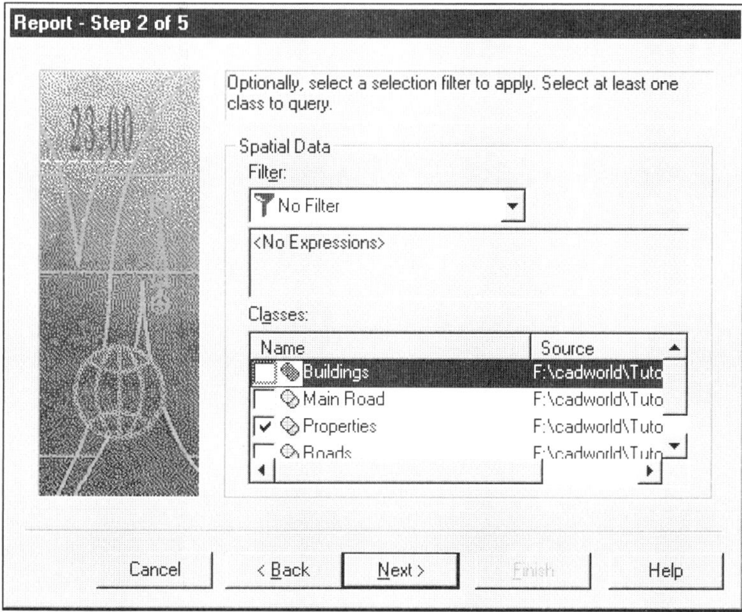

Figure 17–19

At Step 3 of the wizard, select the **Field**(s) to be used. In this case the selection is Property Number. Click on **Next**.

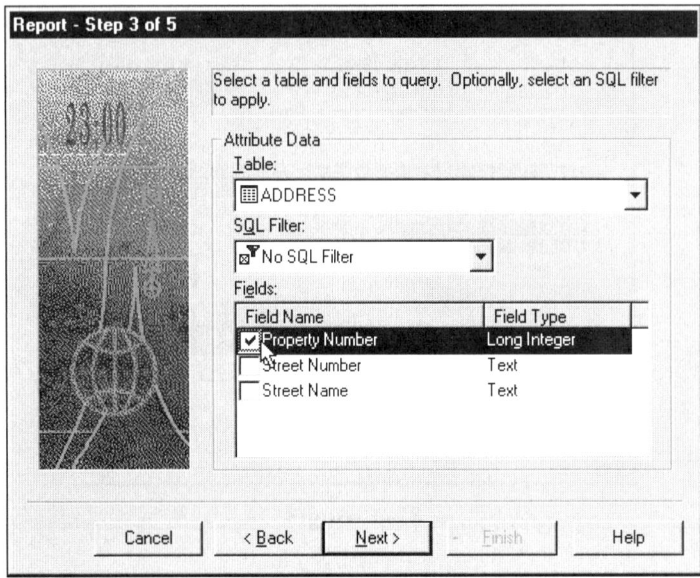

Figure 17–20

At Step 4 of the wizard, enter a name for the report. In this case, the name was testtutorial. The location for the report is also provided. Click on **Next**.

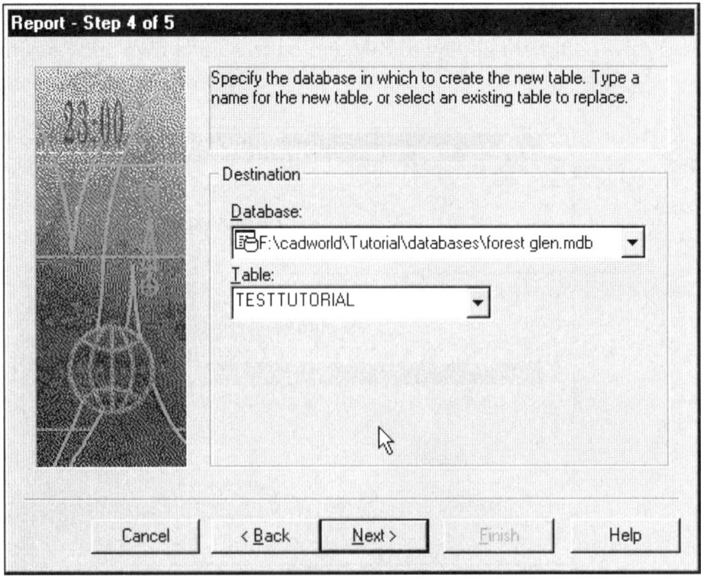

Figure 17–21

At Step 5 of the wizard you can select additional fields of data if required. To complete the report, click on **Finish**, and the tabular report is presented.

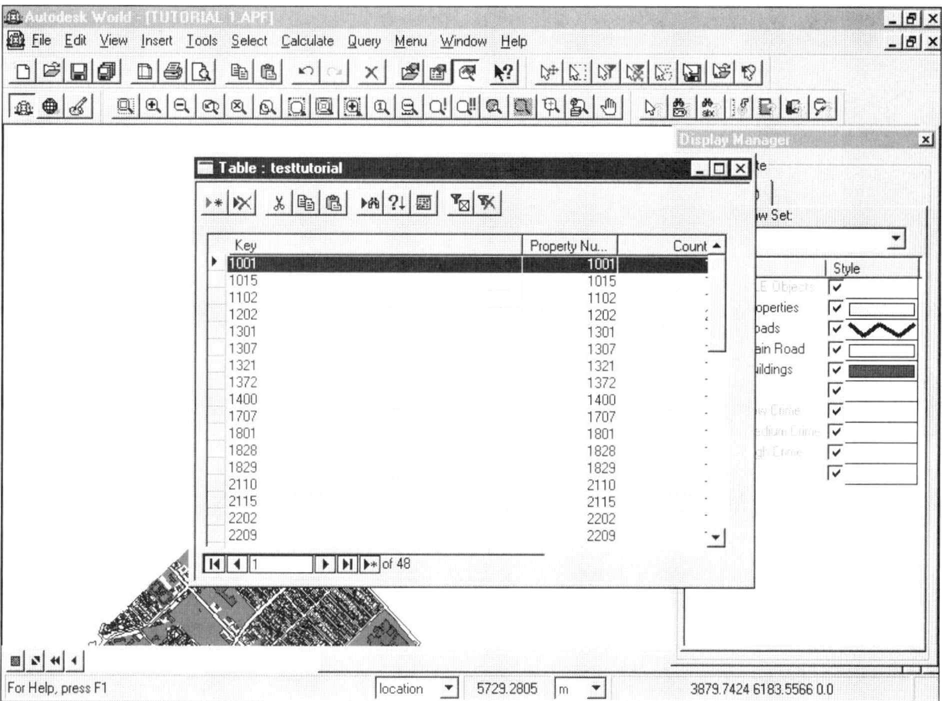

Figure 17–22

Close Autodesk World without saving your work.

The capability of automatically annotating map files as well as being able to quickly query files, generate reports, access a variety of data formats, perform coordinate transformations, and more makes Autodesk World a product worth adding to the software collection of anyone working with GIStechnology.

SUMMARY

Autodesk World is a desktop spatial analysis software package. World can read conventional AutoCAD DWG files, run queries, produce thematic maps and when used in conjunction with AutoCAD Map 2000, the pair of software packages provide the user with unparalleled GIS capabilities and functionalities.

GLOSSARY

AFFINE PROJECTION — One of three types of projection transformations available when calibrating a digitizing tablet. This projection transformation requires at least four ground control points.

BASE DRAWING — The work session drawing on which queries are presented or that is used to attach data in order to build data-enriched maps for data sets.

BUFFER FENCE — A boundary style used in the query process. The query identifies a drawn or user selected "fence" and queries objects within a specified distance from the fence.

BUFFER ZONE — The area on either side of a queried object as specified in the query.

CATALOG — The directory and path for database tables.

CENTROID — A point inside a topology-containing polygon that possesses information about the components of the polygon.

CLOSED POLYGON — A shape drawn using the polyline feature which has the same starting and ending point. The coincident position of the points can be ensured by using the Close function prior to completing the drawing of the shape.

COM PORT — A serial port used to connect peripherals, such as a GPS unit or a graphics tablet, to the computer.

CONFORMAL PROJECTION — A mapping projection that preserves angular accuracy. Area may be distorted.

CONTINUOUS QUERY FUNCTION — A query function used when the data being queried is numeric and distributed over a range of values.

DANGLING OBJECTS — A line segment that appears too short and does not reach another line or node.

DIRECTIVITY	Direct resistance assigned to links in a network topology.
DISCRETE QUERY FUNCTION	A query function used to search for data that is either non-numeric such as text or to search for specific values within the data
DUPLICATE OBJECT	Part of a polygon or link exists in the drawing twice even though it appears as a singular object.
DXF	Drawing Interchange Files, a format for importing and exporting files for AutoCAD and AutoCAD Map 2000.
EDGE MATCHING	A drawing adjustment for accurately aligning two different maps along a common edge.
EMULATION	A mode of operation copied by one device to promote system compatibility. In this text, one brand of graphics or digitizing tablet might "emulate" another brand for which AutoCAD Map 2000 provides built-in drivers.
ENVIRONMENT	The combination of the Catalog, Schema and Table used in linking to an external database.
EQUIVALENT PROJECTION	A mapping projection that preserves area. Angles may be distorted.
EXTENTS	The boundaries of the objects in the drawing.
GEOBASE	A spatial geographic database proprietary to Autodesk World.
GEOREFERENCE	To link a digital map to geographic coordinates from the earth.
GIS	Geographic Information System.
GLOBAL COORDINATE SYSTEM	A system converting the latitude and longitude coordinates from the earth's surface into AutoCAD Map 2000's coordinate system.
GRAPHICS TABLET	A device used for tracing paper maps, thereby converting them into digital maps. Also referred to as a digitizing tablet.
INTERSECTION	Where two lines or links meet.
KEY COLUMN	A field or column on a database used to build the links between the entity in a map and the data in the database.

LIMITS	Two-dimensional drawing coordinates identifying the lower left and upper right corners of the drawing.
LINK	A linear feature represented by a line or polyline used in creating topology.
LINK PATH NAME	The link path name identifies a key column for a current table used as an external database. It serves as an "alias" for the Catalogue, Schema, Table and Key Column.
MAPBOOK	The printed output resulting when a highly detailed map is printed over a series of several small pages rather than one large sheet.
NETWORK	A collection of lines or polylines linked together to model or represent a feature such as a road system.
NODE	The endpoint of a link (line) or an individual, isolated point.
OBJECT DATA	Data in either text, numeric or block form that is attached to features in a drawing.
OBJECT SNAP	A drawing tool used to eliminate digitizing and drawing errors.
OBJECT THEMATIC QUERY	An query based on data embedded in drawings used to graphically present nongraphic data.
ODBC	Database drivers for accessing external databases that are installed through the Control Panel.
ORTHOGONAL PROJECTION	A projection transformation used when digitizing paper maps based on two ground control points.
OVERLAY	The draping of one map over another to study the relationship of the variables from each map.
PIXEL	A picture element. The number of pixels in an image determines its resolution. Typically computer screens can show up to 1,024 pixels in the horizontal and 768 pixels in the vertical direction.
POLYGON	A geometric shape formed by a series of lines or line segments.
PROJECT DRAWING	The first drawing opened in AutoCAD Map 2000.
PROJECTIVE PROJECTION	A projection transformation offered as an option when digitizing and using at least three ground control points.

PROPERTY CONDITION	A condition set during a query. This specifies the nature of the information in the drawing to be queried.
QUERY	A formula or statement used to search for specific data.
RASTER	A dot- or pixel-based image with each pixel having its own color value.
RESISTANCE	The property used in network topology to indicate variations in the ease of movement along a portion of the network.
RUBBER SHEETING	A map modification process that adjusts maps by stretching points on the map to known ground control points.
SAVE SET	The collection of changes, new data, queried data and any other update made to a collection of drawings in the project that is to be saved back to the source drawings.
SCHEMA	The subdirectory of the Catalog that holds the database table.
SOURCE DRAWING	The attached drawing file and data that you wish to query.
SQL	Structured Query Language, used to build database inquiries.
THEMATIC MAPPING	The process of preparing maps that graphically display a theme or attribute that is generally nongraphic data.
TIF	One form of a raster image. Others include jpg, pcx, bmp, etc.
TOPOLOGY	The property that allows the drawing to keep track of what is beside what. It is a set of definite relationships that exist between polygons, points, and lines.
WORK SESSION	The term used to describe the collection of data being used at any one time to study a project including base drawings, source drawings, and attached databases.

INDEX

A

Access, external database connections, 221-223
Access data, AutoCAD map 2000, 224-234
Acid rain, environmental application, 246-255
Affine projection, 182
AutoCAD Map 2000, importing GPS text files as DXF files, 350-359
Autodesk
  map guide viewer, downloading/installing from Internet, 342-347
  world, Microsoft Office and, 362-377

B

Block attribute date, described, 5
Buffer zone analysis, 115-120
Building layout drawings, use of, 288-299

C

Catalog, defined, 202
Centroid, defined, 83
Civil engineering, application, 240-244
Conformal projections, 7
Continuous functions, discrete query functions versus, 64-65
Coordinate systems, assigning, 7-12
Cost estimates, preparing with raster images, 330-335

D

Data set(s)
  establishing drawings as, 268-276
  preparation of, 239-244
Database connections
  external,
    access, 221-223
    DBASE3, 203-205
Data-enriched drawings, preparing, 257-258
DBASE3, external connections to, 203-205
Delivery routes, programming, 303-327
Digital maps
  adding object data, while digitizing, 199
  created, with graphics tablet and UTM coordinates, 183-198
  using UTM coordinates with, 181-182
Digitize, map, explained, 183
Direction, establishing, 107
Discrete query functions, continuous functions versus, 64-65
Drawing assignment, continuing interrupted, 198
Drawings
  accessing multiple, 246-255
  attaching external reference, 280-285
  cleanup of, 70-77
  as data sets, 268-276
  data-enriched, preparing, 257-258
  deleting layers from, 268-270, 276
  linking, 12-64
  multiple,
    querying, 258-267
    viewing simultaneously, 257-258
  preparing for use in queries, 30-40
  using building layout, 288-299
DXF files
  importing, 174-176
  importing from GPS text files, 350-359

E

Edge matching, 142-152
  defined, 142
Environmental application, acid rain, 246-255
Equivalent projections, 7
External database queries, network topology connection and, 302-327
External references, attaching to a drawing, 280-285

F

Facilities management, GIS technology and, 298-299
Flood plane properties, 240-244

G

Geographic calculator, using to import GPS text files as DXF files, 350-359
Georeferencing
  explained, 139
  raster image, 136
GIS
  software, exporting files for AutoCAD 2000, 174-176
    technology, facilities management and, 298-299
Glossary, 379-382
GPS (Global positioning system) text files, importing as DXF files, 350-359
Graphics tables, creating digital maps with, 183-198
Graphics tablet, creating digital maps with UTM coordinates and, 183-198
Ground control points, 182

I

Internet, downloading/installing Autodesk mapguide viewer from, 342-347
Interrupted work session, continuing, 198

K

Key column, defined, 202

L

Layers, deleting from drawings, 268-270, 276
Link objects, defined, 83
Link template, 202
Linking drawings, 12-64
Links, defined, 83

M

Map projection, 179
  defined, 7
Mapbook printing, 157-163
  defined, 157
Mapping, thematic, 13-29
Mercator projection, 7
Microsoft Office, Autodesk world and, 362-377
Multiple drawings
  querying, 258-267
  viewing simultaneously, 257-258

N

Network(s)
  defined, 67
  flood traces, explained, 114
  (line) topology, **69**
    creating, 91-97
  topology, external database queries and, 302-327
  traces, using topology, 100-114
  tracing, defined, 100
Node topology, **68**
  creating, 98-99
Nodes, defined, 67, 83
Non-DWG files, importing, 174-176

O

Object data
  adding while digitizing map, 199
  described, 5
ODBC, use of, 203-235
Orthogonal projection, 182
Overlay analysis, 121-127
Overlays, topology, explained, 121

P

Polygon queries, conducting, 239-244
Polygon topology, **69**
  creating, 78-90
Polygons, defined, 67, 68
Project file, described, 5
Projection
  formula, function of, 179
  map, defined, 7, 179
  selection of, 180
Projective projection, 182
Puck, **183**, 184

Q

Query(ies)
  building a, 12-64
  defined, 6
  determining area of, 19
  functions, continuous versus discrete, 64-65
  multiple drawings, 258-267
  preparing drawings for use in, 30-40
  setting up drawing, 5
Quickview, using, 283-285

R

Raster images
  approximating dimensions of image objects with, 330
  described, 129
  georeferencing, 136
  viewing, 130-139
Resistance, explained, 107
Reverse resistance, explained, 107
Routes, delivery, programming, 303-327
Rubber sheeted, explained, 141
Rubber sheeting, explained with drawbacks noted, 152-156

S

Site selection, 280-285
Source drawings, described, 5
Special analysis
  performing, 246-255
  software, 362-377

T

Table, defined, 202
Thematic data query, when to use a, 4
Thematic mapping, 13-29
Thematic maps
  defined, 12
  function of, 3
Topology
  buffer zone analysis, 115-120
  creating, 83
    network, 91-97
    node, 98-99
    polygon, 78-90
  defined, 67
  drawing, cleanup, 70-77
  network traces using, 100-114
  overlay analysis, 121-127

U

Universal transverse mercator. *See* UTM
UTM
  coordinates,
    creating digital maps with graphics tablet and, 183-198
    using, 181-182
  zone, explained, 180-181

W

Work session
  continuing interrupted, 198
  described, 5

LICENSE AGREEMENT

for Autodesk Press, an imprint of Thomson Learning™

Educational Software/Data

You the customer, and Autodesk Press incur certain benefits, rights, and obligations to each other when you open this package and use the software/data it contains. BE SURE YOU READ THE LICENSE AGREEMENT CAREFULLY, SINCE BY USING THE SOFTWARE/DATA YOU INDICATE YOU HAVE READ, UNDERSTOOD, AND ACCEPTED THE TERMS OF THIS AGREEMENT.

Your rights:

1. You enjoy a non-exclusive license to use the enclosed software/data on a single microcomputer that is not part of a network or multi-machine system in consideration for payment of the required license fee, (which may be included in the purchase price of an accompanying print component), or receipt of this software/data, and your acceptance of the terms and conditions of this agreement.
2. You own the media on which the software/data is recorded, but you acknowledge that you do not own the software/data recorded on them. You also acknowledge that the software/data is furnished "as is," and contains copyrighted and/or proprietary and confidential information of Autodesk Press or its licensors.
3. If you do not accept the terms of this license agreement you may return the media within 30 days. However, you may not use the software during this period.

There are limitations on your rights:

1. You may not copy or print the software/data for any reason whatsoever, except to install it on a hard drive on a single microcomputer and to make one archival copy, unless copying or printing is expressly permitted in writing or statements recorded on the diskette(s).
2. You may not revise, translate, convert, disassemble or otherwise reverse engineer the software/data except that you may add to or rearrange any data recorded on the media as part of the normal use of the software/data.
3. You may not sell, license, lease, rent, loan, or otherwise distribute or network the software/data except that you may give the software/data to a student or and instructor for use at school or, temporarily at home.

Should you fail to abide by the Copyright Law of the United States as it applies to this software/data your license to use it will become invalid. You agree to erase or otherwise destroy the software/data immediately after receiving note of Autodesk Press' termination of this agreement for violation of its provisions.

Autodesk Press gives you a LIMITED WARRANTY covering the enclosed software/data. The LIMITED WARRANTY can be found in this product and/or the instructor's manual that accompanies it.

This license is the entire agreement between you and Autodesk Press interpreted and enforced under New York law.

Limited Warranty

Autodesk Press warrants to the original licensee/ purchaser of this copy of microcomputer software/ data and the media on which it is recorded that the media will be free from defects in material and workmanship for ninety (90) days from the date of original purchase. All implied warranties are limited in duration to this ninety (90) day period. THEREAFTER, ANY IMPLIED WARRANTIES, INCLUDING IMPLIED WARRANTIES OF MERCHANTABILITY AND FITNESS FOR A PARTICULAR PURPOSE ARE EXCLUDED. THIS WARRANTY IS IN LIEU OF ALL OTHER WARRANTIES, WHETHER ORAL OR WRITTEN, EXPRESSED OR IMPLIED.

If you believe the media is defective, please return it during the ninety day period to the address shown below. A defective diskette will be replaced without charge provided that it has not been subjected to misuse or damage.

This warranty does not extend to the software or information recorded on the media. The software and information are provided "AS IS." Any statements made about the utility of the software or information are not to be considered as express or implied warranties. AutoDesk Press will not be liable for incidental or consequential damages of any kind incurred by you, the consumer, or any other user.

Some states do not allow the exclusion or limitation of incidental or consequential damages, or limitations on the duration of implied warranties, so the above limitation or exclusion may not apply to you. This warranty gives you specific legal rights, and you may also have other rights which vary from state to state. Address all correspondence to:

AutodeskPress
3 Columbia Circle
P. O. Box 15015
Albany, NY 12212-5015